# PRINCIPLES OF ORGANIC FARMING

NIPA GENX ELECTRONIC RESOURCES & SOLUTIONS P. LTD.
New Delhi-110 034

## About the Authors

**Dr. S.S. Walia**, is working as Principal Agronomist in the School of Organic Farming, Punjab Agricultural University, Ludhiana. Started organic farming research as a pioneer work at PAU, Ludhiana as Ph.D. student (2001-04) and then worked on organic farming and integrated farming systems in the All India Co-ordinated Research Project on Integrated Farming Systems and Network Project on Organic Farming. Played foremost role in developing infrastructure, state of art model of an integrated farming system, and other research facilities in AICRP-Integrated Farming System project. Handled 19 projects and currently 5 projects are in hand, Involved in 35 product testing. Published 495 research and extension papers in reputed journals/magazines/newspapers. Co-author of eight books namely Science of Agronomy; Crop Management (1st ed), Farming System and Sustainable Agriculture Contingent Crop Planning, Waste recycling, Moringa, Principles of Weed Management, Organic Crop Production; 21 book chapters, ten practical manuals, one research bulletin, four extension bulletins, four extension booklets, delivered 51 TV & Radio talks and guided eleven M.Sc. and four Ph.D. students. Participated in 44 National and International conferences/meetings. Recipient of Best organic farming centre award (2019) by ICAR; Dr M S Randhawa best book award (2017); Fellow, Indian Ecological Society (2016); Gold medal from Society of Recent Development in Agriculture at International Conference (2013); ISA Best Paper Award alongwith cash prize of Rs 5000/- from Indian Society of Agronomy (2011) for paper entitled," Alternate cropping systems to rice-wheat for Punjab"; ISA P. S. Deshmukh Young Agronomist Award (2005) by Indian Society of Agronomy; Fellow, Society of Environmental Sciences (2004). The AICRP on Integrated Farming System, Ludhiana centre received the Best Centre Award (2014) from ICAR- Indian Institute of Farming Systems Research, Modipuram, Meerut. Performance of All India Co-ordinated Research Project on Integrated Farming Systems, Ludhiana centre during QRT review (2007 to 2012 and 2012 to 2017) was rated outstanding and also received four best poster awards. Received Dr. Gurbaksh Singh Gill Gold Medal, Merit Certificate for M.Sc. (Agronomy); S.S. Labh Singh Gold Medal and Merit Certificate for B.Sc. Agri. (Hons.). The chapter on organic farming and integrated farming system in package of practices was included during 2004 and 2017-18, respectively. Sixty six recommendations have been included in the package of practices for mass adoption by the Punjab Farmers especially, resource conservative cropping systems, nine organic farming based cropping systems, production technology of cultivation of direct seeded rice and consortium biofertilizer application in sugarcane, maize, wheat, potato, onion and turmeric. Developed technologies regarding application of consortium in sugarcane, turmeric, potato, onion, maize, wheat crops; integrated nutrient management in maize/ soybean, highly productive alternative cropping systems, direct seeded rice and basmati rice, rice residue management for the mass scale adoption under Punjab conditions. Developed Integrated Farming System Research Model comprising dairy, fishery, horticulture, vegetables, agro-forestry, vermi-composting and chapter on Integrated Farming System included in the Package of Practices.

**Dr. R.K. Nanwal**, Professor (Retd), CCS Haryana Agricultural University, Hisar-125004, India. For being B.Sc. (Hons) Ag and M.Sc. (Ag.) in Agronomy from HAU, Hisar in 1979 and 1981, respectively, he was awarded Ph.D. in Agronomy by IARI, New Delhi in 1992 with merit/ICAR scholarship throughout. He did Post Doctorate specialized training in water resource management in Rainfed Agriculture at University of Georgia, USA on FAO fellowship. He has teaching experience of more than 35 years in the university. Developed new courses on Farming Systems, Sustainable Agriculture and Rainfed Agriculture. Guided 7 M.Sc. and 8 Ph.D. students as Major Advisor. Played foremost role in developing infrastructure, state of art model of an integrated farming system and other research facilities in AICRP-Integrated Farming System project at CCS HAU, Hisar. Developed Integrated Farming System Research Model integrating field crops with dairy, fishery, horticulture, vegetables, agro-forestry, vermi-composting, mushroom, biogas etc. which is very popular among Haryana farmers. He has published 12 books, 10 manuals, 7 book chapters, 153 full research papers, 75 general articles, 35 research notes and presented 60 papers in various national/international seminars/conferences. Received "Indian Council of Agricultural Research, New Delhi Award" for outstanding Multidisciplinary team research in agriculture and allied sciences for the triennium 1994-96, awarded on July 16, 1997 by Union Minister of Agriculture, Govt of India. Awarded CCS Haryana Agricultural University "Best Teacher Award" (Indian Council of Agricultural Research sponsored) for the year 2005-2006. He was awarded "Rashtriya Gaurav Award" by India International Friendship Society on Nov. 11, 2011. He was also awarded Best Citizens of India Award by International Publishing House, New Delhi on Dec. 28, 2011. Received Appreciation Certificate for excellent work by three Vice Chancellors of the university. Superannuated from active service from CCS HAU, Hisar on Feb. 28, 2017. After superannuation re-employed for three years as Officer Incharge, Students' Centre for Career Excellence in CCS HAU, Hisar.

# PRINCIPLES OF ORGANIC FARMING

**Sohan Singh Walia**
Principal Agronomist
School of Organic Farming
Punjab Agricultural University
Ludhiana, Punjab

**R.K. Nanwal**
Professor (Retd)
CCS Haryana Agricultural University
Hisar-125004, Haryana

**NIPA GENX ELECTRONIC RESOURCES & SOLUTIONS P. LTD.**
New Delhi-110 034

**NIPA GENX ELECTRONIC**
**RESOURCES & SOLUTIONS P. LTD.**

101,103, Vikas Surya Plaza, CU Block
L.S.C.Market, Pitam Pura, New Delhi-110 034
Ph : +91 11 27341616, 27341717, 27341718
E-mail:newindiapublishingagency@gmail.com
www: www.nipabooks.com

***For customer assistance, please contact***
Phone: + 91-11-27 34 17 17 Fax: + 91-11- 27 34 16 16
E-Mail: feedbacks@nipabooks.com



ISBN: 978-81-19235-56-8



Composed and Designed by NIPA.

# Preface

Organic agriculture follows the principles and logic of a living organism, in which all elements (soil, plant, farm animals, insects, the farmer and local conditions) are closely linked to each other. This is accomplished by using possible, agronomic biological and mechanical methods, following the principles of these interactions, using natural ecosystem as a model. Looking to the importance of organic food, pulses, oilseeds, vegetables constitute an important ingredient of the vegetarian diet in the Indian sub-continent and play major role in supplying quality proteins and minerals. Organic products therefore, have potential market, especially in India. Organic farming even if practiced on a small scale could be a step in the right direction because it is based on the principle that soil is a living system and not an inert bowl for unloading chemical and it uses the nature as the best role model for rejuvenating the soil health. Moreover, organic farming interlinks the soil health with other resources with least interference to raise healthy plants and animals; by sustaining the resource equilibrium. Keeping in view the present status of Indian agriculture, it seems pertinent to give a rethought to chemical farming at present practiced on large scale. The one option of organic farming might be to replace the chemical farming in coming years more as an inevitable measure than as any new practice to boost the agricultural production. Though organic farming is indeed conceptually not new, a fresh look on the principles of organic farming in context of the present day is necessary, to make it a viable practical preposition in a commercial high yield- high quality scenario.

The book contains detailed and comprehensive information basically to fulfill the requirements of under graduate students of agriculture stream for organic farming considering the syllabus of 5th Dean's Committee of ICAR. We hope that the book will serve as a good reference source for those interested in organic farming. During the preparation of the book, we have consulted various international and national books, periodicals, journals, internet from which we have drawn literatures liberally in this book. We heartily acknowledge each of the scientist and author. In some cases, where it has not been possible to contact and obtain permission for such reproduction, the authors and publisher offer their sincere apologies.

It is beyond our access to acknowledge in words, the ever encourging moral support and cooperation from our family members including parents and spouse during the course of writing the book. We acknowledge the love and affection of our children which always keep us cheerful. The author welcomes constructive comments and suggestions from readers for further improvement of the book. We sincerely thank NIPA, New Delhi for publishing this book well in time.

**Authors**

# Contents

# Glossary

**Accreditation:** accreditation means registration by the national accreditation body for certifying organic farms, products and processes as per the national standards for organic products and as per the guidelines of the national accreditation policy and programme for organic products.

**Accredited programme:** means programme of accrediting inspection and certification agencies which have been accredited by the accreditation agency and which have agreed to comply with the accreditation contract.

**Agriculture:** it is an art, science and business of raising crops and rearing of animals through exploring the natural resources with the coordination of socio economic infrastructure to meet the basic necessities of life . *i.e.* Food, feed, fiber and shelter.

**Agronomy:** the science of crop production and soil management. The name is derived from the greek words *agros*(field) and *nomos* (to manage). It incorporates the basic sciences (biology, chemistry, physics, geology, microbiology, etc.) into an applied science which is the foundation for most agriculture.

**Allelopathy:** phenomenon involving the release of certain chemicals from plant parts into the environment which may when present in sufficient amounts, inhibit or suppress the germination or growth of the plants in the neighbourhood.

**Annual report:** means the report on operators, products and processors submitted annually to the accreditation agency by the accredited inspection and certification agencies.

**Appeal:** shall be the process by which an inspection and certification agency can request reconsideration of a decision taken by the accreditation agency or an operator can request reconsideration of a decision by the certification agency.

**Applicant:** shall be the inspection and certification agency that has applied for accreditation to the accreditation agency.

**Arboriculture:** intensive cultivation of individual trees possibly for fruits gums and resins.

**Arid/aridity:** a term applied to land that is dry also means deficient in rainfall.

**Autoecology:** it deals with the ecology of an individual species and its population including the effect of other organisms and environmental conditions on every stage of its life cycle.

**Autotrophs:** organism able to manufacture their own food from inorganic materials. Using energy from outside sources. Most green plants are completely autotrophs.

**Ayurveda:** ayurveda is a traditional naturopath system of medicines and health care of India.

**Band placement:** the placement of fertilizer in the soil in continuous bands or narrow ribbons in between the plant rows or around each plant. The fertilizer applied in bands is covered by the soil but is not mixed with it.

**Basic seed:** is the progeny of pre – basic seed produced so as to maintain genetic purity and identity.

**Bed:** a narrow flat topped ridge on which crops are grown with a furrow on each side for the drainage of water.

**Biofertilizers:** microbial inoculants probably known as biofertilizers refer to living or latent cells of efficient strains of microorganisms capable of fixing atmospheric nitrogen .e.g., Rhizobium, Consortium.

**Biological yield:** it is the total dry matter produced by a plant as a result of photosynthesis and nutrient uptake minus that lost by respiration.

**Blind hoeing:** hoeing before the crop emerges, usually done in sugarcane.

**Broadcasting:** manual spreading of seed in the field and mixing of the spread seed by ploughing or planking the field.

**Buffer zone:** a clearly defined and identifiable boundary area bordering an organic production site that is established to limit application of, or contact with, prohibited substances from an adjacent area.

**Bulk density ($d_b$):** it is defined as mass of a unit volume of dry soil. This volume includes both solids and pores.

**Cereal crops**: a cereal is defined as crop grown for its edible seed. These crops are also known as grain crops e.g. wheat, rice, maize etc.

**Certificate of registration:** shall mean the document issued by the inspection and certification agency, declaring that the operator is licensed to use the certificate on specified products.

**Certificate:** would mean a document issued by an accredited agency declaring that the operator is carrying out the activities or the stated products have been produced in accordance with the specified requirements in accordance with the national standards for organic products.

**Certification mark:** shall mean certification programme's sign, symbol or logo which identifies the products as being certified according to the national standards for organic products.

**Certification programme:** shall mean the system operated by an inspection and certification agency in accordance with the criteria for carrying out certification of conformity as laid down herein.

**Certification transference:** the formal recognition by an inspection and certification agency of another certification programme or agency or projects or products certified by that programme or agency, for the purpose of permitting its own certified operators to trade or process under the programme's own certification mark, the products which are certified by the other programme.

**Certification:** shall be the procedure by which a written assurance is given by the certification agency that a clearly identified production or processing system has been methodically assessed and confirms to the specified requirements.

**Certified seed:** it is the progeny of basic seed and is produced by registered growers of seed producing agencies.

**Chain of custody:** all relevant steps in the production chain including growing, harvesting, processing, handling and related activities of the accreditation criteria that have been inspected and certified.

**Clayey soil:** a soil is known as clayey which contain at least 30% clay particles and in most cases not less than 40% ,usually it contain 45% clay , 30% silt and 25% sand.

**Climate:** aggregate of atmospheric condition over a long period of time.

**Commercial farming system:** in this type of farming system, crops are raised on a commercial scale for marketing.

**Companion crops:** the two crops grown together are called companion crops. E.g. Berseem and barley.

**Competent authority:** shall mean the official government agency for accreditation.

**Component crop:** it is used to refer to either of the individual crops making up the intercropping situation.

**Composite variety:** a mixture of genotype from general sources maintained by normal pollination.

**Composting:** composting is a biological process in which micro-organisms of both types namely, aerobic and anaerobic, decompose the organic matter and lower the carbon nitrogen ratio of the refuse. The final product of the composting is the well rotten manure known as compost.

**Conservation**: enhance conditions for existing natural enemy survival and reproduction.

**Consultancy:** shall mean the advisory service for organic operations, independent from inspection and certification procedures.

**Contamination:** pollution of organic product or land; or contact with any material that would render the product unsuitable for organic certification.

**Conventional:** farming systems dependent on input of artificial fertilizers and/or chemicals and pesticides or which are not in conformity with the basic standards of organic production.

**Conversion period:** the time between the start of organic management, and the certification of crops as organic. This is also called transition.

**Cover crops:** the crops, which are planted to cover the ground and to reduce the soil erosion and nutrients losses by leaching. e.g., grasses and rye.

**Crop ecology:** it is the study of the interrelationships between crop plants and the environment. Factors controlling the habitat of crop during its entire life cycle such as climate, physiography, soil and biota are considered.

**Crop growth rate (CGR):** CGR represents total dry matter productivity of the community per unit land area over a certain time span.

**Crop rotation:** it is the process of growing different crops in succession on piece of land in a specific period of time, with an object to get maximum profit from least investment without impairing the soil fertility.

**Crop system:** an arrangement of crop populations that transform solar energy, nutrients, water and other inputs into useful biomass. The crop system is a subsystem of cropping system. For example, in maize crop system, maize is the dominant crop which is grown in association with other crops.

**Crop:** a crop is a community of plants grown under field condition for its economic value.

**Cropping index (cropping intensity):** the number of crops grown per annum on a given area of land multiplied by 100.

**Cropping pattern:** the yearly sequence and spatial arrangement of crops or of crops and fallow on a given area.

**Cropping scheme:** the cropping scheme is a plan according to which crops are grown on individual plots of a farm during a given period of time with the object of obtaining maximum return from each crop without impairing the soil fertility.

**Cropping system:** cropping system is a land use unit comprising soils, crop, weed, pathogen and insect subsystems that transform solar energy, water, nutrients, labour and other inputs into food, feed, fuel and fibre. The cropping system is a subsystem of the farming system.

**Cultivator:** which only cut and stir the soil.

**Declaration of interest:** declaration of no personal / commercial conflict of interest by all concerned involved in the process of inspection and certification.

**Disinfectant:** a product that minimizes by physical or accepted chemical means, the number of microorganisms in the environment, to a level that does not compromise food safety and suitability.

**Diversified farming:** this is an expanded type of farming system in which varieties of crops are produced and many types of animals are reared.

**Drainage:** it is the removal of excess surface or ground water from the root zone of a crop by means of surface or sub – surface drains.

**Dry farming:** in which crops and livestock are raised on land which does not receive sufficient rainfall for water intensive crops and no irrigation facilities are available fall into this category.

**Earthing up:** is the operation of pulling up soil from the center of crop rows to the bottom of the plants, which helps to uproot weeds and provide to plants.

**Ecological recycling**: is common in organic farming, where nutrient management is fundamentally different compared to agri-business styles of soil management. Organic farms that employ ecosystem recycling to a greater extent support more species (increased levels of biodiversity) and have a different food web structure. Organic agricultural ecosystems rely on the services of biodiversity for the recycling of nutrients through soils instead of relying on the supplementation of synthetic fertilizers.

**Ecology:** it is defined as the study of plants and animals in reciprocal relationship with their environment or external world.

**Economic yield:** that part of the crop for which crop is grown e.g. Stem in sugarcane root in the case of sugar beet.

**Ecosystem:** any collection of organisms that interact or have the potential to interact along with the physical environment in which they live, form an ecological system or ecosystem.

**Effective rainfall:** precipitation which falls during the growing period of the crop and is available to meet the evapotranspiration requirements of crops.

**Evaluation:** shall be the process of systematic examination of the performance of an inspection and certification agency to meet the specific requirements under the national accreditation programme.

**Facultative weed:** a weed found growing both wild and in association with crops.

**Farm unit:** an agricultural farm, area or production unit managed organically, by a farmer or a group of farmers.

**Farm yard manure (FYM):** it refers to the decomposed mixture of dung and urine of farm animals along with the litter and left over material from fodder fed to cattle. Well rotten FYM contains about 0.5 % N, 0.3 % $P_2O_5$ and 0.5% $K_2O$.

**Farming system:** farming system is a decision making and land use unit comprising the farm household, cropping and livestock systems that produces crop and animal product for consumption and sale. The farming system is a subsystem of a higher level land use system, such as village or water shed, that, in turn, forms a component of the agricultural sector of the regional system.

**Food additive:** food additive is an external permissible ingredient added to improve the keeping quality, consistency, colour and other physico chemical, sensory properties, wholesomeness and safety of food

**Genetic diversity:** genetic diversity means the variability among living organisms from agricultural, forest and aquatic ecosystem. This includes diversity within species and between species.

**Green manure crops:** some crops are grown and ploughed in the soil in green form in order to improve soil fertility e.g. Berseem, guara, *dhaincha*, sunnhemp, cowpea etc

**Green manure:** manure consisting of fresh green plant matter, which is ploughed in or turned into the soil for the purpose of soil improvement.

**Green manuring:** it is a practice of ploughing or turning into the soil undecomposed green plant material for improving the physical condition of the soil or for adding nitrogen when the green manure is a legume.

**Group certification:** certification of an organized group of producers, processors and exporters with similar farming and production systems and which are in geographical proximity.

**Growth analysis:** growth analysis is a technique for quantifying components of plant growth from a series of mathematical equations.

**Growth:** the term growth is applied to quantitative changes occurring during development. It may be defined as an irreversible change in the size of cell, organ or whole organism.

**Guidelines for organic production and processing:** standards for organic production and processing established by the accreditation agencies for specific crops in accordance with the national standards for organic products.

**Habitat:** the area in which a plant or animal species naturally exists.

**Harvest index (HI):** it is the ratio of economic yield to biological yield. It is expressed in percentage

H.I. = economic yield x 100 / biological yield.

**Hazard analysis critical control point :** a systematic process that identifies food safety hazards, critical control points, critical limits, corrective actions and documentation and integrates monitoring procedures to ensure food safety.

**Herd system:** the arrangement of (domesticated) animals populations and their feed sources.

**Homeopathic treatment:** treatment of disease based on administration of remedies prepared through successive dilution of a substance that in larger amounts produces symptoms in healthy subjects similar to those of the disease itself.

**Homeopathy:** homeopathy is a system of medicine based on the principle of *"similia, similibus, curentur* (let likes be treated by likes)".

**Importation**: involves introduction of pest's natural enemies to a new locale where they do not occur naturally.

**Ingredient:** shall mean any substance, including a food additive, used in the manufacture or preparation of a food and present in the final product although possibly in a modified form.

**Inoculant:** the bacteria containing material used to introduce N- fixing rhizobium bacteria into soil.

**Input manufacturing:** shall mean the manufacturing of organic production or processing inputs.

**Inputs banned:** those items, the use of which is prohibited in organic farming.

**Inputs permitted:** those items that can be used in organic farming.

**Inputs restricted:** those items that are allowed in organic farming, in a restricted manner, after a careful assessment of contamination risk, natural imbalance and other factors arising out of their use. Farmers should consult the certifying agency.

**Inspection agency:** shall mean the agency that performs inspection services as per the national accreditation policy and programme.

**Inspection and certification agency:** shall be the organization responsible for inspection and certification.

**Inspection:** shall include the site visit to verify that the performance of an operation is in accordance with the production or processing standards.

**Inspector:** shall be the person appointed by the inspection and certification agency to undertake the inspection of an operator.

**Integrated organic farming system (IOFS):** is an innovative concept of on-farm resource management strategy to achieve economic and sustained agricultural production to meet the diverse requirement of the farm household while preserving the resource base and maintaining high environmental quality.

**Integrated weed management:** the concept of IWM involves the planned use of all possible direct and indirect measures rather than relying on a single method to combat weeds.

**Intercrop yield:** it is a yield of the component crop when grown in intercropping and expressed over the total intercropped area (*i.e.* Area occupied by both the crops).

**Intercropping:** intercropping is the cultivation of two or more crops simultaneously on the same field, with or without row arrangement (row intercropping or mixed cropping).

**Inter-cultivation:** inter cultivation is a process involving operations in the rows of standing crops. Weeding, tiling and cultivating are the examples for the same.

**Internal control system:** a documented quality assurance system that allows the external certification body to delegate the inspection of individual group members to a body identified from within the operators of the group.

**Internal review:** shall mean an assessment of the objectives and performance of a programme by the certification or the accreditation agency itself.

**Irradiation:** high energy emissions from substances for the purpose of controlling microbial, pathogens, parasites and pests in food, preserving the food or inhibiting physiological processes such as sprouting or ripening.

**Irrigated soil:** the soil, which receives irrigation water from well and tube wells, are known as irrigated or chahi soil.

**Irrigation efficiency:** it is a term used to indicate how efficiently the available water supply is being used for crop production.

**Labelling:** means any written, printed or graphic matter that is present on the label accompanies the food, or is displayed near the food, including that for the purpose of promoting its sale or disposal.

**Leaching:** removal of plant nutrients in solution by the passage of water through soil. Among the major nutrients, N is lost in large quantities by leaching.

**Licence:** shall be the accreditation contract that grants a certifier the rights associated with its accredited status in line with the national program for organic production.

**Line:** a cultivar propagated by seed is called a line.

**Livestock and poultry farming:** this category includes farming system in which various kinds of livestock are reared for meat, milk, wool and eggs.

**Livestock system:** a land use unit comprising pastures and herbs and auxiliary feed sources transforming plant biomass into animal products.

**Loamy soil:** the soil is more or less than midway clay and sandy soil and ideal loam soil may defined as a mixture of sand , silt and clay particles which shows the properties of sandy, silt and clayey in equal proportion.

**Localized placement of fertilizers:** this refers to the application of fertilizers into the soil close to the seed or plant. This method is specifically adopted with phosphatic fertilizers.

**Lodging:** the bending or breaking over of a plant before harvesting.

**Main crop/base crop:** this is the crop which is planted at its optimum sole crop population in an intercrop situation and the second crop is planted in between rows of the main or base crop, with a view to obtain some extra intercrop yield without sacrificing the main crop yield.

**Marketing:** means holding for sale or displaying for sale, offering for sale, selling, delivering or placing on the market in any other form.

**Mixed intercropping:** growing two or more crops simultaneously with no distinct row arrangement. Also referred to as mixed cropping.

**Monocropping:** a cropping system in which the populations of crops, other than the main crop, grown in relay, sequential or intercropping with the main crop are greatly reduced.

**Monoculture:** Monoculture is the repetitive growing of the same crop on the same land.

**Mulch:** any material or practice which is used to check the loss of water by evaporation is called mulch.

**Multiple cropping:** the intensification of cropping in time and space dimensions, growing two or more crops on the same field in a year.

**Multiplication:** the growing of seed / stock / plant material to supply for future production.

**National Accreditation Body (NAB):** shall be the agency set up by the steering committee for national programme for organic production for accrediting inspection and certification agencies.

**Natural enemies**: are living organisms that kill pests, decrease pest reproductive potential and compete with pest organisms of plants/crops.

**Natural fibre:** a filament of plant or animal origin.

**No tillage crop (zero tillage crop):** crop grown without any tillage to prepare seed bed or row.

**Nutrient cycle:** the regeneration/cycling of nutrients.

**Nutrient Recovery (NR):** it refers to the actual amount of nutrient taken up from the fertilizers.

**Nutrient recycling:** the term bio-recycling of organic carbon in oceans: "following the actualistic assumption, then, that biological activity is responsible for the source of dissolved organic material in the oceans, but is not important for its activities after death of the organisms and subsequent chemical changes which prevent its bio-recycling, we can see no major difference in the behavior of dissolved organic matter between the pre-biotic and post-biotic oceans."

**Nutrients budget:** a quantitative data of the major nutrients flowing to retained within the discharge from the system.

**Nutrients:** the food for microbial and plant life mainly composed of nitrogen and phosphorous but also of potassium, Mg, Fe, Ca, Co, Cu, Zn and others elements.

**Operator:** shall mean an individual or a business enterprise practicing organic farming or organic processing.

**Organic agriculture:** it is a system of farm design and management to create an eco-system, which can achieve sustainable productivity without the use of artificial external inputs such as chemical fertilizers and pesticides.

**Organic farming**: is a form of agriculture, which excludes the use of synthetic fertilizers and pesticides, plant growth regulators, livestock feed additives and genetically modified organisms. Organic farmers rely on crop rotation, green manure, compost, biological pest control and mechanical cultivation to maintain soil productivity and control pests.

**Organic production unit:** shall mean a unit / holding or stock farm complying with the rules of NPOP regulations.

**Organic seeds and planting material:** seed and planting material produced under certified organic system.

**Organic:** refers to a particular farming system as described in these standards and not to the term used in chemistry.

**Organically-produced feeding stuffs / feed materials:** shall mean feeding stuffs / feed material produced in accordance with the rules of production laid down in NPOP regulations.

**Package of practices:** guidelines for organic production and processing established by the accreditation agencies for specific crops, specific to the region.

**Parallel production:** shall mean any production where the same unit is growing, breeding, handling or processing the same products both in a certified organic and a non-certified organic system. Similarly, a situation with "organic" and "in conversion" production of the same product is also parallel production.

**Parasitoids**: lay their eggs on or in body of an insect host, which then used as a food for developing larvae. The host is generally killed. Most insect parasitoids are wasps, flies and usually have narrow host range.

**Part conversion:** shall be when part of a conventional farm or unit has already been converted to organic production or processing and a part is in the process of conversion.

**Patch cropping:** the component crops are planted in patches.

**Pathogen**: pathogenic micro-organisms include bacteria, fungus and viruses. They kill their host and are relatively host specific.

**Plant protection product:** shall mean any substance intended for preventing, destroying, attracting, repelling, or controlling any pest or disease including unwanted species of plants or animals during the production, storage, transport, distribution and processing of food, agricultural commodities, or animal feeds.

**Plough-sole placement:** with the plough-sole method, the fertilizer is placed in a continuous band on the bottom of a furrow in the process of ploughing. Each band is covered as the next furrow is turned.

**Pore space:** it is a space between soil particles occupied by air and water; it is largely controlled by the texture of soil.

**Potential evapotranspiration (PET):** potential evapotranspiration is the ET that occurs when the ground is completely covered by actively growing vegetation and there is no limitation of soil moisture.

**Potential soil moisture deficit:** it is the difference between a crop potential evapotranspiration and the amount of rainfall received by a crop plus the quantity of water delivered to it in irrigation.

**Pre–basic seed:** it is the progeny of nucleus seed, and is handled so as to maintain specific genetic purity and identity as completely as possible.

**Predators**: predators are mainly free living species that directly consume a large number of prey during their whole life time.

**Preparation:** shall mean the operations of slaughtering, processing, preserving and packaging of agricultural and animal products and also alterations made to the labeling concerning the presentation of the organic production method.

**Pressure potential:** it is produced by actual hydrostatic pressure.

**Processing aids:** a substance or material not consumed as a food ingredient by itself but used in the processing of raw materials, food or its ingredients to fulfil a certain technological purpose during treatment or processing and which may result in unintentional but unavoidable presence of residues or derivatives in the final product.

**Pulses or grain legumes:** the crops belonging to leguminoseae family are grown for their edible seed e.g chick pea, lobia.

**Quality system:** documented procedures, which are established, implemented, and periodically audited to ensure that production, processing, handling, management, certification, accreditation and other systems meet the specified requirements and outcomes by following standardized protocols.

**Raw materials:** all ingredients other than food additives.

**Relay crops:** a relay crops is one which is planted as a second crop after the first crop has reached its reproductive stage of growth but before it is ready for harvest. E.g. Planting of sugar cane in sugar beet.

**Relay intercropping:** growing two or more crops simultaneously during the part of life cycle of each. A second crop is planted after first crop has reached its reproductive stage of growth much before it is ready for harvest. Often referred to as relay cropping.

**Relief:** when rising ground or mountains running at right angle to the prevailing wind.

**Row intercropping:** growing two or more crops simultaneously where one or more crops are planted in a row. Often referred to as intercropping.

**Rumber:** the process in which planker or leveler is used to conserve moisture at watter condition before the preparation of land after rauni.

**Rural compost:** this is prepared by using weeds, crop stubble, cotton stalks, groundnut husk, sugarcane trash, urine soaked earth and litter from cattle shed, waste fodder etc. It contains about 0.5 % N, 0.15 % $P_2O_5$ and 0.5% $K_2O$.

**Sandy soil:** the soil which contains 2.00 to 0.2 mm diameter soil particles, it contain 85% sand and >15% silt and clay. These soils are poor in plant material.

**Sanitize:** to adequately treat the produce or food-contact surfaces by a process that is effective in destroying or substantially reducing the numbers of vegetative cells of micro organisms of public health concern, and other undesirable micro organisms, but without adversely affecting the safety and quality of the product.

**Seed certification:** is the process to secure, maintain and make available high quality seed and propagating materials of superior crop varieties, so grown and distributed as to ensure desirable standards of genetic identity, physical purity and quality attributes.

**Seed:** seed may be defined as a fertilized ovule consisting of intact embryo, stored food and seed coat which is viable and has got a capacity to germinate. This is also known as true seed because in some of the crops, vegetative parts are used for planting and are also called as seeds.

**Seedling**: a young plant that has grown from a seed.

**Seeling:** mechanical weeding process carried out by ploughing with bullocks or tractors.

**Seepage:** losses through the beds and banks of canals and water channels take place mainly by unlined canals.

**Sequential cropping:** growing two or more crops in sequence in the same field in a farming year. The succeeding crop is planted after the preceding crop has been harvested.

**Silage crops:** those crops, which are harvested when still green and succulent and are fed directly to animals without curing, are called silage crops. e.g., Berseem, shaftal, maize.

**Silage:** forage preserved in a succulent condition by partial fermentation in a tight container.

**Soil fertility:** it is defined as the inherent capacity of soil to supply amount of adequate nutrients to plants in balanced form.

**Soil productivity:** it is defined as the capacity of the soil to yield crops per unit area or it can be defined as the total crop production from the field. A soil can be fertile but may not be productive.

**Soil:** a natural body developed from variable mixtures of broken and weathered minerals and decaying organic matter which covers the earth crust in thin layers and supplies proper amount of nutrients and air water and mechanical support to plants.

**Soiling crops:** crops harvested when green and succulent condition is fed to animals shortly after cutting, it is neither dried nor stored for future feeding.

**Sole cropping:** one crop/variety grown alone in pure stands at normal density.

**Split production:** where only part of the farm or processing unit is certified as organic. The remainder of the property can be (a) non-organic, (b) in conversion or (c) organic but not certified.

**Standards:** shall mean the standards for national organic products established by the steering committee for national programme for organic production.

**Surveillance:** the measures undertaken to provide monitoring of an operator's / certification body's compliance with the standards / criteria for meeting the certification / accreditation requirements.

**Synecology:** it deals with the ecology of plant communities. It involves the study of structure, nature, organization and development of plant communities.

**System:** by system we mean a unified whole made of regularly interacting and interdependent components.

**Tillage:** it is usually defined as the mechanical manipulation of the soil aimed at improving its physical condition.

**Tiller:** to put forth shoots other than the mother shoot from roots axis.

**Tilth:** it is the final condition of soil when all the tillage operations are carried.

**Town compost:** this utilizes night soil, street and dustbin refuse. Town compost is rich in nutrients (1.4 % n, 1.0 % $p_2o_5$ and 1.4% $k_2o$) as compared to FYM and rural compost.

**Transaction / import certificate:** document issued by a certification body declaring that the specified lot or consignment of goods is derived from production and / or processing system that has been certified.

**Truck gardening/farming:** it refers to the system in which the bulk of the outputs are produced and marketed. Growing of crops like potato, onion and cabbage on large scale for distinct market.

**Use of GMO and GMOS derivatives:** a plant, animal, microbe or their derivatives that are transformed through genetic engineering.

**Variety:** in general the term variety has been used to refer to a group of similar plant with in a particular species that is distinguished by one or more then one character and given the name.

**Veterinary drug:** means any substance applied or administered to any food-producing animal, such as meat or milk-producing animals, poultry, fish or bees, whether used for therapeutic, prophylactic or diagnostic purposes or for modification of physiological functions or behaviour.

**Watter:** describes the condition when soil moisture level is suitable for cultivation.

**Weed control:** weed control aims at checking the growth or reducing the number of weed to a minimum so that they do not affect crop yields.

**Weeds:** weeds are that species of plant which grow unwanted or are not useful, often prolific and persistent, interfere with agricultural operations, increase labour cost and reduce the crop yields.

# 1

# Organic Farming Introduction

During the last decade organic agriculture has emerged as a dynamic alternate farming system. This has been necessitated as a consequence of resource degradation because of to achieving high yields in various crops to meet the increasing demand of food by the burgeoning population. These have resulted in change in soil structure, imbalance in soil micro-organisms and organic matter, resistance among various weeds to weedicides, increased salinity, sodicity and have affected the water holding capacity of the soil and brought in health and environmental problems. Organic farming is a specified form of diversified agriculture that not only aims to produce nutritious food of high quality but also lead to the generation of more income and employment opportunities for rural population. In true sense, organic farming is a production system which avoids or largely excludes the use of synthetic fertilizers, growth regulators, livestock feed additives and pesticides. Moreover, it relies on agronomic management practices like crop rotation, crop residues, animal manures, legumes, green manures, off-farming organic wastes and aspect of biological pest control to maintain soil productivity.

The inability of Indian agriculture to meet the demand for food in the country during the two and half decades immediately after independence had been a matter of concern at those times. The system of our agriculture based on the traditional knowledge and practices handed down from generation to generation could not produce enough to feed the increasing population. The ignominy of our dependence for food on the western developed nations and the politics of food aid practiced by them added to our determination to be self-sufficient in food production by modernizing agriculture. The green revolution fulfilled our aspirations by changing India from a food importing to a food exporting nation. However, the achievement was at the expense of ecology and environment and to the detriment of the well-being of the people. The agriculture system adopted from the west has started showing increasing unsustainability and once again the need for an appropriate method suitable to our requirements is being felt (Table 1).

In the pre-independence period, Indian agriculture was usually described as a gamble with monsoons. There used to be a great deal of uncertainty about crop prospects, as monsoons played a decisive role in determining agricultural output and their failures resulted in widespread famine and misery. In the last few years, Indian agriculture has made impressive progress and so is more resilient to the vagaries of the monsoon, although the country's population increased from 361 million to more than 130 million. Under the stalward leadership of globally acknowledged scientist Dr M S Swaminathan the green revolution programme was launched in India during sixties, which was based on high yielding fertilizer responsive improved varieties. The Indian farmers adopted these improved technologies and shifted from organic farming to modern agriculture. The green revolution led the country self sufficient in food grain production but later on it was realized that the increase in food grain production was on the cost of emerging second generation problems viz., depletion of natural resources, deterioration of soil health, environmental pollution and health hazards etc, which again realized to rethink about organic and integrated approach. The practice of organic farming, said to the best known alternative to the conventional method, also originated in the west, which suffered from the ill effects of chemical agriculture. However, organic farming is based on the similar principles underlying our traditional agriculture. Organic agriculture aims at the human welfare without any harm to the environment which is the foundation of human life itself.

**Table 1:** Ideological differences between organic agriculture and conventional agriculture.

| Organic Agriculture | Conventional (chemical) Agriculture |
|---|---|
| Holistic approach: Any technology applied considering the system as a whole- No Imbalance | Reductionist approach: Targeted approach for one commodity or one pest or deficiency of nutrient- creates imbalance in system |
| Decentralize production: Most of the inputs e.g. seed, manure, biopesticides etc. produced at farm/village level- suitable to local environment+ generate employment+ low cost of production | Centralize production: Produced in factories/farms, away from the place of use- no proper use of local resources+ least employment+ increase cost of production |
| Harmony with NATURE: Harness the benefit of natural resources, flora and fauna by using or giving favorable environment to them- sustained productivity of natural resources | Domination on NATURE: Agriculture system is forced to produce more- Regenerative capacity of natural resources decreased+ decreased productivity in long term. |
| Diversity: Includes all possible organisms complimentary in a system. Work as mutual service providers for nutrient and pest management. Least cost and time required of system owner. | Specialization: Only one crop or tree or animal. All cost and time of nutrient and pest management has to be borne by system owner/farmer. |
| Input optimization: best use/recycling of available resources. System regenerative capacity and owners economic capacity maintained/enhanced. | Output maximization: Over use of resources disturbs system and resources productivity in long term- Increasing cost. |

| | |
|---|---|
| Knowledge Intensive: Only few resources but need how timely and best integrated. Least dependency on experts/imported technologies, once farmer trained- possible in remotest area. | Input Intensive: Comprehensive list of chemicals with time and method. Needs experts for timely updating. Only possible in resources sufficient areas. |
| Preventive, protective and proactive approach: All the actions/ applications are done in anticipation of system requirement-least use of inputs. | Cause and control approach: Most of the actions/ applications are done to control the damage to system-heavy use of inputs. |
| Decreasing input use: As the system reaching at perfection it conserve/generate its own resources e.g. for nutrition and protection- decreasing requirement of inputs | Increasing input use: Target and action approach that rather, deteriorate systems regenerative capacity- increasing requirement of inputs. |

## Important points about Indian agriculture

- Current Indian population: 139 crores, it will reach to 166 crores 2050 AD.
- We need to produce food grain of 320 and 350 m t by 2025 and 2050 A D, respectively. Current food grain production is 303 m t.
- India has 17% of the world's population and 15 % of livestock but, has only 2.4 %of the geographical area, 4 % of the fresh water, 1 % of rainfall, 0.5 % of forest and 0.5% of grazing land of the world.
- 17 % more water is required by 2025 AD.
- By 2050 AD global population will be 9.4 billion (42 % increase).
- Growth rate of agriculture (1.5 %) is much lower than population growth (2%).
- Net cultivated area in India: 142.8 m ha. It will decline to 137 m ha by 2050 AD.
- 65 % of the population still relies on agriculture for employment and livelihood.
- Agriculture support 58 % of the total working force.
- Agriculture contributes 18 % to GDP.
- 973 million people are undernourished in the world.
- 800 m people living below the poverty line in the world and half of them in South Asia and majority of them in India (FAO).
- 270 m ha area in the world is irrigated area, which consumes 70 % of the world's total human-induced water consumption.
- Before 1970's –ship to mouth stage (PL-480: wheat import during 1960).
- Annual agril. production, which was 0.3 % during 1900 to 1950, jumped to 3.3 % during 1960-70 but the growth became reduced to 1.43 % during 1990-98 compared to the population growth of 18 % during the same period.

- India became the 4th largest fertilizer producing country.
- Fertilizer consumption increased from 0.52 kg/ha during 1950-51 to 94.72 kg/ha during 1999-2000.
- During 2001-02: 18 m t fertilizer was used but the crop removed 28 m t.
- During 2006 total fertilizer consumption from chemical fertilizer was 20.4 m t.
- Country's nutrient requirement will be 45 m t by 2020 AD to produce the food grain required by the population.
- Before 1960 none of the micronutrient (MN) deficiencies were found.
- During 1960: Only Fe was found defiant.
- During 1999: Five MN (Mo, Bo, Mn, Zn and Fe) noticed deficient.
- During 2003: Eight MN's were found deficient.
- Now: Ten MN's are found deficient.
- Karnataka stood first in Zn (78%), Fe (39%) and Bo (32%) deficiencies.
- Combined effect of high yielding var., fertilizers, irrigation and monocropping mined more nutrients from the soil.
- In the beginning of the green revolution a kilo of chemical fertilizer yielded 15-16 kg of grains. The same is now declined to 4-5 kg.
- Nations wealth is measured based on the fertility of its soil.
- According to Indian standards nitrates in soil should not exceed 45 mg/litre (carcinogenic & affects central nervous system, methaemoglobinaemia: blue baby disease).
- Phosphates concentration in the water encourages weed growth (Eutrophication).
- Consumption of pesticide: 1954 =434 tonnes.
- 1984=64,290 tonnes.
- 2018-19= 53453 M. tonnes
- But, pests and diseases multiplied four fold.
- The problem of pests and diseases became ten times more severe compared to 1950, after the synthetic pesticides became in use.
- In 1990, cotton crop occupied only 5% of the area but consumed 54% of the total pesticides. Rice received highest pesticide next only to cotton.
- Use of pesticide was more popular in agriculture with the invention of DDT (Paul Muller who got Noble prize in 1939).

- Each year 25 million people in developing countries suffer from pesticide poisoning.

*"Green Revolution provided food for millions of hungry people but left many unseen miseries of fauna and flora on the land and in water"*

***Eutrophication***: It is the process of enrichment of surface water bodies like lakes, reservoirs and streams with nutrients. Nutrient enrichment of water bodies results in intense proliferation and accumulation of algae and higher aquatic plants in excessive quantities that can result in detrimental changes in water quality and can significantly interfere with the use of water resources.

- About 60 % of our agricultural land currently under cultivation suffers from indiscriminate use of irrigation water and chemical fertilizers.
- The gravity of environmental degradation resulting from faulty agricultural practices has caused alarming concern among the farmers, scientists and conservationists.

## The ills of green revolution on environment

The role of agriculture in economic development in an agrarian country like India is a pre-dominant one. Agriculture provides food for more than 1 billion people and yields raw materials for agro-based industries. Agricultural exports earn foreign exchange. Modernization of Indian agriculture began during the mid-sixties which resulted in the green revolution making the country a foodgrain surplus nation from a deficit one depending on food imports. Modern agriculture is based on the use of high yielding varieties of seeds, chemical fertilizers, irrigation water, pesticides, etc., and also on the adoption of multiple cropping systems with the extension of area under cultivation. But it also put severe pressure on natural resources like, land and water. However, given the continuous growth of modern technology along with the intensive use of natural resources, many of them of non renewable, it is felt that agriculture cannot be sustainable in future because of the adverse changes being caused to the environment and the ecosystem. The environmental non-degradable nature of the agricultural development and its ecological balance have been studied in relation to the modem Indian farming system by experts which shows exploitation of land and water for agriculture.

**Some more ill effects of this modern agriculture on environment are as follow:**

1. Reduction in natural fertility of the soil
2. Destruction of soil structure, aeration and water holding capacity

3. Susceptibility to soil erosion by water and wind
4. Diminishing returns on inputs (the ratio of energy input to output halves every 10 years)
5. Indiscriminate killing useful insects, microorganisms and predators that naturally check excess crop damage by insects pests
6. Breeding more virulent and resistant species of insects
7. Reducing genetic diversity of plant species
8. Pollution with toxic chemicals from the agrochemicals and their production units
9. Endangering the health of the farmers using chemicals and the workers who produce them
10. Poisoning the food with highly toxic pesticide residues
11. Cash crops displacing nutritious food crops
12. Chemicals changing the natural taste of food
13. High inputs increasing the agricultural expenses
14. Increasing the farmer's work burden and tension
15. Depleting the fossil fuel resources
16. Increasing the irrigation needs of the land
17. Big irrigation projects often resulting in soil salinity and poor drainage
18. Depleting the ground water reserves
19. Lowering the drought tolerance of crops
20. Appearance of 'difficult' weeds
21. Heightening the socio-economic disparities and land holding concentration
22. High input subsidies leading to inflationary spiral
23. Increasing the political and bureaucratic corruption
24. Destroying the local culture (commercialization and consumerization displacing self-reliance)
25. Throwing financial institutions into disarray (as impoverished farmers demand write-off of loans)
26. Agricultural and economic problems sparking of social and political turmoil resulting in violence.

**Problems of modern day agriculture**

- High Fertilizer use / Contaminated water
- High Pesticide use/ Contaminated food products
- Soil health / multi nutrient deficiencies
- Heavy metal contamination (Pb, Hg, Ni)
- Loss of biodiversity
- Loss of natural checks and balances (birds, insects, snakes)

**Organic farming**: Organic farming is a type of agriculture that benefits from the recycling and use of natural products. Extensive use of dried foliage and kitchen compost not only reduces the investment overheads, but also ensures the growth of crops that are devoid of synthetic interference. The technique is characterized by the use of green manure, biological pest control methods and special cultivation techniques to maintain soil productivity. Limiting the use of synthetic fertilizers or completely doing away with them reduces the risk of exposure to ailments that arise on account of synthetic pesticides, ingrained plant growth regulators and the presence of genetically modified organisms in organic food products. Today, organic farming caters to a major and preferred industry, worldwide. Organically grown food products have a huge market catered to, by farmlands covering approximately 10% of the total world-farmland cover. The initial effort made by Sir Albert Howard, the Father of Organic Farming, has paid off, metamorphosing into a practice that sustains soil health and ecosystems, by relying on biodiversity, ecological processes and the progress of innovative health sciences.

The organic farming is integration of biological, cultural and natural inputs including integrated disease and pest management practices.

**Definition**, Organic farming is a method of farming system, which primarily aims at cultivating the land and raising crops in such a way, so as to keep the soil alive and in good health.

It is the use of organic wastes (crop, animal and farm wastes, aquatic wastes) and other biological materials, mostly produced in situ along with beneficial microbes (biofertilizers) to release nutrients to crops, which connotes the 'organic' nature of organic farming. It is also termed as organic agriculture. In the Indian context it is also termed as 'Javik Krishi'.

**As per FAO's definition** "Organic agriculture is a unique production management system which promotes and enhances agro-ecosystem health, including biodiversity, biological cycles and soil biological activity, and this is accomplished by using on-farm agronomic, biological and mechanical methods in exclusion of all synthetic off-farm inputs".

Organic agriculture is a production system which avoids or largely excludes the use of synthetic compounded fertilizers, pesticides, growth regulators and livestock additives. To the maximum extent feasible organic farming system relies on crop rotations, crop residues, animal manures, off-farm organic wastes and aspect of biological pest control to maintain soil productivity and tilth, to supply plant nutrients and to control insects, weeds and other pests (Lampkin, 1990).

The USDA, implemented the organic food production Act of 1990, describes organic production as one which will respond to site-specific conditions by integrating cultural, biological and mechanical practices that foster cycling of resources to promote ecological balance and conserve biodiversity. Organic agriculture promotes linkages and connects between land and water, plants and people.

The FAO suggested organic agriculture is a unique production management system which promotes and enhances agro-ecosystem health including bio-diversity, biological cycles and soil biological activity and this is accomplished by using on-farm agronomic, biological and mechanical methods in exclusion of all synthetic off-farm inputs.

Organic agriculture is a system of farm design and management to create an ecosystem which can achieve sustainable productivity without the use of artificial external inputs such as chemical fertilizers and pesticides (NPOP).

## Organic farming help in

- Surviving harsh environmental conditions.
- Meeting family's food and nutritional requirements.
- Meeting the local/ regional market requirement.
- Replacing the external inputs with on-farm inputs.
- Replacing the cost of agricultural practices.
- Optimizing production and improving soil and water quality and the ecology that support agricultural.
- Socio -economic improvement and better food security.
- The economics of organic farming is characterized by increasing profits via reduced water use, nutrient-contamination by pesticides, reduced soil erosion and carbon emissions and increased biodiversity.

- Organic farming produces the same crop variants as those produced via conventional farming methods, but incurs 50% lower expenditure on fertilizer and energy, and retains 40% more topsoil.
- This type of farming effectively addresses soil management. Even damaged soil, subject to erosion and salinity, are able to feed on micro-nutrients via crop rotation, inter-cropping techniques and the extensive use of green manure.
- The use of green pesticides such as neem, compost tea and spinosad is environmentally friendly and non-toxic. These pesticides help in identifying and removing diseased and dying plants in time and subsequently, increasing crop defense systems.

## Area and Status

- World area: 57.8 m ha (2/3 is grasslands) in 186 countries
- Global market in 2016: 89.7 billion USD
- Maximum area under organic farming: Australia
- Highest no of organic producers : India
- Area under organic farming in India: 3.56 mha
- Area under organic cultivation in India: 1.78 mha (In 2003-04: 42000 ha)
- Organic production: 1.7 mt
- Madhya Pradesh > Himachal Pradesh > Rajasthan.

## Concept

Organic farming is the farming/agricultural system which aims at cultivation of the land in a way so that the soil kept the dynamic activity and in good health. At the same time, keeping the environment clean, maintain the ecological balance and providing stability to the production level without polluting the soil, water and air. It is concerned with the best use of natural resources and maintains their balance. In this system, soil is not only considered to be physical medium for supporting crop but also as a shelter of innumerable living entities. Infect, it is only mean to achieve the sustainable target of production. It is feasible by adopting multi-dimensional approach which involves co-ordination and interaction between several major and minor factors. The constituent of organic farming can be integrated plant nutrient supply system, integrated insect-pest and disease management.

- Nature is the best role model for farming
- 'Feed the soil, not the plant'

- The emphasis- improving the general health of soil with minimal use of off-farm inputs.
- Does not believe in mining of the soil of its nutrients and degrading it for present needs.
- Treat Soil-as a living entity
- Soil microbes and other organisms- an important place in sustaining the soil fertility and must be protected and nurtured.
- Waksman, 'Without soil microbes, life on planet would come to standstill'.
- The ultimate aim- to develop economically viable, socially acceptable and ecologically safer agricultural systems by mutual coexistence with the ecosystem rather than dominating it.
- Thus organic farming is not just a chemical free farming; it also covers the sustainability, environmental, economic and social aspects.

## Scope of organic food

In recent years, organic farming experienced tremendous growth in many countries. Organic agriculture is currently practiced in more than 160 countries. It is estimated that worldwide about 37.2 million hectares are organically managed. The share of land area under organic management is highest in European countries. Thus, success of organic agriculture in these countries is highest due to consumer awareness of health and environmental issues.

Organic agriculture is not a new concept in India. At the beginning of the 19$^{th}$ century Sir Albert Howard one of the most important pioneers of organic farming worked in India for many years studied the soil-plant interaction and development of composting methods. In doing so, he capitalized substantially on India's highly sophisticated traditional agricultural system which had long applied many of principles of organic farming.

## Prospects of organic farming in India

Organic farming denotes a holistic system of farming which optimizes productivity in a sustainable manner through creation of interdependent agri-eco systems where annual crop plants (e.g. wheat), perennial trees (e.g. horticulture) and animals (including fishes where relevant) are integrated on a given field or property. It is often confused with certified organic farming or labeled organic produce, which is a commercial quality control and marketing mechanism entailing third party certification or any other form of independent certification for individual commodities and the process of their cultivation. It is important to separate the two.

The first-organic farming is essentially an agricultural activity employing the knowledge/understanding of naturally occurring processes while the second-a certified organic farming is essentially the same as organic farming but there is an addition of checks and balances to maintain soil fertility and control insect-pests and diseases. It is done by involving record keeping on various aspects and is dictated by market forces.

About 74% farmers in India are small and marginal farmers. Organic farming is most relevant to them because they are resource poor to provide costly inputs for enhancing yield. Their only resource viz. land, need to be prevented from degradations. In the organic farming system approach, a piece of land is used optimally and to its fullest potential to produce a range of nutritious and healthy food as well as other required commodities in a manner which can healthily feed a small family, and maintain soil health and productivity by agricultural practices based on principles of nature. Pests (both insects and diseases) are also controlled and managed by the selection of crop mixes and using biological control measures.

The organic farming is not a technology by itself. But it consists of set of practices evolved around several principles based on widely character or resources in different ecosystems. The totality of each eco-system and mechanism of delicate balance available between components of each ecosystem would ultimately decide which principle would weigh more or less and ultimately which practice should be predominant in organic farming. In this background, the assessment of scope of organic farming should adopt widely varying techniques for different ecosystems and for different principles of organic farming. Further, the assessment should also include the harmony achieved between new set of organic farming practice and components of ecosystem. The scope of organic farming cannot be assessed on a uniform scale at global level. The criteria of assessment and in turn the extent of scope of organic farming vary according to soil, climate, cropping system, vegetation, irrigation, allied activities (animal husbandry, sheep rearing, poultry, piggery, fisheries etc.) as well as harmony achieved between different components of each ecosystem. Many examples can be cited to justify this.

1. Scope for preparing farm yard manure by a farmer with large number of livestock is certainly brighter than that of a farmer maintaining a few cattle.

2. An irrigated farmer is able to easily adopt green manuring practice or the practice of returning the crop residues for enrichment of soil, as the availability of moisture can facilities the decomposition but a dryland farmer is not able to do so easily.

3. Temperate climate does not favour faster decompositions of organic residues as compared to tropical climate due to cold temperatures prevailing in temperate climate for most part of the year.
4. A farm nearer urban agglomerations has advantage of obtaining decomposed compost or poultry at cheaper rates compared to interior rural areas.
5. Regions of better rainfall distribution support two crops and result in sufficient biomass for incorporation. As compared to this, poorer distribution of rainfall may offer little scope for biomass incorporation.

Scope of organic farming exists in every situation with varying degree. Within the meaning of broad concepts and varying principles of organic farming, it can be adopted for every crop and in every situation. Each farming situation offers unique scope of adopting some principles of organic farming and evolving a set of practices that may be specific to each situation.

Total area under cultivation : 142 million ha 65% area under rain fed cultivation

- Fertilizer use very low due to erratic rainfall.
- Farming systems- highly diversified with crops, trees, animals
- No much effect on national food security.
- Labour intensive nature of organic farming will help in optimizing the
- Human resource utilization-usually remain under utilized.
- Spice crops of these areas like fennel, fenugreek are in demand in organic food market.
- Almost organic by default and conversion to organic is not difficult.
- Thus, these areas should be priority areas for promotion of organic farming.

The need for organic farming in India arises from the unsustainability of agriculture production and the damage caused to ecology through the conventional farming practices. The present system of agriculture which we call 'conventional' and practiced the world over evolved in the western nations as a product of their socio-economic environment which promoted an over riding quest for accumulation of wealth. This method of farming adopted by other countries is inherently self destructive and unsustainable. Intensive agriculture with use of agro-chemicals in huge amount has resulted in manifold increase in the productivity od farm commodities but the adverse affects of those chemicals are clearly visible on soil structure, soil micro-flora, quality of water, food and fodder material. Pesticides and nitrates from fertilizers have been detected in ground water in many agricultural regions. High concentration of nitrates in drinking water affects human health, particularly infants and can prove fatal in some cases. Traces of banned DDT and BHC isomers have been found in soil, cereals, vegetable oil, bovine and human milk, butter fat, fish, meat, egg and drinking water. Thus organic farming is an answer for making available safe food and clear environment.

## Organic farming in Punjab

- As a state policy is neither desirable nor possible.
- The total organic sources are not adequate to meet the nutrient demand of the current area under crops to exploit the potential yields.
- The estimated total organic sources (FYM) are hardly sufficient for 15 % of wheat and 20% of rice area.
- On an average, there are 2 animals per hectare and to apply full dose of recommended nutrients (N) to wheat and rice it would require about 17 animals per hectare.

## Relevance of organic farming

The relevance and need for an eco-friendly alternative farming system arose from the ill effects of the chemical farming practices adopted worldwide during the second half of the last century. The methods of farming evolved and adopted by our forefathers for centuries were less injurious to the environment. People began to think of various alternative farming systems based on the protection of environment which in turn would increase the welfare of the humankind by various ways like clean and healthy foods, an ecology which is condusive to the survival of all the living and non-living things, low use of the non-renewable energy sources, etc. Many systems of farming came out of the efforts of many experts and laymen. However, organic farming is considered to be the best among all of them because of its scientific approach and wider acceptance all over the world.

## Aims of organic food production

- To produce sufficient quantities of high quality of food, fibre and other products.
- To work compatibility with natural cycle and living systems through the soil, plant and animal in entire production system.
- To maintain and increase long term feasibility and biological activity of soil using locally adopted cultural, biological and mechanical methods as opposed to reliance on inputs.

- To maintain and encourage agricultural and natural biodiversity on the farm and surrounding through the use of production systems and the protection of plant and world life habitat.
- To maintain and conserve genetic biodiversity through alteration to on-farm management of genetic resource.
- To promote the responsible use and conservation of water and all life therein.
- To use, as far as possible, renewable resources in production and processing systems and avoid pollution and waste.
- To foster local and regional production and distribution.
- To create harmonious balance between crop production and animal husbandry.
- To provide living condition that allow animal to express the basic aspect of their innate behavior.
- To utilize biodegradable, recyclable and recycled packaging material.
- To support the establishment of an entire production, processing and distribution chain which is both socially and ecologically responsible.

## Principles of organic agriculture

- Guiding principles for producing quality food in an environment friendly and sustained manner.
- These are the roots from which organic farming grows and develops.
- They express the contribution that organic farming can make to the world.

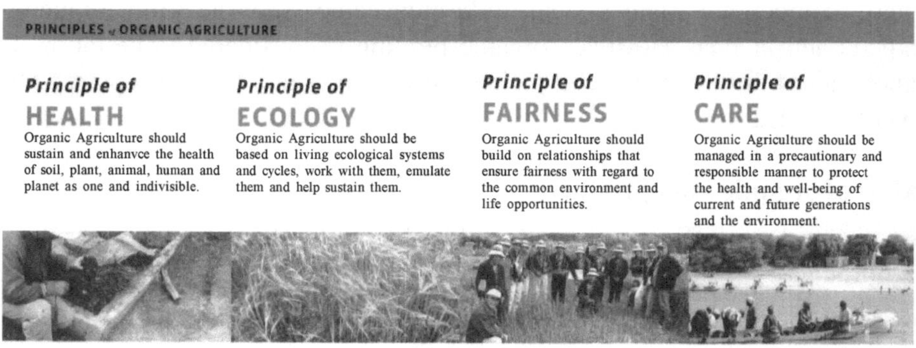


| | | |
|---|---|---|
| Healthy Soil | Agro-ecology | Ecological & Social |
| Healthy Crops | Diversity & Recycling | Precaution |
| Healthy Livestock & | | Justice & Fare Trade? |
| Healthy People | | |

### The principle of health

- Should sustain and enhance the health of soil, plants, animals, humans and planet as one and indivisible.
- Health of individuals and communities cannot be separated from the health of ecosystems.
- Only a healthy soil can produce healthy crops that further contribute to the health of animals and human beings.

  The health encompasses the physical, mental, social and ecological health.
- Should sustain and enhance the health of ecosystems in the entire chain from production to the ultimate consumer and health of organisms from the smallest in the soil to the human beings.
- The use of fertilizers, pesticides, animal drugs and food additives are prohibited in organic farming in the perspective of this principle.

The principle of **health-**Organic Agriculture should sustain and enhances the health of soil, plant, animal, human and planet as one and indivisible. This principle points out that the health of individuals and communities cannot be separated from the health of ecosystems-healthy soils produce healthy crops that foster the health of animals and people. Health is the wholeness and integrity of living systems. It is not simply the absence of illness, but the maintenance of physical, mental, social and ecological well-being. Immunity, resilience and regeneration are key characteristics of health. The role of organic agriculture, whether in farming, processing, distribution, or consumption, is to sustain and enhance the health of ecosystems and organisms from the smallest in the soil to human beings. In particular, organic agriculture is intended to produce high quality, nutritious food that contributes to preventive health care and well-being. In view of this it should avoid the use of fertilizers, pesticides, animal drugs and food additives that may have adverse health effects.

### The principle of ecology

- Organic farming should be based on living ecological systems and cycles, work with them, emulate them and help sustain them.
- Production should be based on ecological processes and recycling.
- Should have harmony with the natural ecosystem and should not try to dominate the natural system.
- It must be adapted to local conditions, ecology, culture and scale.

- Input use should be reduced by reuse, recycling and efficient management of materials and energy in order to maintain and improve the environment quality and conserve resources.
- Ecosystem should be maintained through appropriate system of farm design, establishment of habitats and maintenance of biodiversity.

The principle of ecology-Organic Agriculture should be based on living ecological systems and cycles, work with them, emulate them and help sustain them. This principle roots organic agriculture within living ecological systems. It states that production is to be based on ecological processes, and recycling. Nourishment and well-being are achieved through the ecology of the specific production environment. For example, in the case of crops this is the living soil; for animals it is the farm ecosystem; for fish and marine organisms, the aquatic environment. Organic farming, pastoral and wild harvest systems should fit the cycles and ecological balances in nature. These cycles are universal but their operation is site-specific. Organic management must be adapted to local conditions, ecology, culture and scale. Inputs should be reduced by reuse, recycling and efficient management of materials and energy in order to maintain and improve environmental quality and conserve resources. Organic agriculture should attain ecological balance through the design of farming systems, establishment of habitats and maintenance of genetic and agricultural diversity. Those who produce, process, trade, or consume organic products should protect and benefit the common environment including landscapes, climate, habitats, biodiversity, air and water.

### The principle of fairness

- Should be based on relationships that ensure fairness with regard to the common environment and life opportunities.
- It indicates equity, respect and justice both among people and their relations to other living beings.
- All the stakeholders in organic farming should conduct human relationships in a manner that ensure fairness at all levels and to all the involved in organic farming.
- Natural and environmental resources that are to be used for production and consumption should be managed in a way that is socially and ecologically just and should be held in trust for future generations.
- Animals should be provided with living conditions so as to express their natural behaviour.

The principle of fairness-Organic Agriculture should build on relationships that ensure fairness with regard to the common environment and life opportunities.

Fairness is characterized by equity, respect, justice and stewardship of the shared world, both among people and in their relations to other living beings. This principle emphasizes that those involved in organic agriculture should conduct human relationships in a manner that ensures fairness at all levels and to all parties - farmers, workers, processors, distributors, traders and consumers. Organic agriculture should provide everyone involved with a good quality of life, and contribute to food sovereignty and reduction of poverty. It aims to produce a sufficient supply of good quality food and other products. This principle insists that animals should be provided with the conditions and opportunities of life that accord with their physiology, natural behavior and wellbeing. Natural and environmental resources that are used for production and consumption should be managed in a way that is socially and ecologically just and should be held in trust for future generations. Fairness requires systems of production, distribution and trade that are open and equitable and account for real environmental and social costs.

**The principle of care**

- Should be managed in a precautionary and responsible manner to protect the health and well being of current and future generations and their environment.
- Increased productivity should not be at the cost of human and environmental health.
- Precaution and responsibility are the key concerns in management, development and technology choices.
- Scientific interventions combined with practical experience, accumulated wisdom and traditional and indigenous knowledge should offer valid solutions.
- This principle restricts the adoption of unpredictable technologies such as genetic engineering.

The principle of care - Organic Agriculture should be managed in a precautionary and responsible manner to protect the health and well being of current and future generations and the environment. Organic agriculture is a living and dynamic system that responds to internal and external demands and conditions. Practitioners of organic agriculture can enhance efficiency and increase productivity, but this should not be at the risk of jeopardizing health and well-being. Consequently, new technologies need to be assessed and existing methods reviewed. Given the incomplete understanding of ecosystems and agriculture, care must be taken. This principle states that precaution and responsibility are the key concerns in management, development and technology choices in organic agriculture. Science is necessary to ensure that organic agriculture is healthy, safe and ecologically sound. However, scientific knowledge alone is not sufficient. Practical experience, accumulated wisdom and traditional and indigenous knowledge offer valid solutions, tested by time. Organic agriculture should prevent significant risks by adopting appropriate technologies and rejecting unpredictable ones, such as genetic engineering. Decisions should reflect the values and needs of all who might be affected, through transparent and participatory processes.

## Characteristics of organic farming

Organic farming systems, in the light of principles of organic farming, should:

- Encourage and enhance the biological cycles within the farming system.
- Enhance and maintain the long term soil fertility.
- Mobilize organic matter and nutrient elements locally within closed systems.
- Use, as far as possible, renewable resources in locally organized agricultural systems.
- Avoid all forms of pollution.
- Maintain biodiversity at the regional and local level.
- Allow the livestock to express their innate behaviour.
- Produce qualitative food in sufficient quantity which is acceptable socially and economically.

## Benefits of organic farming

i) Conservation of natural resources: Non-traditional crops avoided, Balanced supply of macro and micronutrients, improved soil health, Better retention of soil moisture and nutrients.

ii) Healthy environment: Reduced pollution, Reduces the $CO_2$ emission by eliminating synthetic fertilizers and storage in the soil, Burning of crop residues restricted.

iii) Reduced health hazards: Prevents pesticide residues in environment and human food chain. The prohibition of use of highly soluble fertilizers- nitrate contamination.

iv) Employment generation.

v) Sustainable agricultural production: Based on the concept of feeding the soil and not the crop plants.

vi) Reduced cost of product: Inputs are on-farm

vii) Higher profitability: Premium/ reduced costs

Organic manure produces optimal conditions in the soil for high yields and good quality crops. They supply all nutrients required by crops (N,P,K, Secondary and micro nutrients). They improve plant growth and physiological activities of plants. Organic farming helps to prevent environmental degradation and can be used to regenerate degraded areas. Organic farming helps to avoid chain reaction in the environment from chemical sprays and dusts. There is increasing consumer demand for agricultural products which are free from toxic chemical residues. In developed countries consumers are willing to pay more for organic food. Organically grown plants are more resistant to diseases and insects and hence only a few chemical sprays are required. They reduce need for purchased inputs. They improve the soil chemical properties such as supply and retention of soil nutrients and promote favourable chemical reactions.

### Constraints

Land resources can move freely from organic farming to conventional farming, they don't move in reverse direction. In changing over to organic farming an initial crop loss generally occur during transitional period. Moreover, the biological control have been weakened or destroyed by chemicals. Vermi-compost or non-edible cakes are costly over chemical fertilizers. Very few bio-insecticides are available and the efficiency of bio-pesticides is low as compared to synthetic pesticides.

- Threat to national food security: 276 million tonnes of foodgrains by 2020.
- Low organic carbon soils: Desired organic carbon > 1.2%
- Lack of awareness: among the consumers and the farmers.
- Lack of market for organic products
- Complex and costly certification
- Limited availability of organic manures
- Competitive use of crop residues
- High input costs
- Lack of quality biofertilizers and biopesticides
- Lack of subsidy on organic inputs
- Lack of organic premium
- Labour and cost intensive
- Lower crop yields
- Few myco-herbicides are available that too very costly.

## Methods of organic farming systems

- Bio-dynamic Farming
- Natural Farming
- Rishi Krishi
- Homa Farming

## Biodynamic farming

- Greek words: *Bios*- life and *Dynamics*- energies.
- Working with the energies which create and maintain life.
- Developed by - Rudolf Steiner in Germany
- It is based on the principle that natural phenomena like cosmic and terrestrial forces play an important role in crop production. Agricultural operations are performed and biodynamic preparations are used according to the positions of the earth, the sun and the moon. Biodynamic calendar is followed for planning the agricultural operations.
- BD preparations: BD 500 - 508
- Derived from a variety of herbs, silica and cow manure.

Biodynamic agriculture is a method of farming that aims to treat the farm as a living system which interacts the environment, to build healthy, living soil and to produce food that nourishes and vitalizes and helps to develop mankind. The underlying principle of biodynamics is making life-giving compost out of dead material. The methods are derived from the teachings of Rudolf Stainer and subsequent practitioners.

## The important components of biodynamic farming are as follows

a) Turning in plant materials such as green crops and straw

b) Not using chemical fertilizers and pesticides

c) Avoiding soil compaction by machinery or animals, particularly in wet weather

d) Keeping soil covered by pasture, crops or mulch not destroying the soil structure by poor farming practices such as excessive use of rotary hoe or cultivation in unsuitable weather (too wet or too dry)

e) Fallowing the land by planting deep-rooting permanent pasture species or using green crops

f) Use of preparations BD-500 and BD-501

g) Compost made with preparations BD-502 – BD-507

h) Liquid manure made with preparations BD-502-BD-507

i) Cowpat pit manure made with preparations BD-502-BD-507

These biodynamic preparations named BD-500 to BD-507 are not food for the plants, but they facilitate the effective functioning of etheric forces. They are also not the usual compost starters, but can stimulate compost organisms in various ways. In short they are biologically active dynamic preparations which help in harvesting the potential of astral and etheral powers for the benefit of the soil and various biological cycles in the soil.

So, far 9 biodynamic preparations have been developed, named as formulation 500 to 508. Out of these, formulation- 500 (cow horn compost) and formulation- 501 (horn-silica) are very popular and are being used by large number of organic farmers. Formulations-502 to 507 are compost enrichers and promoters, while formulation 508 is of prophylactic in nature and helps in control of fungal diseases.

### *Rishi Krishi*

*Rishi krishi* system- based on the *Vedas*

Generally practiced in Maharashtra and Madhya Pradesh

On-farm sources of nutrition are used for soil enrichment through *Rishi krishi* formulation known as- *amritpani* and soil collected from under a banyan tree (*Ficus bengalensis*).

- *Amritpani- ghee* (butter) + cow dung fb mixing of honey and diluting the mixture with water.
- Seed treatment (*beej sanskar*), enrichment of soil (*bhumi sanskar*) and foliar spray on plants (*padap sanskar*).

Drawn from Vedas, the Rishi Krishi method of natural farming has been mastered by farmers of Maharashtra and Madhya Pradesh. In this method, all on-farm sources of nutrients including composts, cattle dung manure, green leaf manure and crop biomass for mulching are exploited to their best potential with continuous soil enrichment through the use of Rishi Krishi formulation known as "*Amritpani*" and virgin soil. 15 kg of virgin rhizosperic soil collected from beneath of Banyan tree (*Ficus bengalensis*) is spread over one acre and the soil is enriched with 200 litres Amritpani. It is prepared by mixing 250 g ghee into 10 kg of cow dung followed by 500 g honey and diluted with 200 litres of water. This formulation is utilized for seed treatment (*beej sanskar*), enrichment of soil (*bhumi sanskar*) and foliar spray on plants (*padap sanskar*). For soil treatment it need to be applied through irrigation water as fertigation. The system has been demonstrated on a wide range of crops i.e. fruits, vegetables, cereals, pulses, oilseeds, sugarcane and cotton.

### ***HOMA Farming***

- *Homa* farming- origin from the *Vedas*
- Based on the principle that, 'you heal the atmosphere and the healed atmosphere will heal you'.
- The basic aspect of *homa* farming is *Agnihotra puja* with chanting of Sanskrit *mantras* at specific times.
- The ash that results from the *puja* is used to energize the composts, plants, animals etc.

Homa farming has its origin from Vedas and is based on the principle that "you heal the atmosphere and the healed atmosphere will heal you" The practitioners and propagators of homa farming call it a "revealed science". It is an entirely spiritual practice that dates from the Vedic period. The basic aspect of homa farming is the chanting of Sanskrit mantras (*Agnihotra puja*) at specific times in the day before a holy fire. The timing is extremely important. While there is no specific agricultural practice associated with homa farming, the farm and household it is practiced in, is energised and "awakened". The ash that results from the *puja* is used to energise composts, plants, animals, etc. Homa Organic

Farming is holistic healing for agriculture and can be used in conjunction with any good organic farming system. It is obviously extremely inexpensive and simple to undertake but requires discipline and regularity.

*Agnihotra* is the basic Homa fire technique, based on the bio-rhythm of sunrise and sunset, and can be found in the ancient sciences of the Vedas. Agnihotra has been simplified and adapted to modern times, so anybody can perform it. During *Agnihotra*, dried cow dung, ghee (clarified butter) and brown rice are burned in an inverted, pyramid- shaped copper vessel, along with which a special mantra (word-tone combination) is sung. It is widely believed that through burning organic substances in a pyramid-formed copper vessel, valuable purifying and harmonizing energies arise. These are directed into the atmosphere and are also contained in the remaining ash. This highly energized ash can successfully be used as organic fertilizer in organic farming.

Besides the practice of *Agnihotra* and the frugal distribution of *Homa* ash on beds and fields, a variety of further applications have also been recommended. Here are some examples:

***Impregnation of Seeds and Bulbs-*** Before planting/sowing, seeds and bulbs are treated i.e., impregnated with a mixture of *Agnihotra* ash and cow urine. It is recommended to prepare a mixture of cow urine and water in a ratio of 50:50, to which up to 4 tablespoons of *Agnihotra* ash per 5 liters of solution are added and stirred. Seeds and bulbs should soak in this solution for 30-40 minutes. This strengthens the germinating plant and makes it more resistant to pests. Like cow dung, cow urine has antibacterial effects and provides a protective coating around the seeds and bulbs. After this time of treatment, seeds are spread on filter paper, or other absorbent paper, to dry. They should be dry enough to spread, but moist enough so that the core of the seed doesn't dry out. Through the impregnation, germination is started-which would be ended if the seeds completely dried out. Bulbs may be planted immediately after being treated with the solution.

***Fertilizers-*** In addition, plants can be fertilized with a mixture of *Agnihotra* ash, stinging nettles, and water. This special liquid fertilizer strengthens plants. The stinging nettles are fermented *i.e.* decomposed in the water for 7-14 days, depending on weather conditions and the amount of nettles needed. This mixture should then be diluted to a solution with a ratio of 1:9. In other words, 1 part stinging nettle solution is mixed with 9 parts water and filtered with a fine screen (sieve) into a spraying container or watering can.

***Plant Nutrient Solution -*** To make an *Agnihotra* plant nutrient solution, up to 4 tablespoons of *Agnihotra* ash and up to 4 tablespoons of pulverized, dried cow dung are stirred in approximately 5 liters of water and then applied to plants. This may be repeated every 14 days, depending on how much it is needed.

***Spray Solution*** **-** A nutrient solution to be sprayed can be made by mixing up to 4 tablespoons of *Agnihotra* ash with 5 liters of water. This spray solution is left standing for 3 days and then filtered through a fine screen before it is used to protect plants against pests and diseases. A spray solution can also be made from certain fern blossoms, in which approximately 150 grams of the blossoms, mixed with 2 liters of water and 2 tablespoons of *Agnihotra* ash, are left standing to ferment for 7-10 days. Filtered through a narrow-meshed screen and then finally distributed on the plants with a sprayer, this helps keep away pests such as snails.

### Gloria Biosol an effective homa biofertilizer

Gloria Biosol is a very effective bio-fertilizer which can be produced simply in Homa atmosphere. Biosol liquid can be used for foliar application to nourish plants and soil. Biosol is superior to vermiwash as it contains high numbers of beneficial microorganisms and energy of homa process. *Agnihotra* Ash has a significant positive effect on all the materials used and makes the Biosol rich in macronutrients.

### Materials required to make the Biosol are

- Fresh cow dung
- Vermicompost
- Cow urine
- Agnihotra ash
- Water

Materials are mixed in a large tank (200, 500 or 1000 litre). One copper Shree Yantra disc is placed in the tank. The tank is then sealed and kept for 20 to 30 days. After digestion is complete, the slurry can be removed. Biosol is used diluted with Agnihotra ash water solution in the ratio of 1:10. For one hectare of agricultural area, 200 liters of Biosol in solution are required. Biosol in solution can be sprayed on any type of crop at an interval of fifteen days. The application of the Biosol solution should be made before sunrise or after sunset. If we preserve Biosol liquid in air tight cans it will last longer, say about six months. Left over solid Biosol which is having maximum macro nutrients should be mixed with any type of organic manure at a ratio of 1:5.

## Natural Farming

- Developed by Masanobu Fukuoka who wrote 'The One Straw Revolution'.
- Based on the understanding of nature and stress is laid on no-till farming.
- Pruning of fruit trees is prohibited.
- All synthesized chemicals, untreated animal manures and urban sewage are prohibited.
- Composts fermented using plant organic materials are used to increase soil fertility and improve the soil physical properties.

## Natural farming

Natural farming emphasizes on efficient use of on-farm biological resources and enrichment of soil with the use of Jivamruta to ensure high soil biological activity. Use of Bijamruta for seed/ planting material treatment and Jivamruta for soil treatment and foliar spray are important components. The use of both these ingredients have been incorporated in the package described above.

Jivamruta has been found to be rich in various beneficial microorganisms. As per the studies conducted by Bio Centre Bangalore the Jivamruta contains following microorganisms:

- Azospirillum $2 \times 10^6$
- PSM $2 \times 10^6$
- Pseudomonas $2 \times 10^2$
- Trichoderma $2 \times 10^6$
- Yeasts and moulds $2 \times 10^7$

200 litres of jivamruta is needed for one application in one acre. It can be applied through irrigation water by flow, by drip or sprinkler or even by drenching of mulches spread over the field or under the tree basin.

## Natueco farming

**The Natueco farming** follows the principles of eco-system networking of nature in the farming system. It is different from organic or natural farming both in philosophy and practice. It offers an alternative to the commercial, heavily inorganic modern techniques of farming. It emphasizes harvesting the sun through a critical application of scientific inquiries and methods that are rooted in the neighborhood resources. It depends on developing a thorough understanding of plant physiology, plant geometry of growth, plant fertility and plant biochemistry.

Natueco Farming emphasizes 'Neighborhood Resource Enrichment' by 'Additive Regeneration' rather than through dependence on external, commercial inputs. The four relevant aspects of Natueco Farming are as below:

**Soil:** Enrichment of soil by recycling the biomass and by establishing a proper energy chain

***Roots:*** Development and maintenance of white root zones for efficient absorption of nutrients

***Canopy:*** Harvesting the sun through proper canopy management for efficient photosynthesis

**External Resources**: Focus on minimizing the use of external resources including water.

The fundamental belief system in Natueco concept is that it is a science of life and life is all about Energy.

### No work method

This method is very similar to the Natural Farming method. Ruth Stout published a book in 1971 called The Ruth Stout No-Work Garden Book" (Rodale Press, Emmaus, Penna, 1971). The emphasis is on lots of organic matter for mulch to control weeds and provide nutrients. Worms and other organisms are allowed to do the work of getting nutrients to the roots.

### Permaculture

Permaculture is a holistic method of farming popularized by Bill Mollison and others in 1980's. This concept has originated in Australia or New Zealand. Permaculture goes even farther to evaluate and take into account the entire landscape and physical features of a growing area for farming & getting maximum potential of crops. Plant and animal raising are considered together to make the best use of the land. Sustainability of practices and design and care for the earth are key components of Permaculture.

### Veganics

Some people have chosen to use the term *veganics* as opposed to *organics*. The main difference is basing nutrient management on only plant derived nutrients and organic matter as opposed to making use of manures and other animal products.

All these are not widely used term but do point out that there are different degrees of organic farming/gardening and what people consider important.

### Background

- Agriculture: 7500 BC (Organic Farming)
- 1900- Albert Howard- Father of modern organic farming-'An Agricultural Testament'
- 1922- Rudolph Steiner-biodynamic agriculture - 'Demeter'
- 1940- The term 'organic' in relation to farming- Northbourne Lord- 'Look to the Land'.
- 1962- Rachel Carson-'Silent Spring'-ill effects of pesticides on the environment.
- 1972- IFOAM- International Federation of Organic Agriculture Movements-France.
- 1975- Masanobu Fukoka- One Straw Revolution- Natural Farming
- 1991- EU Regulation
- 1999- Codex Guidelines
- 2000- NPOP- National Programme on Organic Production
- 2004- NSOP- National Standards for Organic Production

### Production technology

- **Sir Albert Howard**-Father of Modern Organic Agriculture

I regard Indian Farmers as my professors. I learnt from them how to grow healthy crops without the slightest help from artificial manures or insecticides.

### Crops & varieties in organic farming

When organic seed and plant material is available, they shall be used. When certified organic seed and plant material are not available, chemically untreated, conventional seed and planting materials shall be used; that are not chemically treated. The use of genetically engineered seeds, pollen, transgenic plants material is not allowed.

- Crops- Locally adapted
- Intercropping: Legume
- Crop rotation: Legume
- Varieties - Bt

### Choice of crops and varieties

1. All seeds and plant material should be certified organic.
2. Species and varieties cultivated should, as far as possible, be adapted to the soil and climatic conditions and be resistant to pests and diseases.

3. A moratorium on all transgenic food and agricultural applications should be imposed.
4. In the choice of varieties genetic diversity should be taken into consideration.

### Crop rotations

Rotations should be as varied as possible and aim to:

1. Maintain soil fertility
2. Reduce nitrate leaching
3. Reduce weed, pest and disease problems.

It is recommended that specific rotations including legumes are instead upon by the certification programmes.

## Organic weed management

### Weeds

Weeds are the plants growing at a place where they are not required or are harmful to the interest of man. Under field conditions these are the undesirable plants that compete with the crop plants for light, nutrients, space and moisture.

The management of weeds in organic farming is aimed at minimizing the weed competition with the crop.

It is achieved by integrating all the method of weed management excluding the herbicides.

### Weed management options

- Herbicides
- Manual/Mechanical control
- Cultural control
- Allelopathy (Sorghab)
- Biological control

### Preventive methods

- These are the methods and practices that check the spread of weeds. This can be achieved by careful management of agricultural machinery, using weed free seed and proper irrigation methods.
- Sanitation and Composting

- Manual method
- Mechanical method
- Allelochemicals
- Eucalyptus leaves/rice straw/Sorghum.

## Cultural methods

i) **Selection of crop cultivars:** quick growing and early canopy cover providing cultivar

ii) **Crop rotation**: breaking association of crop and weeds. Wild oat can be completely eliminated by growing rice and *Phalaris minor* by rotating wheat with other *rabi* crops such as berseem, potato, raya or gobhi sarson etc.

iii) **Date of sowing:** unfavourable for germination of weed seeds but crop performance is not adversely affected. *Phalaris minor* infestation is reduced in October sown wheat.

iv) **Planting pattern:** minimum space should be left for the weeds

**Seed rate/crop density:** to attain more number of crop plants per unit area in order to boost initial smothering potential on weeds. High seed rate in wheat/higher plant population in rice.

vi) **Method of sowing:** bed planted and transplanted crops.

vii) **Stale seed bed/Dab system**

viii) **Water management**: submergence in transplanted rice/ Drip irrigation in wider row crops/ Alternate furrow system.

ix) **Method of ploughing:** Summer ploughings- *Cyperus rotundus*/Puddling.

x) **Mulching**

xi) **Intercropping**: Intercropping of rapid growing, early maturing and good canopy structured crops like cowpea or green gram.

## Bioherbicides

Bioherbicides are microbial plant pathogens which are applied as sprays to kill or supress the weeds.

De-Vine (*Phytophthora palmivora*) -1$^{st}$ registered bioherbicide in 1981 for control of milkweed in Florida citrus. 2$^{nd}$ commercially developed micoherbicide- Collego (cga) in USA for control of Northern joinvetch, a problem BLW in rice and soybean.

### Advantages

- High degree of specificity
- No effect on non target organisms
- No problem of weed resistance
- No residues
- Long lasting
- Environmental friendly

### Disadvantages

- Lengthy registration procedure
- Slow in action
- In-consistence results
- Production of bioherbicides is expensive

## Organic farming: Standards & Certification

### Organic standards

These are the minimum requirements that must be met to get the organic status.

The Demeter certification trademark set up in Germany in 1928 was the 1$^{st}$ production standard to be set up in the world.

The first international standards for organic products were published by IFOAM in 1980 and are known as the mother of standards. India formulated NSOP in 2004.

### National Standards for Organic Production

NSOP are grouped under six categories

i) Conversion
ii) Crop production
iii) Animal husbandry
iv) Food processing and handling
v) Labeling
vi) Storage and transport

### Conversion

The minimum conversion period for plant products produced annually is 24 months prior to the sowing of crop. For perennial plants the conversion period is 36 months prior to harvesting. This period can be extended or reduced depending upon the previous land use but it can not be less than 12 months.

- Transition Period: 2 year before sowing
- Inputs- Prohibited/Restricted/Permitted
- Seedbed -
- Varieties - Bt not allowed
- Seed - Organic
- Seed Treatment - Fungicides
- Nutrient supply - Fertilizers (not allowed), FYM, Human excreta (not allowed)
- Weed Management - Weedicides (not allowed)
- Insect - Pest/Disease Management: Pesticides (not allowed)
- Irrigation: Sewage water (not allowed)
- * One legume crop in rotation

**Organic animal husbandry standards**

i) Natural behaviour

ii) Free movement

iii) Natural day light

iv) Poultry and rabbits are not to be kept in cages

v) Herd animals are not to be kept individually

vi) Conversion period is 12 months

vii) All animals are to be raised on organic farm and if not available can be brought in: 2 days old chicks for meat, 18 week old hen for eggs, Calves upto 4 week

viii) AI is allowed

ix) Embryo transfer techniques not allowed

x) Feed should be organic

xi) Medicines- homeopathic, ayurvedic, unani

xii) If conventional medicine used - 2 year withholding period.

xiii) Legally required vaccinations -allowed

**Pest and disease control during post harvest management**

i) General cleanliness and hygiene.

ii) The practices that can be used for managing the pests are - physical barriers, sound, ultra-sound, light, UV-light, traps, temperature control, humidity control, mechanical, physical and biological methods and pesticidal substances recommended for organic farming.

iii) There should never be direct or indirect contact between organic products and the prohibited substances.

iv) Irradiation is prohibited.

### Processing

i) No co-mingling with non-organic products.

ii) There should not be any contamination of the produce.

iii) The processing methods should not cause pollution of any form.

iv) Processing methods should be based on mechanical, physical and biological processes like smoking, extraction, precipitation, filtration etc.

v) Extraction should be done only with water, ethanol, plant and animal oils, vinegar, carbon dioxide, nitrogen or carboxylic acids and should be of food grade quality.

vi) Irradiation is not allowed.

### Packaging

i) Biodegradable and eco-friendly materials should be used for the packaging.

ii) The materials that affect the organic nature of the contents should be avoided.

iii) Use of PVC materials is prohibited.

iv) Laminates and aluminium should be avoided.

v) Reusable systems should be used wherever possible.

### Storage & transport

i) Bulk stores for organic products should be separate from conventional product stores and clearly labeled.

ii) Organic and non-organic products should not be stored and transported together except when labelled or physically separated.

iii) The stores and containers should be decontaminated, cleaned or disinfected as per organic standards.

iv) Cooling, Freezing, Drying and Humidity regulation are allowed.

v) Fumigation (methyl bromide) is not allowed.

## Organic certification

- It is a procedure to verify the compliance of organic standards.
- It is a third party guarantee to the consumer about the genuineness of the organic product.

## Certification agencies

| | |
|---|---|
| • BVQI | Mumbai (022-66956300) |
| • Ecocert | Aurangabad (0240-2377120) |
| • IMO Control Pvt. Ltd. | Bangalore (080-25285883) |
| • INDOCERT | Cochin (0484-2630908) |
| • AOCA | Bangalore (080-65369888) |
| • LACON | Thiruvalla (Kerala) (0469-2606447) |
| • NOCA | Pune (020-25457869) |
| • OneCert | Jaipur (0141-2701882) |
| • ROCA | Jaipur (0141-2227104) |
| • SGS | Gurgaon (0124-2399990) |
| • Control Union Certifications | Mumbai (022-67255396) |
| • USOCA | Dehradun (0135-2760861) |
| • Aditi Organic Certifications Pvt. Ltd. | Bangalore (080-32537879) |
| • Food Cert India Pvt. Ltd. | Hyderabad (040-23301618) |
| • ISCOP | Coimbatore (0422-2544199) |
| • Vedic Organic Certification Agency | Hyderabad (040-65276784) |
| • Chhattisgarh Certification Agency (CGCERT) | Chhattisgarh (0771-2283249) |
| • Tamil Nadu Organic Certification Deptt | Coimbatore (TN) (0422-2405080) |

## Substances permitted, restricted and prohibited for use in soil fertilization

### Permitted

Farmyard manure produced on organic unit, Crop residues and green manure, Straw and other mulches, vermicompost, Biodynamic preparations, oilseed cake meal, biofertilizers, botanical extracts, rock phosphate, Magnesium sulphate, Calcium chloride, Calcium sulphate (gypsum), Sodium chloride

**Restricted**

Composted mixture of vegetable matter, Farmyard manure (produced by conventional means), Blood meal, meat meal, bone meal, Seaweeds and seaweed products, Basic slag, Potassium sulphate, Elemental sulphur.

**Prohibited**

Dried farmyard manure and dehydrated poultry manure, Human excrement.

**Products permitted, restricted and prohibited for plant pest and disease control**

**Permitted**

Plant based extracts (neem, garlic, pongamia etc.), Sodium bicarbonate, Clay, Bacterial preparations (*Bacillus* species etc.), Viral preparations (granulosis viruses, nuclear polyhydrosis viruses etc.), Fungal preparations (*Trichoderma* species etc.), Parasites, predators and sterilized insects, Homeopathic and Ayurvedic preparations, Herbal and Biodynamic preparations, Use of predator insects, Physical methods (chromatic traps, mechanical traps, light traps, sticky traps and pheromones), Physical barriers, sound, mulches, nets.

**Restricted**

Natural acids (e.g. vinegar), Natural plant preparations (excluding tobacco), Copper salts/inorganic salts (bordeaux mixture, copper hydroxide, copper oxychloride), Copper sulphate, Sulphur, Lime sulphur, Hydrated lime, Mineral powders (stone meal, silicates), Mineral oils, Potassium permanganate, Silicates (sodium silicate, quartz), Sterilised male insects

**Prohibited**

Mineral powders (stone meal, silicates), Tobacco tea, Ethyl alcohol

**Certification status:** any written, printed or graphic matter that is present on the label, accompanies the food or is displayed near the food, including that for the purpose of promoting its sale or disposal.

**Certification status**

| | |
|---|---|
| 1st year produce | Nil |
| 2nd year produce | In conversion to organic |
| 3rd year produce | Certified organic |

### Labelling guidelines

No claim about

- Healthier
- Tastier
- Nutritious

### Participatory Guarantee System (PGS)

- It is a locally focused quality assurance system in which people in similar situations assess, inspect and verify the production practices of each other and collectively declare the entire holding of the group as organic.

### Advantages of PGS

- Only basic documentation.
- Better surveillance by group members.
- Very low cost of certification.
- Every farmer gets individual certificate and each farmer is free to market its own produce independent of group.

### Limitations of PGS

The certification is valid for domestic market only.

### Certification status

- Farmers which have one or more major non-compliance or are under conversion period will be declared- 'PGS-INDIA Green'
- After the full compliance and conversion period they will be declared - 'PGS-INDIA Organic'
    - APEDA (Agril. and processed food products export development authority)
    - Coffee Board
    - Spices Board
    - Tea Board
    - Coconut Development Board
    - Directorate of cashew and cocoa development.

## Problems of organic farming

- Package of organic farming technology has to be location specific.
- Bulky nature of organic manure leading to higher transport and application cost.
- Small farm holdings reduce the scope for in-situ organic manure production.
- Lack of awareness among farmers about scientific procedures of compost making and application.
- Lack of scientific data on the long – term benefits and limitations of organic farming.
- Limited domestic market for organically grown horticultural produce.

During this period, the size of farm holdings and the per capita availability of agricultural land have also been decreasing and they are expected to be around 1.4 and 0.14 hectares respectively, by the turn of this century. With competing demands on land for other sectors of development, this decline is likely to aggravate further.

India is now one of the 39 countries where organic farming regulations are fully implemented. The total area under organic farming in the country is around 37050 ha covering about 5347 organic farms as against the 24 m ha area under organic farming worldwide. Internationally, maximum organic food retail sales takes place in Europe and USA estimated at US $ 10000-11000 million during 2002. Organic farming is receiving acceptance in all the states but particularly states like Uttranchal, Chhattisgarh, Nagaland, Meghalaya, Sikkim, Rajasthan and Madhya Pradesh are the real area where it is being adopted very fast because these states already using negligible quantity of fertilizers as well as pesticides. To compete in the international market India has a great scope on account of diverse agro-climatic conditions available with respect to agricultural commodities like fruits, vegetables, herbs, aromatic plants, plantation crops like tea, coffee, beside cereals like rice, wheat and cotton.

The northern parts of the country have made great strides to boost total foodgrains production of the country (297.4 m tonnes). But still the foodgrains production is not enough to match with burgeoning population. A over- view demonstrate that India has compulsive need to raise food production by 5 m tonnes every year and this was the reason that modern science based agriculture do not encourage organic farming. Moreover, in organic farming the crop yield may come down during the transitional period. But to enable our agriculture to be competitive in the global market, putting 5-6 per cent area will hardly impact the total production. In addition, when the organic resources (livestock and crop residues) are visualized then a ray of hope emerges as India has around 193 m cattle and 79 m buffaloes (Paroda 1999) and 493 m tonnes of crop residues. Out of this Punjab alone produce residue of wheat, rice and cotton of the order of 19, 18 and 1 m tonnes, respectively (Beri 2003). In addition urban compost, vegetable and fruit wastes etc. are also available. Total nutrient requirement (NPKS) from chemical fertilizer is about 422.85 LMT. To substitute the same from the organic sources which are available would suffice to meet the entire requirement of organic agriculture. But there is a need to streamline the things and create awareness among the peoples involved in agriculture.

At present, concerns are voiced both in respect of environment and health quality. Eutrophication of water bodies due to high nitrate and phosphate concentration, increasing levels of nitrates in drinking water sources, accumulation of heavy metals such as lead, chromium and cadmium in soil and water resources are the principal causes of environmental deterioration in agriculture. Dhaliwal *et al* (2000) reported the high concentration of $NO_3^-N$ (mg/litre) at different locations in Punjab, Haryana and UP which was much higher than the safe limit of 45 mg $NO_3$/litre (Table 2). Likewise, pesticide contamination of food and feed in Punjab were studied by Dhaliwal *et al.* (2000) and sizeable amount of contamination was found in grains, vegetables, feed and milk (Table 3).

**Table 2:** Nitrate Content of Well Water.

| District | State | $NO_3$-N (mg/l) |
|---|---|---|
| Aligarh | Uttar Pradesh | 61 |
| Meerut | Uttar Pradesh | 157 |
| Mahendergarh | Haryana | 296 |
| Ambala | Haryana | 223 |
| Bathinda | Punjab | 128 |
| Sangrur | Punjab | 98 |

*Source :* Dhaliwal *et al* (2000)

**Table 3:** Pesticide contamination

| Commodity | Number of Samples | |
|---|---|---|
| | Analysed | Contaminated |
| Cereals | 30 | 30 |
| Rice grain | 99 | 99 |
| Vegetable | 96 | 64 |
| Animal feed | 15 | 15 |
| Animal feed + fodder | 105 | 105 |

Nitrate in drinking water is the main reason of blue baby syndrome. Herbicide residues affect the central nervous system, respiratory system and gastro-intestinal system of human beings. These chemical residues also cause depression, insomnia, oral automatism, myoclous and hyper-reflexia of man (Ghosh 2000).

Such harmful effects are now being focused at various forums on account of alarming health hazard situations. To negate such harmful effects organic farming is one of the important options for adoption.

## Benefits of organic farming

- Conservation of natural resources and avoid pollution hazards
- Appropriate use of crop and livestock residues.
- Encourage export potential.
- Create employment opportunities.
- Restore original ecosystem (production, consumption and decomposition)
- Reduce the cost of production
- Recycling of biomass as a renewable energy source.

## Current status

The organic foods particularly in the northern part are produced on a limited scale at few farms. The main reasons identified discouraging organic farming are as under:

- Limited certification avenues
- Lack of awareness
- Marketing of organic produce
- Large volume of organic manure is required
- Fermented compost, vermicompost, green manuring-labour and cost intensive.
- Divergent nutritional value of each organic sources
- Lack of fixed price premium
- No fixed price premium and committed procurement
- Non-availability of bio-pesticides in number and packing size

## Policies to support organic farming

Yield plateau both in irrigated and rain fed areas of country have promoted agricultural scientists and policy markers to re-evaluate to current agricultural practices and investigate alternative protection methods. Presently, up to 30% area is default organic in India. About 10% of total agriculture land in India can be brought under organic production method. There are niche area is terms of potential medicinal herbs, crops, fruit and vegetable which can up-scaled for organic farming through technological backstopping, mass awareness, cooperative management and group certification.

Limited availability and high prices could be considered a barrier related to the immature character of organic markets in India. Thus, supply-rather than the demand- is the larger issue today, even though, on a quick note, one also hears complaints of non-availability of markets. Strategies to achieve this include supporting the conversion to organic production through reduced cost of certification and organic input production, arranging supplies from different regions of the country in order to encourage the pooling of produce from small farmers into larger consignments and providing vital institutional and technical support to this organic production in niche and potential areas of the country. Moreover, knowledge and ecological awareness are particularly important not only for organic farmers but for consumers, processors, teenagers and children as well.

**Research/knowledge gaps**

- Only crop based componential technologies have been developed and issues of strong/weak sustainability of organic farming is to be addressed.
- Nutrient weak farming systems are prevalent (Resource cycling is weak).
- Eco-management practices as a base of cropping or farming systems are missing.
- Bio-intensive technologies at micro level don't exist.
- Organic livestock knowledge is completely lacking.
- Market led integrated organic farming systems or trade-off based economic integrated organic farming systems is still an issue at micro-level.
- On-line data bank of organic growers, their soil resources, input & output flow and real time changes in beneficial microflora & fauna need to be established.
- Standardization of local inputs (manures, biopesticides, biodynamic manures, liquid bio- sprays) with quality parameters and operational guidelines.
- Nutrient management schedule for mitigating nutrient deficiency in standing crops.
- Research on standardization of organic ITKS
- Establishment of scientific model organic farms

**Future challenges**

Organic farming aims for human welfare without harming the environment and follows the principles of health, ecology, fairness and care for all including soil. The modern concept of organic farming combines the tradition, innovation and science. Although, history states that the movement for organic way of life recognized in 1905, it could gain ground after realizing the ill effects of modern agriculture in the late 1990's. In 1905, the British botanist Sir Albert Howard, often referred to as the father of modern organic agriculture, documented traditional Indian farming practices, and came to regard them as superior to conventional agriculture science. During 1940, In Japan, Masanobu Fukuoka, a microbiologist working in soil science and plant pathology, quit job as a research scientist, returned to his family's farm, and devoted the next 30 years to develop a radical no-till organic method for growing grain, now known as "Fukuoka farming". Many other practices such as Rishi krishi, Natueco farming, homa farming, panchagavya krishi and bio dynamic farming are associated with organic agriculture. The reports indicates organic farming can minimize energy consumption by 30.7% per unit of land by eliminating the energy required to manufacture synthetic fertilizers and pesticides and by using internal farm inputs, thus reducing fuel used for transportation. India can emerge as global leader due to the presence of large number of organic producers (almost 6.5 lakh producers) and they needs to be supported with technical knowledge and inputs besides marketing infrastructure. The research results available for little over a decade confirms the yield advantage in many crops such as basmati rice, maize, cotton, chickpea, soybean, groundnut etc. However, the major impediment for growth of organic farming in India is yield reduction in the initial years due to swift switch over from inorganic to organic, wide gap between availability of organic source of nutrients and requirement and lack of pest and disease management options. Most of the organic growers have expressed that lack of support price for organically grown crops and marketing infrastructure as the major constraint in promotion of organic agriculture. Although, much progress on research in organic farming has been done, the new emerging areas of human health benefits, understanding the economics with environmental markets, climate friendly farms and carbon farming with organic farming system models needs to be addressed in future. The certification systems of grower group, participatory guarantee system, know your farm and know your food should be promoted in large scale.

## Status of Organic Farming in India

- The task force of GOI under the chairmanship of Shri Kunwarji Bhai suggested need for alternative to modern conventional agriculture.
- Ministry of Commerce, GOI, has launched national programme for organic production in March, 2000
- National Standards for organic products (NSOP) have been standardized during May, 2001 and all the products sold under the logo "*India Organic*".

- National Accreditation Policy Programme (NAPP) has been formulated with the accreditation regulations announced in May, 2001.
- This made it mandatory that all certification bodies engaged in inspection and certification of organic crops and products should be accredited by an accreditation agency.
- Accreditation agency: APEDA, Coffee Board, Tea Board, Spice Board etc.,
- National Institute of Organic Farming (NIOF) established at Ghaziabad. The purpose of this institute is to formulate rules, regulations and certification of organic farm products in conformity with International Standards.

The organic food production costs are higher in the developed countries as organic farming is labour intensive and labour is costly in these countries. However, in country like India, where labouris abundant and is cheap. OF is seen as a good cost effective solution to the increasing costs involved in chemical farming. Currently most of the organic farmers in India are still in the transition phase and hence, their costs are high. As these farmers continue with OF, the production costs are expected to reduce, making India as one of the most important producers of organic food.

## Organic manures

- Organic manures such as FYM, biogas slurry, compost, straw or other crop residues, biofertilizers, green manures and cover crops can substitute for inorganic fertilizers to maintain the environmental quality.
- Organic farmers may also use the see weeds, fish manures and some permitted fertilizers like basic slag and rock phosphate.

## Non-Chemical weed control measures

- Mechanical weed control, especially in row crops.
- No herbicides are used. Cultural, biological and physical methods are resorted to.
- Botanical herbicides, mycoherbicides and bioagents are employed to control weeds.

## Biological pest management

- Non-chemical, biological pest management is encouraged.

- Natural enemies of pests are encouraged. Botanical pesticides such as those derived from neem could be used.
- Microbial pesticides like Bacillus thuringiensis is used.

## Nutrient management in organic farming

- The first task in organic farming is to protect the soil fertility and health. Use of organic manures, crop rotation, use of crop residues, green manures, intercropping with legumes, use of biofertilizers etc., are resorted to.

## Limiting nutrient losses

- Better recycling of wastes.
- Handling of organic wastes.
- Application of organic matter at right time, method and quantity.
- Reducing run off by following conservation practices.
- Conservation of organic matter by decreasing burning of crop residues.
- In wet land, deceasing denitrification losses of nutrients.
- Nutrient release and time of uptake must be synchronized.
- Cropping pattern.
- Pumping of nutrient by hedge row planting.
- Minimizing the exports of nutrients from the farm.

## Organic agriculture suits for mitigation of climate change

- A study commissioned by IFOAM discusses the potential of Organic Agriculture both to avoid and to sequester Greenhouse Gases (GHG), and makes comparisons with conventional agriculture (Kotschi & Müller-Sämann, 2004)
- Organic agriculture offers options to reduce GHG significantly. One is to reduce emissions and, thereby, to minimise the production of atmospheric $CO_2$, $CH_4$ and $N_2O$. The second option consists in systematically sequestering carbon dioxide in soils and in plant biomass. It is unique for all types of land use. Although sinks in vegetation and soils have a high potential to mitigate increases of $CO_2$ in the atmosphere, they are not sufficient to compensate for heavy inputs from fossil fuel burning. The long-term solution lies in a reduction of the use of fossil fuel (developing alternatives to fossil fuel, reduce energy consumption etc.). A number of inbuilt practices and functional system mechanism in organic agriculture makes it suitable to mitigate and adapt to climate change. The important processed of

mechanisms which make organic agriculture suitable to mitigate effect of climate change in agriculture are as follows:

- ***Eco-functional intensification:*** Organic agriculture uses fewer external inputs, establishes closed nutrient cycles and fosters the optimal use of natural resources, ecosystem services and biological functions, such as predator prey dynamics (IFOAM EU Group, 2010). The practice of encouraging such processes with targeted interventions is known as eco-functional intensification (TP Organics, 2010). In this way, organic farming responds perfectly to the need for risk-reducing strategies, as it lowers input costs, fosters diversity and invests in healthy organisms
- ***No use of fertilisers and pesticides:*** Organic agriculture uses no inorganic fertilisers and pesticides. Therefore, organic agriculture is looked as one of the solutions for climate change mitigation because it emits much lower levels of greenhouse gases (GHG), and also effectively sequesters carbon in the soil (Panwar et al., 2010; IFOAM, 2009).
- ***Promotes climate resiliency through water and weather adjustment capacity:*** In addition to this organic agriculture also makes farms and people more resilient to climate change, mainly due to its water efficiency, resilience to extreme weather events and lower risk of complete crop failure.
- ***Lower the emission of greenhouse gases:*** It is believed that shifting from conventional crop production systems to organic crop production systems would significantly lower the emission of greenhouse gases because organic production systems produce smaller amount GHG emissions than conventional industrial farming systems (Meredith, 2008).
- ***Reduction of carbon dioxide emissions:*** As a viable alternative for intensive agricultural systems, it uses significantly less fossil fuel in comparison to conventional agriculture and hence reduces $CO_2$ emissions. This is mainly due to the following factors, (i) Soil fertility is maintained mainly through farm internal inputs (organic manures, legume production, wide crop rotations etc.), (ii) Energy-demanding synthetic fertilizers and plant protection agents are rejected, and, (iii) External animal feeds-often with thousands of transportation miles-are limited to a low level. As a consequence, the organic variants have in most cases a more favourable energy and carbon balance.
- ***Reduction of methane emission:*** In avoiding methane, Organic Agriculture has an important though not always superior impact on reduction. Through the promotion of aerobic microorganisms and high biological activity in soils, the oxidation of methane can be increased. Secondly, changes in ruminant diet can reduce methane production considerably. However, technology research on methane reduction in paddy fields – an important source of methane production - is still in its infancy.

- ***Reduction of $NO_2$ emission:*** Nitrous oxides are mainly due to overdoses and losses on nitrogen. These are effectively minimized in Organic Agriculture because:
    - i) No synthetic nitrogen fertilizer is used, which clearly limits the total nitrogen amount and reduces emissions caused during the energy demanding process of fertilizer synthesis.
    - ii) Agricultural production in tight nutrient cycles aims to minimize losses.
    - iii) Animal stocking rates are limited.
    - iv) Dairy diets are lower in protein and higher in fibre, resulting in lower emission values.
    - v) Using biomass as a substitute for fossil fuel represents another emission reduction option.
    - vi) It has the advantage that inorganic N fertilizers are not applied, which cause significant emissions of $N_2O$ and use a lot of energy.
- ***Contribute to carbon sequestration:*** Global adoption of organic agriculture has the potential to sequester up to the equivalent of 32% of all current man-made GHG emissions (Jordan et al., 2009). FAO also stated that organic systems contribute less to GHG emissions due to use of lower energy inputs and sequesters more carbon in the biomass than conventional systems (Ziesemer, 2007).
- Soils are the major sink for atmospheric $CO_2$. Organic farming increases organic carbon through organic manures, crop cover and crop rotation and restores it for the longer duration. It is reported that soil carbon sequestration rates on arable land can range from 200 kg to 2000 kg of carbon per hectare per year above 'business as usual' conventional agriculture depending on the organic agriculture soil management practice (IFOAM, 2009).
- ***Low input and low energy use minimises risks:*** Eyhorn et al. (2007) reported that organic farming is a low-risk farming strategy with reduced input costs, therefore, lower risks with partial or total crop failure due to extreme weather events or changed conditions in the wake of climate change and variability. Aher et al. (2012) reported that organic yields match with the conventional yields and organic farming uses 45% less energy and is more efficient than the conventional farming systems.

## Misconceptions about organic farming

Can we sustain the food security to ever increasing population through organic farming?

- Is it possible to meet the nutrient requirements of the crops entirely through organic sources?
- Are there any significant benefits of organic farming in terms of food quality?
- Is organic farming economically viable?
- Is it possible to manage weeds, insect-pests and diseases in organic farming to sustain yield levels?
- Complex and costly procedure of identification.
- The risk involved in marketing of organic produce at premium rates are not available in domestic markets.

**Lack of active interest among farming community**

There is considerable latent interest among farmers in conversion to organic farming because of the following reasons:

- Perceived high costs of doing so, which is mainly due to incomplete knowledge about principles and practices of organic agriculture among farmers. Farmers often seek for off farm inputs, leading to escalation in production costs, which is against the basic philosophy of organic agriculture. Moreover, very high government subsidies on chemical fertilizers in conventional agriculture are not taken into account while comparing the economic costs.
- Non availability of adequate quantities of organic manures and other organic inputs in the local market from reliable sources. Farmers are, more often than not, forced to recycle major quantities of crop residues as animal fodder and animal dung as source of household energy. Sizable quantities of crop residues are also sold off to paper and cardboard industry to earn cash for household needs.
- Complete knowledge about organic farming principles, practices and advantages accrued to grower as well as mankind, is not filtered down to the small farmers, which should be actual target and potential beneficiary of organic farming. Further, whatsoever information reaches to target groups, is very often not backed by scientifically proven results in tropical/ sub-tropical climatic conditions and different farming situations prevailing in Indian sub-continent (Gill and Prasad, 2009).

# 2

# History, Initiatives Taken for Promotion of Organic Agriculture

## The Evolution of Organic Agriculture

The story of organic or sustainable agriculture over the last 100+ years and as we know it today is perhaps more a story of observation of the natural system and taking the time to look back more carefully at the previous several thousand years of agriculture. It is also hard to say if it is a story of the last 100+ years or the last 10,000 years. It is pretty clear that the up until the last 150 years, the only type of agriculture was organic agriculture as it currently is defined. So in some sense, the experiment in conventional agriculture, which has been in practice perhaps 1% of the time people have had some form of cultivating crops.

In 1880, a fellow by the name of Charles Darwin was watching worms and making some observations about how the organic matter ingested was broken down and might provide nutrients to plants. Often noted as a keen observer, it seems reasonable that anyone looking at worms might wonder what else was in the soil. It was about this time that the development of microbiology and knowledge of microscopic organisms like bacteria were being observed by humans for the first time. Could it be that the soil was alive and that there were other factors at play? In the years that followed, others looked at the soil as being alive and asking questions about the importance of organic matter.

In 1911, *Farmers of Forty Centuries*, was published by F.H. King. He told of his observations in Asia where the same land had been farmed for as long as 4000 years. *The Living Soil* was published in 1927 by Lady Eve Balfour and she reported on what else was going on in the soil. In 1940, *An Agricultural Testament* was by Sir Albert Howard. Howard reported on his work in India with composting and the effect of compost on soils and food quality. And in 1945, *Pay Dirt*, by J.I. Rodale hit the American scene. These are just some examples of those who saw that the soil is alive.

The early part of the 1900's also saw great strides in the understanding of soil chemistry and processes influencing nutrient availability. There was already evidence of a split in thinking about how to best provide the necessary nutrients. The value of legumes was well recognized and at least one university researcher (Hopkins, 1911) asked the question of why farmers would pay for nitrogen when there was many times more than ample nitrogen sitting in the air above their field that could be fixed with legume crops.

Through the 60's and 70's the alternative view to conventional agriculture evolved into what we have know for over 25 years as *sustainable agriculture*. The concepts of sustainable agriculture are based on developing an agriculture system that will protect soil quality and health for the long term.

## History of Organic Farming Pre-World War II

The first 40 years of the 20th century saw simultaneous advances in biochemistry and engineering that rapidly and profoundly changed farming. The introduction of the gasoline-powered internal combustion engine ushered in the era of the tractor and made possible hundreds of mechanized farm implements. Research in plant breeding led to the commercialization of hybrid seed. And a new manufacturing process made nitrogen fertilizer-first synthesized in the mid-19th century-affordably abundant. These factors changed the labor equation. Consciously organic agriculture (as opposed to the agriculture of indigenous cultures, which always employs only organic means) began more or less simultaneously in Central Europe and India. The British botanist **Sir Albert Howard** is often referred to as the **father of modern organic agriculture**. From 1905 to 1924, he worked as an agricultural adviser in Pusa, Bengal, where he documented traditional Indian farming practices and came to regard them as superior to his conventional agriculture science. His research and further development of these methods is recorded in his writings, notably, his 1940 book, *An Agricultural Testament*, which influenced many scientists and farmers of the day.

In Germany, Rudolf Steiner's development, biodynamic agriculture, was probably the first comprehensive organic farming system. This began with a lecture series Steiner presented at a farm in Koberwitz (now in Poland) in 1924. Steiner emphasized the farmer's role in guiding and balancing the interaction of the animals, plants and soil. Healthy animals depended upon healthy plants (for their food), healthy plants upon healthy soil, healthy soil upon healthy animals (for the manure).

In 1909, American agronomist F.H. King toured China, Korea, and Japan, studying traditional fertilization, tillage, and general farming practices. He published his findings in *Farmers of Forty Centuries* (1911, Courier Dover Publication).

King foresaw a "world movement for the introduction of new and improved methods" of agriculture and in later years his book became an important organic reference.

**The term *organic farming* was coined by Lord Northbourne** in his book *Look to the Land* (written in 1939, published 1940). From his conception of "the farm as organism," he described a holistic, ecologically balanced approach to farming.

In 1939, influenced by Sir Albert Howard's work, Lady Eve Balfour launched the Haughley Experiment on farmland in England. It was the first scientific, side-by-side comparison of organic and conventional farming. Four years later, she published *The Living Soil*, based on the initial findings of the Haughley Experiment. Widely read, it led to the formation of a key international organic advocacy group, the Soil Association.

In Japan, Masanobu Fukuoka, a microbiologist working in soil science and plant pathology, began to doubt the modern agricultural movement. In 1937, he quit his job as a research scientist, returned to his family's farm in 1938, and devoted the next 60 years to developing a radical no-till organic method for growing grain and many other crops, now known as Nature Farming (Natural Farming), 'do–nothing' farming or Fukuoka farming.

## Post-World War II

Technological advances during World War II accelerated post-war innovation in all aspects of agriculture, resulting in large advances in mechanization (including large-scale irrigation), fertilization, and pesticides. In particular, two chemicals that had been produced in quantity for warfare, were repurposed for peace-time agricultural uses. Ammonium nitrate, used in munitions, became an abundantly cheap source of nitrogen. And a range of new pesticides appeared: DDT, which had been used to control disease-carrying insects around troops, became a general insecticide, launching the era of widespread pesticide use.

At the same time, increasingly powerful and sophisticated farm machinery allowed a single farmer to work larger areas of land and fields grew bigger.

In 1944, an international campaign called the Green Revolution was launched in Mexico with private funding from the US. It encouraged the development of hybrid plants, chemical controls, large-scale irrigation, and heavy mechanization in agriculture around the world.

During the 1950s, sustainable agriculture was a topic of scientific interest, but research tended to concentrate on developing the new chemical approaches. In the US, J.I. Rodale began to popularize the term and methods of organic growing, particularly to consumers through promotion of organic gardening.

In 1962, Rachel Carson, a prominent scientist and naturalist, published *Silent Spring*, chronicling the effects of DDT and other pesticides on the environment. A bestseller in many countries, including the US, and widely read around the world, *Silent Spring* is widely considered as being a key factor in the US government's 1972 banning of DDT. The book and its author are often credited with launching the worldwide environmental movement.

In the 1970s, global movements concerned with pollution and the environment increased their focus on organic farming. As the distinction between organic and conventional food became clearer, one goal of the organic movement was to encourage consumption of locally grown food, which was promoted through slogans like "Know Your Farmer, Know Your Food".

In 1972, the International Federation of Organic Agriculture Movements (IFOAM) was founded in Versailles, France and dedicated to the diffusion and exchange of information on the principles and practices of organic agriculture of all schools and across national and linguistic boundaries.

In 1975, Fukuoka released his first book, *The One-Straw Revolution*, with a strong impact in certain areas of the agricultural world. His approach to small-scale grain production emphasized a meticulous balance of the local farming ecosystem, and a minimum of human interference and labor.

In the 1980s, around the world, farming and consumer groups began seriously pressuring for government regulation of organic production. This led to legislation and certification standards being enacted through the 1990s and to date. Since the early 1990s, the retail market for organic farming in developed economies has been growing by about 20% annually due to increasing consumer demand. Concern for the quality and safety of food, and the potential for environmental damage from conventional agriculture, are apparently responsible for this trend.

## Twenty-first century

Throughout this history, the focus of agricultural research and the majority of publicized scientific findings have been on chemical, not organic farming. This emphasis has continued to biotechnologies like genetic engineering. One recent survey of the UK's leading government funding agency for bioscience research and training indicated 26 GM crop projects, and only one related to organic agriculture. This imbalance is largely driven by agribusiness in general, which, through research funding and government lobbying, continues to have a predominating effect on agriculture-related science and policy.

Agribusiness is also changing the rules of the organic market. The rise of organic farming was driven by small, independent producers and by consumers. In recent years, explosive organic market growth has encouraged the participation of agribusiness interests. As the volume and variety of "organic" products increases, the viability of the small-scale organic farm is at risk, and the meaning of organic farming as an agricultural method is ever more easily confused with the related but separate areas of organic food and organic certification.

In Havana, Cuba, a unique situation has made organic food production a necessity. Since the collapse of the Soviet Union in 1989 and its economic support, Cuba has had to produce food in creative ways like instituting the world's only state-supported infrastructure to support urban food production.

## Initiative taken by central and state government, NGO's and other organizations for promotion of organic agriculture

Organic farming envisages a comprehensive management approach to improve soil health, the ecosystem of the region and the quality of farm produce. It includes all agricultural systems that promote environmentally sound production of food and fibres.

India's National Programme for Organic Production defines organic farming as, "A system of farm design and management to create an ecosystem, which can achieve sustainable productivity without the use of external inputs such as chemical fertilizers and pesticides".

International Federation of Organic Agriculture Movements (IFOAM) defines organic farming as, "a production system that sustains the health of soil, ecosystems and people. It relies on ecological processes, biodiversity and cycles adapted to local conditions, rather than use of inputs such with adverse effects. Organic farming combines tradition, innovation and science to benefit the shared environment and promotes fair relationships and a good quality of life for all involved".

United States Department of Agriculture: It defines organic farming as, "a production system which avoids or largely excludes the use of synthetically compounded fertilizers, pesticides, growth regulators and livestock additives".

The word organic in organic farming is a process claim and not a product claim. Chemically, every food is organic but here it is the production process which makes the food organic. Therefore, it is a method of production, processing, storage and transportation that makes a product to qualify to be organic.

Some of the best sources of biopesticides, like neem (*Azadirachta indica*), Karanji (*Pongimia pinnata*), and *Calatropis spp*, are abundantly available in drylands (Rajeshwar Rao 1999).

Similarly, minerals like rock phosphate, gypsum and lime are available naturally in large quantity in Rajasthan. These minerals help improve soil conditions and supply plant nutrients.

Finally, properly handled and aged animal and human wastes are among the best organic fertilizers available, traditional dryland farming systems are well supplied with both. Indeed, in such areas animal populations are growing at 1.5 to 2.5% per year (Purohit 1998).

## Potential areas

The net sown area of the country is 141 million hectare. The area having very low level of fertiliser consumption, namely dryland/rainfed areas, hilly areas, and north eastern States are most suitable for conversion to organic farming. The dryland /rainfed agriculture which constitutes about 60% of the net sown area could play an important role for organic farming as average fertiliser use in this area is very low (25-35 kg/ha as compared to national average of 89.8 kg per ha).

Hill agriculture is by and large natural farming except in valleys and high value crops. The hills show medium fertility level and heavy nutrient removal annually with little replenishment. Uttaranchal and Himachal Pradesh show high mining of potassium followed by Nitrogen and Phosphorous. Some of the districts of Uttaranchal and NE States use negligible fertiliser (less than 5 kg/ha fertilisers).

In fact, India is endowed with all types of climatic conditions ranging from arid to humid and tropical to temperate. The rainfall and temperature vary considerably from region to region besides the soil and land topography including the vegetation. Considering the varied agro-climatic conditions the country has been divided into 15 agro-climatic regions which provide the opportunity to cultivate a variety of crops.

## Based on above, priority zones for organic farming have been identified:

Category I: Are those areas which are rainfed and mostly under monocrop and traditionally no chemical input has ever been used. They can easily be classified as organic produce areas. Broadly, these areas exist in the States of N.E. Region, Jharkhand, Uttaranchal and Rajasthan.

Category-II: Are those areas primarily under rainfed farming having little irrigation support. These are normally under monocropping rarely under double cropping. Broadly the States of Orissa, HP, J&K, MP, Chattisgarh and Gujarat and also parts of Maharashtra and Karnataka will fall under this category.

Category-III: Are those areas which have moderate to heavy use of chemical fertilisers as well as pesticides. The areas are mostly under multiple cropping. The conversion of these areas into organic farming will initially cause some loss of productivity. For these areas balanced and conjunctive use of biomass, organic and inorganic fertilisers and controlled use of chemicals through integrated nutrient and pest management (INM & IPM) will be promoted to achieve the sustainable increases in agricultural production.

## Identification of Crops

1. Indian organic agriculture accounts for a very negligible part of total agricultural production in the country. Though very nascent the Indian organic sector is growing very rapidly and has already made inroads into the world organic market in certain key sectors such as tea, coffee, spices, fruits, vegetables, cotton, cereals, oilseeds, pulses, etc. Based on above, thrust will be given to grow the following crops organically.
   i) Major horticultural crops including vegetables. It will include mainly grapes, banana, mango, papaya, pineapple, guava, passion fruits, mausambi, orange, cashewnut, walnut and fresh vegetables.
   ii) Export oriented cereals like basmati rice, and few others like sorghum and pearl millets.
   iii) All pulses, soybean, groundnut and cotton.
   iv) Chillies, garlic, turmeric, coriander, ginger

Presently, the area under pulses, soyabean, groundnut, cotton is 20.03, 6.42, 6.73 and 8.58 million ha (total 41.76 million ha). However, organic farming per se could not be promoted for food basket crops like rice and wheat in the country except in cases of contracted organic farming for export purposes.

## Strategies for promoting organic farming in drylands

As shown above, there is immense potential for adoption of organic farming in India's drylands. Socio-economic and ecological reasons also favor it (Arora and Mohan,1986): many aspects of drylands that would pose constraints for the intensive agriculture system of the Green Revolution can be converted into opportunities for organic farming.

## Specific Steps would go a Long Way Towards Promoting Organic Farming

### 1. Popularize organic farming without compulsion of certification

In drylands, farmers are very poor and are largely unable to afford the cost of certification programs. Promoting certification as a universal requirement of organic farming has thus had a negative impact on its adoption by smallholders in these areas. A more appropriate approach for drylands, at least during a certain initial period until the industry is established, would be to promote non-certified organic farming with a goal of providing local employment and improving local/regional food security rather than developing export crops.

**2. Promote ley farming**

In India's drylands, rotation of grasses and food grains is known as ley farming. This system is traditional; research by the Central Arid Zone Research Insitute (CAZRI) has validated it as one of the best options for fertility restoration in these regions (Muthana et al. 1985).

**3. Integrate efforts of supporting agencies**

Several government agencies and NGOs are working individually to promote organic farming. This individual approach, however, may result either in lack of adequate funding or lack of adequate knowledge of organic farming and/or marketing techniques. For example, rural ministries have a scheme of "margin money" in which they subsidize 25-30% of loans made to establish vermicompost units; but they are unable to ensure a steady market for the produce of such units. Similarly, state agriculture departments and the institutes that make up the Indian Council of Agriculture Research (ICAR) have a wealth of information on organic farming but are unable to provide financial support for initiating such programs. Thus there is need for all related agencies to create integrated programs, linking the storehouses of technologies with financial institutions, in order to effectively promote organic farming. Since funding agencies always have the financial upper hand, they must take the lead to ensure sufficient technology backup and marketing of these products. This could be effected through a coordinating committee with representatives from all concerned organizations.

**4. Encourage decentralized input supply**

Local, decentralized production of all inputs for organic farming should be encouraged not only so that local resources can be utilized but also so that village-level employment can be generated. Locally produced inputs are also much less likely to be adulterated. All of this organic input production should be legally categorized as cottage industry. Subsidies or micro-financing to help set up small-scale input production units that meet local requirements could be provided to village cooperatives and self-help groups (that is, informal groups of people with a joint livelihood system and a common bank account). Banning inter-state transportation of organic inputs could be another effective method for keeping such production local.

## 5. Adopt improved methods of composting

In general, farmers understand the importance of adding organic matter to drylands soils, but the majority apply undecomposed animal and crop wastes. This not only reduces the availability of nutrients to plants but also invites several pests. It would be better to apply these materials after composting them (Durgude et al. 1996) with any of several suitable methods (Table 3). Because these composting methods would help reduce disease and improve local sanitary conditions, they could be popularized and financially supported under India's existing "Clean village scheme." This scheme, sponsored by the government's Ministry of Rural Development and executed by state governments and volunteer organizations, focuses on safe disposal and use of human and animal wastes as a means of improving hygiene and health at the village level.

## 6. Increase public awareness and build capacity

Conferences, seminars, and farmers' fairs may be organized to raise awareness and encourage adoption of organic farming. Programs demonstrating how to establish organic systems and training in how to produce and manage organic inputs, may be started at the village level. Under the NPOF, sufficient provision has been made to train farmers for organic production and internal control and to develop both model organic farms and a nationwide network of organic service providers (to provide guidance, establish farmers' groups and arrange organic inputs). However, most of these programs focus on irrigated areas because of ensured success and export possibilities. A rational allocation policy for drylands is needed.

## 7. Subsidize organic inputs and produce

Subsidies may be provided for organic inputs and produce while the industry is still getting established. In India, subsidies are mainly provided by the national government and channeled through state agriculture departments; the technique is well-tested, having already been used for the synthetic fertilizer and pesticide industry. Indeed, subsidies have been provided for setting up biofertilizer and vermicomposting units under NPOF and for setting up export schemes under NPOP. Additional subsidies could be provided for:

- Setting up organic input production units for composting, biopesticides etc.
- Compensating organic farmers during the period of conversion to organic techniques, to compensate for yield reductions if any.
- Establishing village-level grading and packaging units for organic produce.
- Developing local and regional marketing infrastructure for organic produce in dryland areas, where regional/local food security is more important than crops for export.

### 8. Promote high-value crops

The dryland climate favors quality production of several spices and medicinal plants that are in already in great demand both nationally and internationally. Furthermore, demand for organic versions of these high-value products is increasing. Thus, once the local market is well established and the potential for growing crops for export has increased, they may be promoted to dryland farmers as excellent plants to include in the overall design of their organic system.

### 9. Develop organic farming clusters of villages

Since the drylands are already an area of focus for governmental development programs based on a watershed approach, clusters of villages previously established for such programs (Khan 2002) may be converted into organic clusters of villages by providing technical support. This will be cost-effective and make the eventual certification process of organic produce easier for these villages once the local organic produce market has been well established.

### 10. Develop certification programs and marketing chains

Ultimately, once the drylands organic agriculture industry is strong, cost-effective facilities for sale of certified organic produce in regional and global markets may be created. The cost of certification should be met by the exporters.

### 11. Conduct ongoing research on organic systems

ICAR institutes and State Agriculture Universities have already done much research on various components of organic farming such as vermicompost, biopesticides, biofertilizers, and agroforestry. However, more research is needed to integrate these efforts and assess their effects. Ongoing research is needed on all the following aspects of organic agriculture simultaneously (Sharma 2001):

- Survey and scientific validation of traditional organic farming practices.
- Use of locally available resources for production of manures and biopesticides.
- Assessment of economic and ecological returns from organic vis-à-vis intensive agriculture systems.

Due to low, erratic rainfall and poor soils in India's drylands, farming systems based on synergism with nature, and thus essentially "organic," have been followed for centuries. This opens the door to immense possibilities for improving the soil health and overall environment, and providing sustainable livelihoods in these areas, by updating these age-old systems with modern research and techniques. The key to achieving this is enhanced collaboration and cooperation among all the agencies and programs that have an interest in supporting these goals. By working together, the goals of promoting organic practices, improving soil health and local food security and, ultimately, creating markets for organic produce both regionally and globally, can be realized for the benefit of all of India.

## Organizations involved in organic agriculture

### 1. International guidance on organic agriculture

Several attempts were made to set up standards for organic agricultural products, all over the world. As a result some guidelines were also issued to the member countries to get accredited to organic status. These guidelines are as follows:

1. **Codex Alimen Tarius**: Guidelines heightened consumer interest in food quality and safety as well as increased concern about the potential for food standards to be applied as trade barriers led to the establishment of the Codex Alimentarius Commission (Codex) by a resolution of the governing bodies of the Food and Agriculture Organization of the United Nations (FAO) in 1961 and the World Health Organization (WHO) in 1963. Primary objectives of this commission are to protect consumer health and to ensure fair practices in food trade through the elaboration, harmonization and publication of food standards and other related texts. Codex brings together scientists, technical experts, government regulators and international consumer and industry organizations to develop food standards. Codex operates based on its Procedural Manual, which includes the Codex Statutes, the Rules of Procedure and the Procedures for the Elaboration of Codex Standards and Related Texts, as well as guidelines and other provisions applicable by Codex subsidiary bodies which should be read in conjunction with the Procedures. The Codex Alimentarius includes many maximum residue limits established for pesticides and veterinary drugs in foods and animal feeds, acceptable levels of food additives and maximum levels for contaminants. The Codex Commodity Committee subject matters range from fresh fruits and vegetables to processed fish and fishery products. General subject Committees focus on so-called "horizontal" subjects – such as food hygiene, labelling, additives and contaminants. These committees develop concepts and principles applicable to foods in general (or applicable to specific foods or food groups) review provisions in Codex commodity standards as required Committees addressing food safety issues often rely on expert advice, consulting internationally recognized experts in special subject areas and seeking guidance from independent FAO/WHO expert committees or consultations.
2. **International Federation of Organic Agriculture:** The International Federation of Organic Agriculture Movements (IFOAM), established in

1972 and located in Bonn, Germany, is the international nongovernmental organization providing an umbrella for all organic agriculture organizations. Currently uniting around 700 member organizations in more than 100 countries, its goal is the worldwide adoption of ecologically, socially and economically sound systems based on the principles of organic agriculture. Through international conferences, meetings and other fora, IFOAM facilitates an ongoing dialogue about the status and future of organic agriculture. IFOAM has developed and maintains the Organic Guarantee System, which seeks to provide a common system of standards, verification and market identity for the organic world. IFOAM also implements specific projects that facilitate the adoption of organic agriculture methods, particularly in developing countries, and represents the organic agriculture movement at the United Nations and other intergovernmental agencies. Its mission is to, "Lead, unite and assist the organic movement in its full diversity." and vision is the "worldwide adoption of ecologically, socially and economically sound systems, based on the Principles of Organic Agriculture". Among its wide range of activities, the federation maintains an organic farming standard, and an organic accreditation and certification services.

3. **FAO's involvement with organic agriculture:** FAO's interest in organic agriculture stems from its potential to contribute to the goal of greater food security and rural development. At the 2007 International Conference on Organic Agriculture and Food Security in Rome, FAO member countries noted organic agriculture's ability to increase access to food in rural communities, create a more stable food supply, and sustain natural resources. FAO has been involved in organic agriculture since the late 1990s, when the Committee on Agriculture recommended that an inter-disciplinary program be developed on the topic. Organic agriculture was officially included onto FAO's agenda as a means to promote sustainable development in March 1999. These first FAO activities were mainly concerned with disseminating information on organic agriculture to allow actors; legislators, producers, consumers to make informed decisions. To that end, the FAO website6 was launched in 2000 and is regularly updated with publications and information on organic agriculture. FAO has also tried to harmonize standards and requirements for organic agriculture through partnerships with IFAOM and the United Nations Conference on Trade and Development ("UNCTAD"). To that end, the International Taskforce on Harmonization and Equivalence in Organic Agriculture, compromising representatives from ministries of agriculture and trade as well as actors in the field, was established in 2002 to facilitate international trade. FAO's involvement in organic agriculture has also included providing assistance to member countries, upon their request

in establishing legislative and regulatory frameworks provide information on accessing markets and agricultural expertise on improving quality and performance of organic farms.

## 2. Legal framework on organic agriculture in India

To have consonance with these international standards, India had to take steps to identify organic products. As a result in 1985 Agricultural and Processed Food Products Export Development Authority Act has been passed. Along with that, two authorities are also been established to regulate the organic agriculture field.

1. **The Agricultural and Processed Food Products Export Development Authority (APEDA)**: APEDA was established under the Agricultural and Processed Food Products Export Development Authority Act, passed by the Indian Parliament in December 1985. Pursuant to said Act, its main functions include: promoting the export oriented production and development of scheduled products (including fruits, vegetables, meat, dairy, cereal, cacao products and non-basmati rice) fixing of standards and specifications for the scheduled products for the purpose of exports, carrying out inspection of meat and meat products for quality purposes and other matters. These are three out of the five Statutory Boards under the Ministry of Commerce and Industry (the other two being on rubber and tobacco). Only products certified by accredited bodies as conforming to national organic standards may be marketed as organic in India.
2. **National Programme on Organic Production (NPOP):** In order to promote organic farming in the country in view of its great export potential and also because of its eco-friendly nature, the National Steering Committee under the Chairmanship of Secretary Commerce approved the National Programme for Organic Production at its meeting on 2nd May 2001. NPOP notified under the Foreign Trade Development and Regulation Act and controlled by Agricultural and Processed Food Products Export Development Authority (APEDA) deals with the export requirements. The NPOP notified under APGMC Act and controlled by Agricultural Marketing Advisor, Directorate of standards and Inspection looks after the domestic certification. NPOP deals with the standards for organic production, system criteria and procedures for accreditation of inspection and certification bodies, the organic logo and the regulations governing its use. The institutional mechanism for the implementation of standards for organic production and certifications provided through a National Accreditation Policy and Programme. The major objectives of NPOP are:

- To provide the means of evaluation of certification programmes for organic production and products as per the approved criteria.
- To accredit certification programmes.
- To facilitate certification of organic products in conformity to the standards laid for organic products.
- To encourage the development of organic farming and organic processing.

Under the National Programme for Organic Production (NPOP), documents like National standards, Accreditation Criteria for accrediting, inspection and certification agencies, Accreditation Procedure, Inspection and Citification Procedures have all been prepared and approved by the National Steering Committee. These documents were prepared on the basis of the guidelines evolved by the representative international organization, International Federation for Organic Agricultural Movements (IFOAM), EU Regulations and Codex standards. Theses are available for sale with the Accreditation Agencies.

**Accreditation agencies**

The National Steering Committee has designated the following statutory Boards/ Authority as Accreditation Agencies:

1. Agricultural and Processed Food Products Export Development Authority (APEDA)
2. Coffee Board
3. Spice Board
4. Tea Board
5. Coconut Development Board
6. Directorate of Cashew and Cocoa Development

APEDA has been assisting the Commerce Ministry in preparing the various documents under the National Programme for Organic Production and has also developed the National Organic Logo.

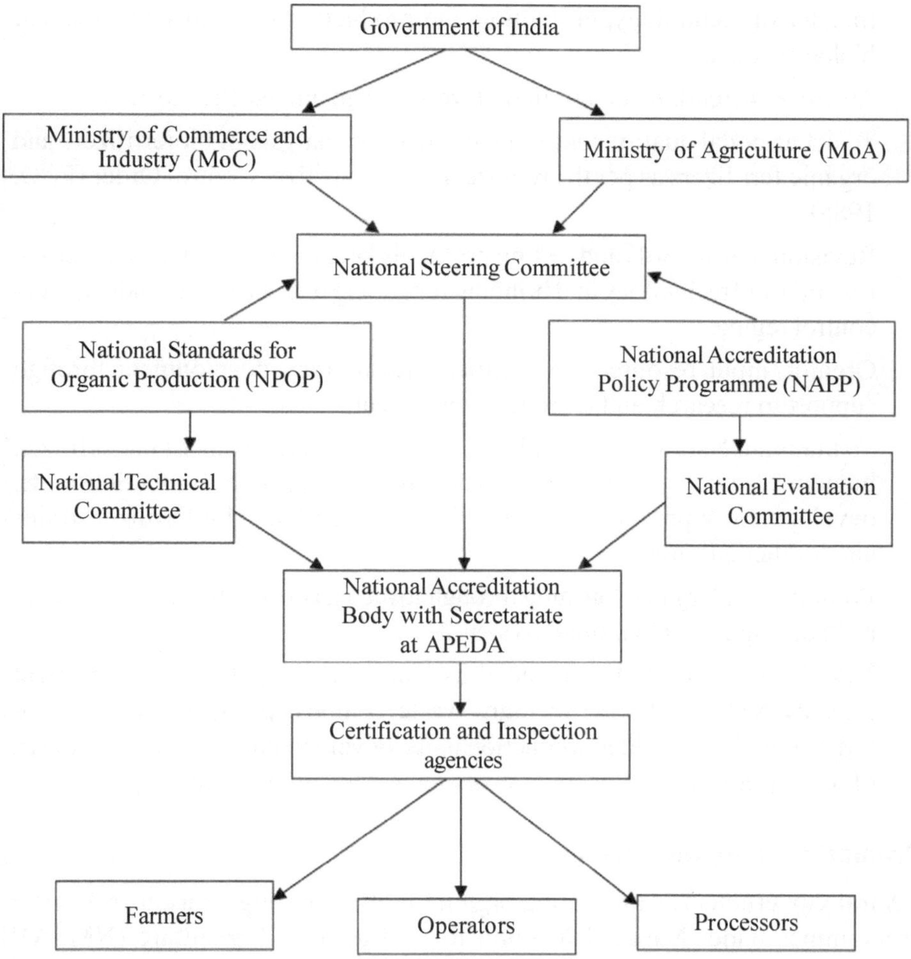


**Fig :** Operational structure of National Programme On Organic Production

## 3. National Project on Organic Farming

National Project on Organic Farming (NPOF) is a continuing central sector scheme since 10th Five Year Plan. NPOF is being implemented by National Centre of Organic Farming at Ghaziabad and its eight Regional Centres at Bangalore, Bhubaneshwar, Panchkula, Ghaziabad, Imphal, Jabalpur, Nagpur and Patna. Besides working for realisation of targets under NPOF, NCOF and RCOFs are also performing specific roles in promotion of organic farming.

### Objectives

- Promotion of organic farming in the country through technical capacity building of all the stakeholders including human resource development, transfer of technology, promotion and production of quality organic and biological inputs.

- Awareness creation and publicity through print and electronic media.
- To act as nodal quality control laboratory for analysis of biofertilizers and organic fertilizers as per the requirement of Fertilizer Control Order (FCO, 1985).
- Revision of standards and testing protocols keeping in view the advances in research and technology and bringing remaining organic inputs under quality control regime.
- Organic input resource management, technology development through support to research and market development.
- To maintain National and Regional culture collection bank of biofertilizer, biocontrol, waste decomposer organisms for supply to production units, development & procurement and efficacy evaluation of biofertilizer strains and mother cultures.
- Promotion of Organic Farming through low cost certification system known as "Participatory Guarantee System".
- NPOF scheme provides financial assistance through Capital Investment Subsidy Scheme (CISS) for agro-waste compost production units, bio-fertilizers/bio-pesticides production units, development and implementation of quality control regime, human resource development, etc.

## Promotion of organic farming

Central Government is promoting organic farming through various schemes/ programmes under National Mission for Sustainable Agriculture (NMSA)/ Paramapragat Krishi Vikas Yojana (PKVY), Rashtriya Krishi Vikas Yojana (RKVY), Mission for Integrated Development of Horticulture (MIDH), National Mission on Oilseeds & Oil Palm (NMOOP), Network Project on Organic Farming of ICAR.

In addition to this, Government is implementing a Cluster based programme to encourage the farmer for promoting organic farming called Paramparagat Krishi Vikas Yojana (PKVY):

- Groups of farmers would be motivated to take up organic farming under Paramparagat Krishi Vikas Yojana (PKVY). Fifty or more farmers will form a cluster having 50 acre land to take up the organic farming under the scheme.
- Organic farming will be promoted by using traditional resources and the organic products will be linked with the market.

- It will increase domestic production and certification of organic produce by involving farmers.

## Role of Punjab Government in promotion of organic farming

There are several things going on in Punjab related to organic farming. The Punjab Agro Export Corporation is involved in promotion of organic farming, appointing consultants, roping over 1200 farmers and on the other hand, several large companies are taking-up organic farming projects in state, taking large chunks of lands on lease from government. Punjab government is also going to establish Food Park at Fatehgarh Sahib. Bhoomi Vardaan Foundation established in collaboration with Prince Charles for promotion of organic farming. More over state council for organic farming is constituted.

Organic Farming Council of Punjab (OFCP) was created to promote and develop organic farming, and to advise the State Government regarding policy initiative on organic farming and value addition of crops produced under the programme. It was started in February 2006. Under the project initiated by OFCP to promote organic farming, five districts of Punjab – Mohali, Fatehgarh Sahib, Patiala, Gurdaspur and Sangrur were chosen. Covering an area of 1144 acres and assisting 547 farmers.

## Objectives of Organic Farming Council of Punjab

- To encourage diversification in agriculture through organic farming.
- To advise the state government regarding the policy initiatives on organic farming and value addition of the crops.
- To catalyse agro-industrial growth related to organic farming.
- To make arrangements for packing, storage, marketing, processing, preservation, transport and export of organic produce.
- To formulate policy/action plan for promotion of organic farming.
- To undertake or assist in undertaking programmes for employment generation, growth and diversification of agriculture and industries based on organic produce.
- To organize technology transfer through training and extension and carrying out research and development related to organic farming.
- To build a cadre of skilled managers for managing the affairs of various units being established as a part of the development process and hire the services of experts in related field.
- To purchase or take land on lease for the promotion of organic farming programme in the state.
- To raise funds/grants from government of India, industry and individuals.

**Services Offered Organic Farming Council of Punjab (OFCP)**: OFCP promotes and implements organic farming practices in an organized way. It registers the farmers in target districts preferably in cluster form to facilitate communication, operation and the certification process. It is accomplished through awareness camps in target districts and by the registration of farmers.

## NGO's involved in organic agriculture

Rural Development organization-it is registered NGO in India, working on number of organic farming projects.

Sanjeevani NGO is implementing two programme with Uttarakhand Organic Commodity Board Dehradun. Encouraging farmers to adopt the organic systems and giving support in technical of trainings, capacity building exposures etc.

Navdanya's - Navdanya is a movement for Earth Democracy based on the philosophy of 'Vasudhaiv Kutumbakam'. Navdanya have so far trained above 400,000 men and women farmers, students, govt. officials, representatives of national as well as international NGO's Voluntary Organizations on biodiversity conservation and organic farming.

Some organizations have undertaken the training of personnel towards organic farming.

- Institute for Integrated Rural Development at Aurangabad
- The Society for Equitable Voluntary Actions (SEVA) in West Bengal
- The Indian Agency for Organic Agriculture (IAOA)
- The Ecological Development Society in Pondicherry
- Peekay Tree Crops Development Foundation (PTCDF) at Cochin

All India Federation of Organic Farmers also promotes organic farming in the country. The following crops are cultivated under organic farming methods in our country.

- Cereals: wheat, paddy, jowar , bajra, maize
- Pulses: pigeonpea, chickpea, greengram, blackgram, chana
- Oilseeds: groundnut, castor, mustard, sesame
- Commodities: cotton, sugarcane, particularly for sugarcandy (gur)
- Spices: ginger, turmeric, chillies, cumin
- Plantation crops: tea, coffee, cardamom
- Fruits: banana, sapota, custard apple and papaya
- Vegetables: tomato, brinjal, cucurbits, cole crops, leafy vegetables

India is home to 30 per cent of the total organic producers in the world, but accounts for just 2.59 per cent (1.5 million hectares) of the total organic cultivation area of 57.8 million hectares, according to the World of Organic Agriculture 2018 report.

At the same time, most organic farmers are struggling due to poor policy measures, rising input costs and limited market, says a study by the Associated Chambers of Commerce and Industry of India (ASSOCHAM) and global consultancy firm Ernst & Young.

### Organic farming is yet to taste success

Problems are evident even in Sikkim, which was recognised as the country's first organic state in 2018. A survey by Delhi-based Centre for Science and Environment showed that the state's transition to organic farming is yet to become a true success. The survey found that the phasing out of chemicals in Sikkim was not complemented by a simultaneous increase in availability of and access to organic manure.

Farmers also complained of low productivity during the transition from conventional chemical farming to organic farming. Pest attack on organic crops is another reason cited by the farmers for low productivity and demanded education and training to deal with it. The problem of pest attacks increased after the conversion to organic farming, but the state is yet to maintain data on this, which is needed for plant disease management.

Similarly, nearly 98 per cent farmers in Rajasthan are aware of ecological hazards of conventional chemical-based farming, but fear of decline in production and unavailability of organic inputs in the market discourage them from switching to organic farming (study conducted by the Consumer Unity and Trust Society).

### Organic farming fraught with hurdles

The organic produce on an average dips by 6.7 per cent in the first year during the transition. To ensure marketing of organic produce, connecting farmers with the domestic and global supply chain is extremely important. However, the supply chain is underdeveloped and as a result, the small and mid-sized farmers located in hilly regions and tribal belts find it extremely difficult to access the market.

Expensive organic produce discourages customers, affects sales even after the produce reaches the markets, consumers find them expensive and discouraging. The organic produce are usually more expensive due to higher labour cost and comparatively lower yields. According to the ASSOCHAM report, post-harvest handling of relatively small quantities of organic foods also results in higher costs because of the mandatory segregation of organic and conventional produce, especially for processing and transportation. "Specialised farmer training costs, processing and inventory holding costs (without chemical additives), and increased packaging, logistics and distribution costs (due to low volumes), contribute to the high price of organic food products.

### Multiple Certification Systems

Even as farmers are struggling to find a better market, the existing certification systems for organic food are making things difficult for them. The certification systems are not only cumbersome and time-consuming, but also expensive. It is important to eliminate confusion over multiple certification systems and multiple ministries regulating organic production and sales.

**Paramparagat Krishi Vikas Yojana (PKVY)**-the Centre's free certification programme for organic farmers-is an example. However, corrective measures are needed to ensure that the states become responsible and contribute toward "organic India".

# 3

# Organic Ecosystem and Their Concepts

Organic farming endorses the concept that the soil, plant, animals and human beings are linked. Therefore, its goal is to create an integrated, environmentally sound, safe and economically sustainable agriculture production system. Soil is a living system linked to an organism with different components. Human interact with these natural components (minerals, organic matter, micro-organisms, animals and plants) to achieve harmony with nature and create a sustainable agricultural production. A key feature of organic farming is the primary dependence on natural resource and those developed locally (green manures, crop residues, farm wastes etc.), rather than external inputs (especially synthetics). The farmer manages self-regulating ecological and biological processes for sustainable and economic production of products. Organic farming systems do not use toxic agrochemical inputs (pesticides, fungicides, herbicides and fertilizers). Instead, they are based on development of biological diversity and the maintenance and replenishment of soil productivity.

## Organic Ecosystem

Agriculture is dual a consumer and producer of ecosystem. A number of ecosystems are utilized to produce other ecosystems as food, which is supported by maintenance of soil health, plant protection measures, water regulation etc,. It is the practice of farming using principles of ecology i.e. the study of relationships between organisms and their environment. It has been defined as "an integrated system of plant and animal production practices having a site-specific application that will last over the long term. It is committed to maintain and preserve the agriculture base of soil, water and atmosphere ensuring future generations the capacity to feed themselves with an adequate supply of safe and wholesome food". Organic ecosystem will help for organic food production that:

- Supports profitable production
- Protects environmental quality
- Uses natural resources efficiently
- Provides consumers with affordable, high quality products
- Decreases dependency on non renewable resources

- Enhances the quality of life for farmers and rural communities by reducing input cost and
- Will last for generations to come.

## Organic farming within the sustainability spectrum

Most definitions include both objective and subjective elements. The latter, which may be inserted subconsciously, include biases, judgments and personal agendas that have their roots in psychological distress. Thus, the objectivity of any definition can be limited by individual knowledge, awareness, vision and level of empowerment. An approach that we have found useful is to define such farming systems according to their proximity, along a spectrum, to a set of preferred goals. These should include optimal nourishment, optimal individual and social development, and sustainability.

Organic farming comprises a range of approaches within the broader sustainable agriculture spectrum. In its most developed form, ecologically sustainable agriculture (including organic farming) is both a philosophy and a system of farming. It is based on a set of values that reflects an awareness of both ecological and social realities, and on a level of empowerment that is sufficient to generate responsible action. Efforts to ensure short-term viability are tested against long-term environmental sustainability, and attention to the uniqueness of every operation is considered in relation to ecological and humanistic imperatives, with an awareness of both local and global implications. It emphasizes benign designs and management procedures that work with natural processes and cycles to conserve all resources (including beneficial soil organisms and natural pest controls), and minimize waste and environmental damage, prevent problems, and promote agro-ecosystem resilience, self-regulation, evolution, and sustained production for the nourishment and optimal development of all (including rural communities both here and abroad). Special attention is paid to the relationships between soil conditions, food quality and livestock health; and livestock is cared for in the most humane way possible. In addition, organic farmers tend to avoid the use of synthetically compounded fertilizers, pesticides, growth regulators, and livestock feed additives. Instead, they rely upon crop rotations, crop residues, animal manures, legumes, green manures, off-farm organic wastes, mechanical cultivation, and mineral-bearing rocks to maintain soil fertility and productivity. Insects, weeds and other pests are managed by means of natural, cultural and biological controls. The potential of this approach, however, goes far beyond its present expression, which has largely been limited to the substitution of environmentally benign products and practices. As this new vision of what is ecologically responsible becomes established, significant development can be expected in the science and art of agro-ecosystem design and management.

In the efficiency stage, conventional systems are altered to reduce both consumption of resources and environmental impact, e.g., by banding fertilizers, monitoring pests, and optimal siting and timing of operations. In the substitution phase, finite and environmentally disruptive products are replaced by those that are more environmentally benign, e.g., synthetic nitrogen fertilizers by organic nutrient sources, non-specific pesticides by biocontrol, herbicides by appropriate systems of cultivation. Because neither the efficiency nor the substitution strategies confront the causes of problems, they condemn producers to repeated reliance on externally derived curative solutions and inputs.

In contrast, the redesign stage aims to avoid problems by site and time-specific design and management approaches. The farm is made more ecologically and economically diverse, resource self-reliant and self-regulating. The various approaches to sustainable agriculture, including organic farming, can be arranged along a spectrum using the ESR framework

The challenge for organic farmers worldwide is to build on the positive elements of these approaches. This involves emphasizing at every opportunity both the prevention of problems and the meeting of real needs in sustainable ways, through the appropriate design and management of agro-ecosystems.

## Concept of organic ecosystems

- Nature is the best role model for farming, since it does not use any inputs nor demand unreasonable quantities of water.
- The entire system is based on intimate understanding of nature's ways. The system does not believe in mining of the soil of its nutrients and do not degrade it any way for today's needs.
- The soil in this system is a living entity.
- The soil's living population of microbes and other organisms are significant contributors to its fertility on a sustained basis and must be protected and nurtured at all cost.
- The total environment of the soil, from soil structure to soil cover is more important.

Thus in today's terminology it is a method of farming system which primarily aims at cultivating the land and raising crops in such a way, as to keep the soil alive and in good health by use of organic wastes (crop, animal and farm wastes, aquatic wastes) and other biological materials along with beneficial microbes (biofertilizers) to release nutrients to crops for increased sustainable production in an eco-friendly pollution free environment.

**Organic farming describes two major aspects of alternative agriculture**

- Substitution of manures, farm organic resources and biofertilizers (INM) for inorganic fertilizers.
- Biological and cultural pest, diseases and weed management (IPM, IDM and IWM) instead of chemical control.

**The key characterization of organic farming in relation to soil fertility and crop production includes:**

- Protecting the long-term fertility of soil by maintaining soil organic matter levels, fostering soil and biological activity and careful mechanical inversion,
- Plant nutrients supply through relatively insoluble nutrient sources (organic sources) made available by the action of soil microbes,
- Meeting crop need of nitrogen through nitrogen fixation by leguminous crops in the cropping systems and recycling of farm organic materials including crop residues and livestock wastes,
- Importance of crop rotation, natural predators, resistance varieties and other agronomic manipulations of plant protection including weed management, and
- Biodiversity management, soil and environmental health.

Organic agriculture is viable alternative to conventional agriculture. It protects the soil from erosion, improves natural resource base and sustains production at levels commensurate with carrying capacity of managed agro- ecosystem because of reduced dependence on fertilizers and plant protection chemicals. It minimizes environmental pollution and aids in regeneration of ecosystem.

Organic farming is one of several to sustainable agriculture and many of the techniques used (intercropping, crop rotation, ploughing, mulching, integration of crops and livestock etc) are practices under various agricultural systems. What makes organic farming unique is that almost all synthetic inputs are prohibited and soil health improving agronomic practices are mandated.

Organic farming is the pathway that leads to live in harmony with nature. Organic agriculture is the key to sound development and sustainable environment. It minimizes environmental pollution and the use of non-conventional natural resources (resources other than traditional resources). It conserves soil fertility and soil erosion through implementation of appropriate conservation practices. In philosophical terms organic farming means "farming in sprits of organic relationship". In this system everything is connected with everything else. Since organic farming means placing farming on integral relationship, we should be well aware about the relationship between the soil, water and plants, between soil-soil microbes and waste products, between the vegetable kingdom and the animal kingdom of which the apex animal is the human being, between agriculture and forestry, between soil, water and atmosphere etc. It is the totality of these relationships that is the bedrock of organic farming.

It works on the principle that we must meet the needs of present without compromising the ability of future generation to meet their own needs. Organic agriculture claims to be sustainable in the context of agriculture. Sustainability refers to the successful management of resources of agriculture to satisfy human needs while at the same time maintaining or enhancing the quality of the environment and conserving natural resources. Sustainability in organic farming must therefore be seen in a holistic sense, which includes ecological, economical and social aspects.

**Ecological Sustainability**

- Recycling the nutrients instead of applying external inputs
- No chemical pollution of soil, air and water
- Promote biological diversity
- Improve soil fertility and build up humus
- Prevent soil erosion and compaction
- Animal friendly husbandry
- Using renewable energies

**Social Sustainability**

- Sufficient production of subsistence and income
- A safe nutrition of the family with healthy food
- Good working conditions for both men and women
- Building a local knowledge and traditions

**Economic Sustainability**

- Satisfactory and reliable yields
- Low costs on external inputs and investments
- Crop diversification to improve income safety
- Value addition through quality improvement and on-farm processing
- High efficiency to improve Competitiveness

Thus only if the above three dimensions are fulfilled an agricultural system can be called sustainable.

## Sustainable agriculture

A Sustainable agriculture system is one that can indefinitely meet demands for food and fibre at socially acceptable, economic and environment cost. Sustainable Agriculture refers to an agricultural production and distribution system that:

- Achieves the integration of natural biological cycles
- Protects and renews soil fertility and the natural resource base
- Reduces the use of non renewable resources and purchased ( external or off-farm) production inputs
- Optimizes the management and use of on- farm inputs
- Provides an adequate and dependable farm income
- Promotes opportunity in family farming and farm communities
- Minimizes adverse impacts on health, safety, wildlife, water quality and the environment.

Sustainable agriculture is the successful management of resources for agriculture to satisfy changing human needs while maintaining or enhancing the quality of environment and conserving natural resources.

Sustainable agriculture is also known as eco-farming or organic farming or sometimes as natural farming or permaculture. Some other designating it as regenerative agriculture or alternative farming.

## Concept of sustainability in cropping system

The concept of sustainability applied to agriculture developed mainly as a result of growing awareness of negative impacts of intensive farming systems on the environment and the quality of life of rural and neighbouring communities.

- Protecting the natural resources
- More efficient use of arable lands and water supply
- The sustainability concept has promoted the need to propose major adjustments in conventional agriculture to make it more environmentally, socially and economically viable and compatible. The concept of sustainability is useful
- It captures a set of concerns
- Several possible solutions to the environmental problems.
- The main focus lies on the reduction or elimination of agrochemical inputs.

## Basic principles of sustainable agriculture

- Based on both biological potential and biological diversity, land can be classified into conservation, restoration and sustainable intensification areas.
- Effectiveness in water saving, equity in water sharing and efficiency in water delivery and use are important for sustainable management of available surface and groundwater resources.
- An integrated system of energy management involving the use of renewable and non-renewable resources of energy in an appropriate manner is essential for achieving desired yield levels.
- Soils in India are often not only thirsty but also hungry. There is need for reduction in the use of market purchased inputs and not of inputs *Pre se*. It is in this context integrated systems of nutrient supply assume importance.
- Genetic diversity and location specific varieties are essential for achieving sustainable advances in productivity.
- The control of weeds, insect pests and pathogens is one of the most challenging jobs in agriculture.
- Whole plant utilization methods and preparation of value added products from the available agricultural biomass are important both for enhancing income and for ensuring good nutritional and consumer acceptance properties.
- Recycling of crop waste and livestock management.
- Growing legume crops
- Genetically Engineering crops (GM crops)

## The goal of sustainable agriculture

The goal of sustainable agriculture is to maintain production at levels necessary to meet the increasing demand of an expanding world population without degrading the environment and that sustainability implies concern for the i) generation of income ii) the promotion of appropriate policies iii) the conservation of natural resources iv) enhance efficiency of use of input v) minimize adverse environmental impacts on adjacent and down stream environments vi) minimize the magnitude and rate of soil degradation and to enhance soil quality and resilience so that the crop productivity can be sustained with minimum adverse impact on soils and environment vii) enhance compatibility with social and political conditions.

However, many people use a wider definition judging agriculture to be sustainable if it is

- Ecologically sound: quality of natural resources is maintained

- Economically viable: farmers can produce enough for self-sufficiency
- Socially just: resources and power are distributed in such a way that basic needs of all members of society are met and their rights to land use, adequate capital, technical assistance and market opportunities.
- Humane: all forms of life (plant, animal and human) are respected.
- Adaptable: rural communities are capable of adjusting to the constantly changing for farming, population growth, policies, market demand etc.

**Scope of sustainability in cropping system play role in saving soil resource**

When farmers grow and harvest crops, they remove nutrients from the soil, without replenishment, land suffers from nutrient depletion and becomes either unusable or suffers from reduced yields. The farmer who reduces purchased chemical fertilizer use by relying more heavily on green-manure crops in a rotation to improve soil fertility will likely experience some change in the pattern of profitability over time (Table 1). Under sustainable agriculture possible sources of nutrients that would be available indefinitely, include:

1. Recycling crop waste and livestock manure (FYM, Compost, and Vermicompost).
2. Growing legume crops and forages such as peanuts or alfalfa that form symbioses with nitrogen-fixing bacteria called *rhizobium.*
3. Crops that require high levels of soil nutrients can be cultivated in a more sustainable manner with certain fertilizer management practices.
4. Soil erosion can also be control by no-tillage, wind-breaks to hold the soil, incorporate all nutrients back to soil.
5. Growing of cover crops such as cowpea, sweet potato and stop using chemical fertilizer which have sufficient amount of salts.

**Table 1.** Common green manure crops and their N content

| Crop species | Biomass (t/ha) | N (per cent) |
|---|---|---|
| Sunnhemp | 21.2 | 0.43 |
| Dhaincha | 20.0 | 0.43 |
| Pillipersara | 18.3 | 1.10 |
| Mungbean | 8.0 | 0.53 |
| Cowpea | 15.0 | 0.49 |
| Guar | 20.0 | 0.34 |
| Senji | 28.6 | 0.57 |
| Khesari | 12.3 | 0.54 |

*Source:* Abrol and Palaniappan (1987)

It is generally recognized that any type of agricultural land use will result in a significant loss in topsoil. Even idle land in grass steadily loses topsoil. An environmental goal of a sustainable cropping system might not be to actually increase the quantity of topsoil on the farm, but rather to employ a cropping system that minimizes the amount of topsoil loss over time.

### Play role in saving water rsources

Considering the fact that 80 to 90% of the developed water resources are diverted towards irrigated agriculture, its efficient management is of crucial importance. For irrigation systems to be sustainable they require proper management and must not use more water from their source than is natural replenished, otherwise the water source becomes a non-renewable resource. Improvements in water well drilling technology and submersible pumps combined with the development of drip irrigation and sprinkler system have made it possible to regularly achieve high yield crops as well as by conserving water resources. Sustainable recommendations have been evolved for efficient use of irrigation water.

1. On-farm water management technology such as land shaping, land levelling and efficient design and layout of irrigation methods.
2. Furrow/alternate furrow/skip furrow irrigation for all row crops.
3. Adopting proper cropping systems in place of continuous lowland rice.
4. Conjunctive use of surface and ground water and canal and tank command area for increasing the productivity of crops.
5. Introduction of sprinkler and drip irrigation systems in water scarcity areas.

It is a better strategy to increase irrigated area than to increase irrigation intensity. A close coordination between irrigation engineers and agronomists will result in improvements.

### Play role in management of rain water

There is wide variability in rainfall occurrence over space and time. About 72% of the cultivated area of the country depends entirely upon rainfall. Hence, efficient rain water management is of prime importance for insuring stability and sustainability in agricultural production. Options for sustainable management are:

1. Agro-climatic analysis that gives an index of probability of rainfall.
2. Contingent crop planning for aberrant weather conditions.
3. In-situ soil moisture conservation practices.
4. Runoff harvesting and recycling.

### Play role in conservation of environment

1. Minimizing the use of non-renewable resources such as natural gas (used in converting atmospheric nitrogen into synthetic fertilizer), or mineral ores (phosphate).
2. Judicial use of chemical fertilizers and pesticide which are responsible for the problem of nitrate accumulation in ground water (permissible limit 45 mg/litre nitrate but in Muktsar 600 mg/litre nitrate found).
3. Residue management by incorporating them in soil instead of burning because burning of residues creates air pollution deteriorates fertility status of soil and kills beneficial organisms present in soil.
4. Recycling of crops wastes and animal dung's as manure and treating sewage water for irrigation purposes are sustainable approaches toward conserving the environment.

### Play role in pest management

Pest management in cropping system is carried out by:

1. A number of cultivars resistant to pests and diseases have been developed. Selection of such cultivars can bring down the pest and disease incidence considerably.
2. Ploughing brings about unfavourable conditions for multiplication of pests. Summer ploughing reduces the reserve of several pests.
3. Early sowing in the season considerably reduces the pest and disease incidence.
4. Mechanical or manual inter cultivation suppresses the pests similar to that of preparatory tillage. Microclimate conducive to pest and disease build up will be disturbed by inter cultivation.
5. Many pests prefer feeding on a particular plant to others. Groundnut as component crop in cotton reduces pest load on cotton. Based on the pests infesting a particular crop, companion or border crops can be grown to minimise the incidence of pests.

| Hierarchical level | Sustainability Indicators (economic, social & environmental) |
|---|---|
| Cropping system/ farming system | Non-negative trends in:<br>i) Farm productivity<br>ii) Net farm income<br>iii) Total factor productivity,<br>iv) Nutnent balance<br>v) Soil quality |

| | |
|---|---|
| | vi) Residues in soil plant products<br>vii) Farm water use efficiency<br>viii) Farmer skills education<br>ix) Debt-service ratio<br>x) Health<br>xi) Tune spent in other social cultural activities |
| Aeroecosvstem (watershed, Asroecozone, etc) | Non-negative trends in:<br>i) Regional production<br>ii) Regional income<br>iii) Regional total factor productivity<br>iv) Regional nutnent balance<br>v) Income distribution<br>vi) Species diversity<br>vii) Soil loss<br>viii) Surface water quality<br>ix) Groundwater quality<br>x) Regional social and economic development indicators |
| Global, National, Region al Systems | Indefinitely meet demands at acceptable social, economic, and environmental costs. |

Sustainable agriculture is the practice of farming using principles of ecology, the study of relationships between organisms and their environment. It has been defined as "an integrated system of plant and animal production practices having a site-specific application that will last over the long term:

- Satisfy human food and fibres need
- Enhance environmental quality and the natural resource base upon which the agricultural economy depends
- Make the most efficient use of non-renewable resources and on-farm resources and integrate, where appropriate, natural biological cycles and controls
- Sustain the economic viability of farm operations
- Enhance the quality of life for farmers and society as a whole."

Sustainable agriculture in the United States was addressed by the 1990 farm bill. More recently, as consumer and retail demand for sustainable products has risen, organizations such as Food Alliance and Protected Harvest have started to provide measurement standards and certification programs for what constitutes a sustainably grown crop.

## Implications for Research

The above framework helps define a vision for agricultural research, target it effectively to prioritize investments and set production and productivity goals at various levels that match national goals. But, it calls for a major paradigm shift in agricultural research and education from the current commodity and input-based approach to management of agricultural resources, to and approach that emphasizes a systems framework and process-knowledge based management to increase production. The new emphasis is on alternatives to agrochemical use and increasing the rates of existing biological processes to control nutrient cycling and pests. The concept of economic discounting of future value of natural resources is also altered. All of these will place far greater demands on research capacity and farmer knowledge. They will also require agricultural research to become more grounded in theory than it has been so far. Other major issues for agricultural research policy and design are described below.

- **Characterization of Systems:** Research designs for sustainability will require clear characterization of production systems, agro ecosystems and their boundaries, the marketing systems, and the linkages between them. Setting objectives: An appropriate balance is required between commodity focused research based on intensive use of agrochemicals (which formed the research paradigm up to now and which was responsible for the green revolution) and the resource management focused research (which forms the backbone of research within the sustainability paradigm).
- **Research prioritization:** Higher priority would be needed for research on systems which currently are tending towards unsustainability. And to problems which are contributing most to the degradation of the system.
- **Externalities and measurement of sustainability:** Research will have to be initiated on identification and measurement of externalities and tradeoffs to develop sustainability 7indicators for agricultural systems at different levels. This research will be interdisciplinary and will need interactions with economics and ecology, and between theory and experiment.
- **Farmer response:** The demands on farmer knowledge and responses will be much higher for sustainable agriculture than for traditional agriculture.
- **The organizational challenge:** Incorporating the sustainability perspective into research policy, design, and management will require important organizational changes in the Indian NARS. Major changes in all the three vital organizational components, namely, its structure, systems, and skills will be required.

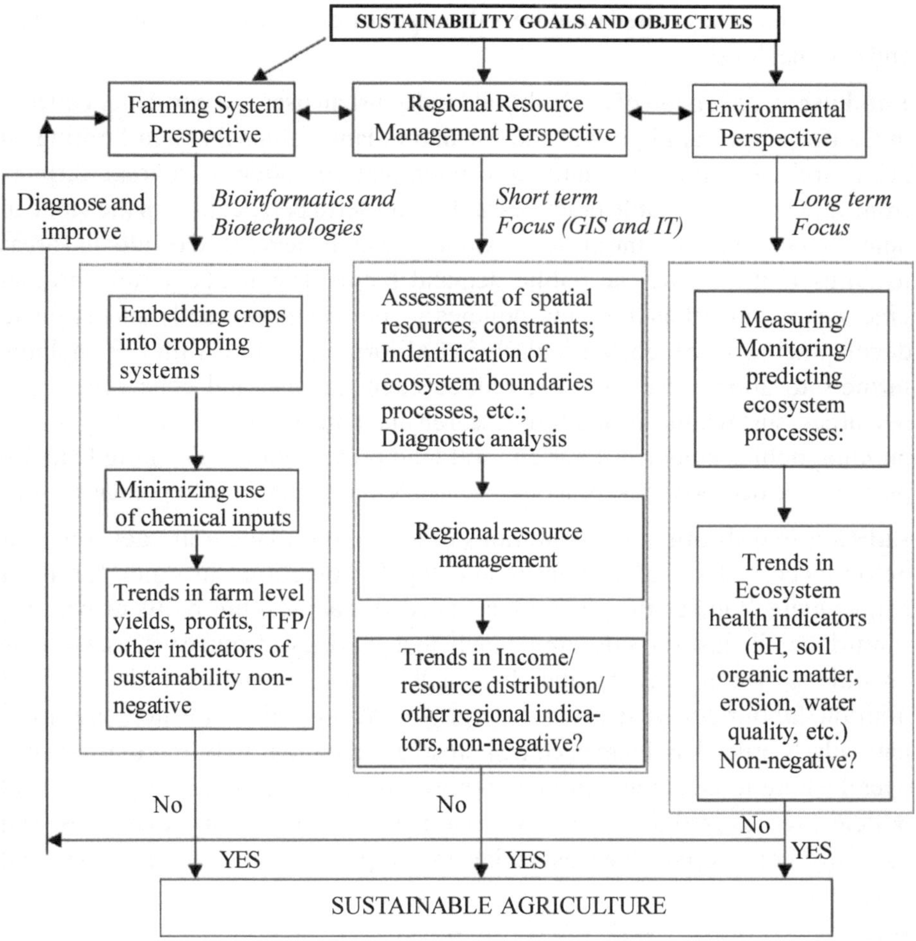


**Fig. 1:** Analytical framework for sustainable agriculture (adopted from Barnett *et al*, 1995)

## Areas of concern where change is most needed include the following:

**Food and agricultural policy:** Existing federal, state and local government policies often impede the goals of sustainable agriculture. New policies are needed to simultaneously promote environmental health, economic profitability, and social and economic equity. For example, commodity and price support programs could be restructured to allow farmers to realize the full benefits of the productivity gains made possible through alternative practices. Tax and credit policies could be modified to encourage a diverse and decentralized system of family farms rather than corporate concentration and absentee ownership. Government and land grant university research policies could be modified to emphasize the development of sustainable alternatives. Marketing orders and cosmetic standards could be amended to encourage reduced pesticide use.

Coalitions must be created to address these policy concerns at the local, regional, and national level.

**Land use:** Conversion of agricultural land to urban uses is a particular concern in California, as rapid growth and escalating land values threaten farming on prime soils. Existing farmland conversion patterns often discourage farmers from adopting sustainable practices and a long-term perspective on the value of land. At the same time, the close proximity of newly developed residential areas to farms is increasing the public demand for environmentally safe farming practices. Comprehensive new policies to protect prime soils and regulate development are needed, particularly in California's Central Valley. By helping farmers to adopt practices that reduce chemical use and conserve scarce resources, sustainable agriculture research and education can play a key role in building public support for agricultural land preservation. Educating land use planners and decision-makers about sustainable agriculture is an important priority.

**Labour:** In California, the conditions of agricultural labour are generally far below accepted social standards and legal protections in other forms of employment. Policies and programs are needed to address this problem, working toward socially just and safe employment that provides adequate wages, working conditions, health benefits, and chances for economic stability. The needs of migrant labour for year-around employment and adequate housing are a particularly crucial problem needing immediate attention. To be more sustainable over the long-term, labour must be acknowledged and supported by government policies, recognized as important constituents of land grant universities, and carefully considered when assessing the impacts of new technologies and practices.

**Rural Community Development:** Rural communities in California are currently characterized by economic and environmental deterioration. Many are among the poorest locations in the nation. The reasons for the decline are complex, but changes in farm structure have played a significant role. Sustainable agriculture presents an opportunity to rethink the importance of family farms and rural communities. Economic development policies are needed that encourage more diversified agricultural production on family farms as a foundation for healthy economies in rural communities. In combination with other strategies, sustainable agriculture practices and policies can help foster community institutions that meet employment, educational, health, cultural and spiritual needs.

**Consumers and the Food System:** Consumers can play a critical role in creating a sustainable food system. Through their purchases, they send strong messages to producers, retailers and others in the system about what they think is important. Food cost and nutritional quality have always influenced consumer choices. The challenge now is to find strategies that broaden consumer perspectives, so that environmental quality, resource use, and social equity issues are also considered in shopping decisions. At the same time, new policies and institutions must be created to enable producers using sustainable practices to market their goods to a wider public. Coalitions organized around improving the food system are one specific method of creating a dialogue among consumers, retailers, producers and others. These coalitions or other public forums can be important vehicles for clarifying issues, suggesting new policies, increasing mutual trust, and encouraging a long-term view of food production, distribution and consumption.

## Indices of sustainability

Quantification of sustainability is essential to objectively assess the impact of management systems on actual and potential productivity and on environment. One can assess sustainability or several indices. Indices may be simple involving one parameter or complex involving several parameters. Although general principle may be the same, there indices must be fine tuned and adapted under local environments. Some indices of sustainability include the following:

1. Productivity **(P):** Production per unit of resource used can be assessed by

   P = P/R

   Where, P is productivity

   P is total production

   R is resource used

2. Total Factor Productivity **(TFP):** It is defined as productivity per unit cost of all factors involved.

3. Coefficient of sustainability **(CS):** It is measure of change in soil properties in relation to production under specific management system.

   Cs = F(Oi, Od, Om)t

   Where, Cs is coefficient of sustainability

   Oi is output per unit that maximizes per capita productivity or profit

   Od is output per unit decline in the most limiting or non-renewable resource

   Om is the minimum assured output

   t is time

   The time scale is important and must be carefully selected.

4. Index of sustainability **(Is):** It is a measure of sustainability relating productivity to change in soil and environmental characteristics (Lal, 1993; Lal and Miller, 1993)

Is = f (Pi*Si*Wi*Ci)t

Where, Is is index of sustainability

Si is alteration in soil properties

Wi is change in water resources and quality

Ci is modification in climatic factor and t is time

5. Agrricultural Sustainability **(Ag):** It is a broad based on several parameters associated with agricultural production.

   As = d (Pt*Sp*Wt*Ct)dt

   Where, As is agricultural sustainability

   Pt is productivity per unit time of the limited or non renewable resources

   Sp is critical property of rooting depth, soil organic matter content

   Wt is available water capacity including water quality

   Ct is climatic factor such as gaseous flux from agricultural activity

   T is time

**Crop productivity as an indicator of sustainability**

A measure of crop productivity is a good integrator of all soil, water, climatic and biotic factors. It is important to assess potential vis-a-vis actual productivity. In a science based management system, actual production exceeds potential production in soils of low inherent fertility and in harsh environments. The potential productivity, soils productive potential within a biome, can be estimated by several models e.g., CERES and Tropical soil productivity are Land use Factor (L), Land Equivalent Ratio (LER) and Area Time Equivalent Ratio (ATER) etc.

Sustainability coefficient (SC): It is complex and a multipurpose index based on a range of parameters and is similar to As. It is defined as:

Sc + F (Pt*Pd*Pm) t

Sc = d (Pd*Wt*Ct) dt

Where, Pt is productivity per unit input of limited resource

Pd is productivity per unit decline in soil property

Sc is critical level of soil property

Wt is soil water regime and quality

Ct is climatic factor

t is time

The Land use factor (L) is defined as the ratio of cropping period C plus fallow period F to cropping period C.

L = C + F/C

The factor L is generally high for low intensity systems e.g., shifting cultivation.

**Input management for sustainable agricultural systems**

1. **Optimizing nutrient availability:** A very important condition for good plant growth and health and, indirectly for good animal and human health is the timely provision of sufficient and balanced quantities of nutrients that can be taken up by the plant roots. Nutrient deficiencies and imbalances are main constraints to crop production, especially in regions with poor or alkaline soils. There is a constant flow of nutrients through the farm. For example, it has been estimated that in Africa nutrient losses through soil erosions and other processes exceed application of artificial fertilizers.
2. **Micronutrient deficiencies:** Due to intensive cropping the micronutrients are removed to a considerable extent, which control various aspects of plant growth. A tudy of Ranchi, India revealed that applying looks NPK (10:25:25) per ha. Led to depletion of Zinc by 0.619/ha and copper by 0.49/ha, this can depress yields by up to 4 t/ha in rice, 2 t/ha in wheat and 3.4 t/ha in maize. Also iron is a limiting factor in rice production in the new rice-wheat rotation evolved in the non-traditional rice growing areas of Punjab. At Punjab Agricultural University, Ludhiana field experimental results proved that application of poultry manure, pig manure and farmyard manure were effective in meeting zinc requirements in a maize-wheat rotation.
3. **Limiting nutrient losses:** Nutrient losses can be limited by:
   - Recycling organic wastes by returning them to the field, either directly or treated (composted, fermented etc.)
   - Applying organic and artificial fertilizers in such a way that nutrients are not leached by excessive rain or volatilized by high temperature or solar radiation.
   - Reducing losses due to run-off and soil erosion.
   - Minimizing nutrient losses due to biomass bussing.
   - Reducing volatilization of nitrogen by denitrification under wet soil conditions.
   - Avoiding leaching by using organic and artificial fertilizers, which release nutrients slowly maintaining high humus content in the soil and intercropping plant species with different rooting depth.

- Limiting nutrient export in products by producing crops with relatively high economic value relative to nutrient content.

4. **Use of chemical fertilizers:** The use of chemical fertilizers is essential for obtaining high crop yields. However, many small landholders and resource poor farmers cannot offer costly fertilizers. Most soils in the tropics are so deficient in primary nutrients that it is imperative that strategies be developed for adding them from outside the ecosystem. There is some potential for enhancing N supply by biological N fixation. Additional fertilizer N and other nutrients must be supplied. The requirements for chemical fertilizers however can be reduced considerably by decreasing losses, recycling nutrients and through biological N fixation.

5. **Nutrient recycling:** Nutrient recycling or regime is an important strategy for sustainable crop production. It involves returning nutrients removed by crops to the soil for further use. In addition, soil fauna (e.g., earth worms, termites) also play an important role in recycling of plant nutrients. Growing deep-rooted crops is important in order to recycle nutrients from the sub soil by returning them through crop residue to the surface where the succeeding shallow rooted crops can use them. Use of mulches, incorporation of crop residues and animal waste, growing legumes as intercrops in cereals etc., can substantially reduce chemical fertilizer requirements.

6. **Use of crop residues:** Crop residue contain substantial quantities of plant nutrients. The beneficial effects of returning crop residues as mulch on crop yields are well known. These benefits are not only to the recycling of plant nutrients but also to improvements in soil moisture and temperature, enhancement of soil structure and soil erosion control.

7. **Biological nitrogen fixation:** Augmenting the nitrogen supply to crops through biological nitrogen fixation is a viable officer for resource poor farmers of the tropics. The amount of N fixed by legumes can range from 20-250 kg/ha/yr depending on the species, soil type, climate and agro-eco-region.

8. **Use of biofertilizers:** Biofertilizers have been recognized as important inputs in integrated plant nutrition systems. The use of legume green manure, blue green algae and Azolla for rice: Azotobacter and Azospirillum for wheat, millets and vegetable crops; Rhizobium for pulses and oil legume crops, Phosphate Solubilizers Vesicular Arbuscular Mycorrhizae) for various crops is well reported, on an average these biofertilizers can minimize the use of inorganic N by 25-50 kg/ha.

9. **Green manuring:** Green manuring is the practice of ploughing or turning into the soil undecomposed green plant tissues for the purpose of improving

physical structure as well as fertility of soil. The plant material used in this way is called a green manure (GM). The short duration legume crops are grown and buried in the same site when they attain the age of 60-80 days after sowing. This system of on-site nutrient resource generation is most prevalent in northern and southern parts of India where rice is the major crop in the existing cropping systems. Green leaves and tender plant parts of the plants are collected from shrubs and trees growing on bunds, degraded lands or nearby forest and they are turned down or mixed into the soil. It is done 15-30 days before sowing of the crops depending on the tenderness of the foliage or plant parts.

*There are now unique opportunities for launching a food-for-sustainable development initiative, in the form of a "grain for green" movement...* Prof. M. S. Swaminathan.

## Advantages of green manuring

i) Improves the soil fertility by the increasing the humus content of the soil

ii) Adds nutrients and organic matters

iii) Improves the soil structure

iv) Improves soil aeration

v) Increases the water holding capacity of the soil

vi) Increases the nutrient holding capacity of the soil

vii) Helps control the soil born insects/ mite (neem, Vitex, etc), nematodes (marigold) and diseases (crucifers)

viii) Used as a soil binder in the sloppy areas

ix) Conserves and recycles plant nutrients

x) Protects the soil from erosion

xi) Promotes habitat for natural enemies

xii) Serves as a good food for the earthworms

xiii) Increases soil biodiversity of beneficial microbes by stimulating their growth

# 4

# Organic Nutrient Management

Organic farmers use animal manure obtained from animals reared with organic standards, certain processed fertilizers such as seed meal and various mineral powders such as rock phosphate and green sand, a naturally occurring form of potash which provides potassium. In some cases pH may need to be amended. Natural pH amendments include lime and sulfur. Composting of all organic waste in general and of farm yard manure or feedlot manure in particular is important in organic farming with the following guidelines:

1. Sufficient quantities of organic material should be returned to the soil after every crop to increase or at least maintain its humus content on long term basis.
2. Accumulation of heavy metals and other pollutants should be avoided.
3. Organic material produced on organic farms should form the basis of the manure/compost programme.
4. Management and handling of manure and compost should minimize nutrient losses.
5. Soil, pH-values, which are appropriate to the soil type and the crops cultivated, should be maintained.
6. The use of human feaces is restricted.
7. Precautions should be taken to prevent the transmission of pests, parasites and other infectious agents. e. g. by high temperature composting.
8. Non-synthetic mineral fertilizers and brought in fertilizer of biological origin should be regarded as supplementary and not a replacement for nutrient recycling.
9. Minimizing the downward movement of nutrients in the soil profile:
    i) The rotation with deep-rooted crops.
    ii) The avoidance of high-solubility nutrient sources.
    iii) The avoidance of use of mould board ploughing in favour of chisel ploughing
    iv) The seasonal use of cover crops in and around the major cash crop.

***Recycling of nutrients**:* Nutrients comprise the 22 or so chemical elements, which are known to be essential for the living organisms growth. The list varies somewhat because some elements are only needed in very specific groups of organisms or specific circumstances. This deals mainly with nitrogen, sulfur, phosphorus, and carbon-all elements needed in relatively large quantities (the so called macronutrients), and with cycles that have been substantially altered by human activities.

Broadly speaking, nutrients can occur in gaseous form (such as $N_2$, $CO_2$), mineral form (such as apatite, the main P-containing mineral), inorganic ionic form ($NH_4^+$, $NO_3^+$, $SO_4^-$, $H_2PO_4$), and organic form (bound into various C-based compounds in living or dead organisms or their products). Nutrients are mostly taken up by plants in the ionic form and by animals in organic forms through consumption of living or dead tissues; microorganisms in general may use nutrients in any mineral or organic form, with sometimes high degrees of specialization at the guild or species level. The inter conversion between forms is mediated by the ecosystem.

Nutrient cycling describes the movement within and between the various biotic or abiotic entities in which nutrients occur in the global environment. These elements can be extracted from their mineral or atmospheric sources or recycled from their organic forms by converting them to the ionic form, enabling uptake to occur and ultimately returning them to the atmosphere or soil. Nutrient cycling is enabled by a great diversity of organisms and leads to creation of a number of physical structures and mechanisms that regulate the fluxes of nutrients among compartments. These structures and processes act as buffers to limit losses and transfers to other ecosystems.

***Nutrient cycle**:* A nutrient cycle (or ecological recycling) is the movement and exchange of organic and inorganic matter back into the production of living matter. The process is regulated by food web pathways that decompose matter into mineral nutrients. It means nutrient cycles occur within ecosystems. Ecosystems are interconnected systems where matter and energy flows and is exchanged as organisms feed, digest, and migrate about. Minerals and nutrients accumulate in varied densities and uneven configurations across the planet. Ecosystems recycle locally, converting the mineral nutrients into the production of biomass, and on a larger scale they participate in a global system of inputs and outputs where matter is exchanged and transported through a larger system of biogeochemical cycles in the ecosystem. Global biogeochemical cycles are the sum product of localized ecological recycling regulated by the action of food webs moving particulate matter from one living generation to the next generation.

## Organic manures

**Manures:** Manures are plant and animal wastes that are used as sources of plant nutrients. They release different nutrients after their decomposition. The art of collecting and using wastes from animal, human and vegetable sources for improving crop productivity is as old as agriculture. Manures are the organic materials derived from animal, human and plant residues that contain plant nutrients in complex organic forms. Naturally occurring or synthetic chemicals containing plant nutrients are called fertilizers. Manures with low nutrient content per unit quantity have longer residual effect besides improving soil physico-chemical and biological properties compared to fertilizer with high nutrient content.

## Major sources of manures are

1. Cattle shed wastes-dung, urine and slurry from biogas plants.
2. Human habitation wastes-night soil, human urine, town refuse, sewage, sludge and sullage.
3. Poultry litter, droppings of sheep and goat.
4. Slaughterhouse wastes-bone meal, meat meal, blood meal, horn and hoof meal, Fish wastes.
5. Byproducts of agro industries-oil cakes, bagasse and press mud, fruit and vegetable processing wastes etc.
6. Crop wastes-sugarcane trash, stubbles and other related material.
7. Water hyacinth, weeds and silt, and
8. Green manure crops and green leaf manuring material.

Manures can also be grouped, into bulky organic manures and concentrated organic manures based on concentration of the nutrients.

***Bulky organic manures:*** Bulky organic manures contain small percentage of nutrients and they are applied in large quantities. Farmyard manure (FYM), compost and green-manure are the most important and widely used bulky organic manures globally. Use of bulky organic manures has several advantages:

1. They supply plant nutrients in synchrony with plant growth including micronutrients.
2. They improve soil physical properties like structure, water holding capacity etc.
3. They increase the availability of nutrients.
4. Carbon dioxide released during decomposition acts as a $CO_2$ fertilizer and
5. Plant parasitic nematodes and fungi are controlled to some extent by altering the balance of microorganisms in the soil.

1. ***Farmyard manure:*** Farmyard manure refers to the decomposed mixture of dung and urine of farm animals along with litter and left over material from roughages or fodder fed to the cattle. On an average well decomposed farmyard manure contains 0.5 per cent N, 0.2 per cent $P_2O_5$ and 0.5 per cent $K_2O$.The present method of preparing farmyard manure by the farmers is defective. Urine, which is wasted, contains one per cent nitrogen and 1.35 per cent potassium. Nitrogen present in urine is mostly in the form of urea which is subjected to volatilization losses. Even during storage, nutrients are lost due to leaching and volatilization. However, it is practically impossible to avoid losses altogether, but the losses can be reduced by following improved method of preparation of farmyard manure. Trenches of size 6.0 m to 7.5 m length, 1.5 m to 2.0 m width and 1.0 m deep are dug. All available litter and refuse is mixed with soil and spread in the shed so as to absorb urine. The next morning, urine soaked refuse along with dung is collected and placed in the trench. A section of the trench from one end should be taken up for filling with daily collection. When the section is filled up to a height of 45 cm to 60 cm above the ground level, the top of the heap is made into a dome and plastered with cow dung earth slurry. The process is continued and when the first trench is completely filled, second trench is prepared. The manure becomes ready for use in about four to five months after plastering. If urine is not collected in the bedding, it can be collected along with washings of the cattle shed in a cemented pit from which it is later can be added to the farmyard manure pit. Partially rotten farmyard manure has to be applied three to four weeks before sowing while well rotten manure can be applied immediately before sowing. Generally 10 to 20 t/ha is applied depending upon nutrient requirement of crop, but more than 20 t/ha is applied to fodder grasses and vegetables. In such cases farmyard manure should be applied at least 15 days in advance to avoid immobilization of nitrogen. The existing practice of leaving manure in small heaps scattered in the field for a very long period leads to loss of nutrients. These losses can be reduced by spreading the manure and incorporating by ploughing immediately after its application in the field.

   The vegetable crops like potato, tomato, sweet-potato, carrot, raddish, onion etc., respond well to the farmyard manure. The other responsive crops are sugarcane, rice, maize, turmeric, napier grass and orchard crops like oranges, banana, mango and plantation crop like coconut. The entire amount of nutrients present in farmyard manure is not available immediately. About 30 per cent of nitrogen, 60 to 70 per cent of phosphorus and 70 per cent of potassium are available to the first crop.

2. ***Sheep and Goat Manure:*** The droppings of sheep and goat contain higher nutrients than farmyard manure and compost. On an average, the manure

contains 3 per cent N, 1 per cent $P_2O_5$ and 2 per cent $K_2O$. It is applied to the field in two ways. The sweeping of sheep or goat sheds are placed in pits for decomposition and it is applied later to the field. The nutrients present in the urine are wasted in this method. The second method is sheep penning, wherein sheep and goats are kept overnight in the field and urine and fecal matter added to the soil is incorporated to a shallow depth by working blade harrow or cultivator or cultivator.

3. ***Poultry Manure:*** The excreta of birds ferment very quickly. If left exposed, 50 percent of its nitrogen is lost within one month. Poultry manure contains higher nitrogen and phosphorus as compared to other bulky organic manures. The average nutrient content is 3.03 per cent N; 2.63 per cent $P_2O_5$ and 1.4 per cent $K_2O$.

4. ***Concentrated organic manures:*** Concentrated organic manures have higher nutrient content than bulky organic manure. The important concentrated organic manures are oilcakes, blood meal, fish manure etc. These are also known as organic nitrogen fertilizer. Before their organic nitrogen is used by the crops, it is converted through bacterial action into readily usable ammonical nitrogen and nitrate nitrogen. These organic fertilizers are, therefore, relatively slow acting, but they supply available nitrogen for a longer period.

5. ***Oil cakes:*** After oil is extracted from oilseeds, the remaining solid portion is dried as cake which can, be used as manure. The oil cakes are of two types:

    1. Edible oil cakes which can be safely fed to livestock; e.g. Groundnut cake, mustard cake, cotton seed cake, Coconut cake etc., and

    2. Non edible oil cakes which are not fit for feeding livestock; e.g.: Castor cake, Neem cake, Mahua cake etc.,

Both edible and non-edible oil cakes can be used as manures. However, edible oil cakes are fed to cattle and non-edible oil cakes are used as manures especially for horticultural crops. Nutrients present in oil cakes, after mineralization, are made available to crops 7 to 10 days after application. Oilcakes need to be well powdered before application for even distribution and quicker decomposition. The average nutrient content of different oil-cakes is presented in the following table.

| Oil-cakes | Nutrient content (%) | | |
|---|---|---|---|
| | N | $P_2O_5$ | $K_2O$ |
| **Non edible oil-cakes** | | | |
| Castor cake | 4.3 | 1.8 | 1.3 |
| Cotton seed cake (undecorticated) | 3.9 | 1.8 | 1.6 |
| Karanj cake | 3.9 | 0.9 | 1.2 |
| Mahua cake | 2.5 | 0. | 1.2 |
| Safflower cake (undecorticated) | 4.9 | 1.4 | 1.2 |
| **Edible oil-cakes** | | | |
| Coconut cake | 3.0 | 1.9 | 1.8 |
| Cotton seed cake (decorticated) | 6.4 | 2.9 | 2.2 |
| Groundnut cake | 7.3 | 1.5 | 1.3 |
| Linseed cake | 4.9 | 1.4 | 1.3 |
| Niger cake | 4.7 | 1.8 | 1.3 |
| Rape seed cake | 5.2 | 1.8 | 1.2 |

6. ***Other Concentrated Organic Manures:*** Blood meal when dried and powdered can be used as manure. The meat of dead animals is dried and converted into meat meal which is a good source of nitrogen. Average nutrient content of animal based concentrated organic manures is given as follows.

| Nutrient content (%) | N | $P_2O_5$ | $K_2O$ |
|---|---|---|---|
| Blood meal | 10 - 12 | 1 - 2 | 1.0 |
| Meat meal | 10.5 | 2.5 | 0.5 |
| Fish meal | 4 - 10 | 3 - 9 | 0.3 - 1.5 |
| Horn and Hoof meal | 13 | - | - |
| Raw bone meal | 3 – 4 | 20 - 25 | - |
| Steamed bone meal | 1 - 2 | 25 - 30 | - |

## Composting

Composting is the natural process of 'rotting' or decomposition of organic matter by microorganisms under controlled conditions. Raw organic materials such as crop residues, animal wastes, food garbage, some municipal wastes and suitable industrial wastes, enhance their suitability for application to the soil as a fertilizing resource, after having undergone composting. Compost is a rich source of soil organic matter. Soil organic matter plays an important role in sustaining soil fertility, and hence in sustainable agricultural production. In addition to being a source of plant nutrient, it also improves the physico-chemical and biological properties of the soil. As a result of these improvements, the soil system becomes more resistant to stresses such as drought, diseases and toxicity, helps the crop in improved uptake of plant nutrients and possesses an active nutrient cycling capacity because of vigorous microbial activity.

## Composting is necessary

- The rejected biological materials contain complex chemical compounds such as lignin, cellulose, hemicellulose, polysaccharides, proteins, lipids etc.
- These complex materials cannot be used as such as resource materials.
- The complex materials should be converted into simple available nutrient.
- The material put into soil without conversion will undergo conversion inside the soil.
- This conversion process takes away all energy and available nutrients from the soil affecting the crop. Hence conversion period is mandatory before using it as organic manure for nutrient source.

***Different Stages of Composting**:* Organic manure during its decomposition passes through different stages (Figure 1) and some of the stages are as follows:

i) ***Mesophilic stage***- This is initial stage of composting. During this stage sugars and readily available microbial food sources are readily metabolized, causing the temperature in compost pile to rise from ambient to over 40$^0$C.

ii) ***Thermophilic stage***- This stage occurs during the next week or two. Temperature in pile rises to 50 to 75$^0$C. Thermophilic organisms decompose cellulose and other more resistant materials. Frequent mixing is essential to maintain $O_2$ supplies in pile and assure even heating of all material. Humus like compounds is formed.

iii) ***Second mesophilic or curing stage***- This stage follows the temperature falls back to near ambient. The material is re-colonized by mesophilic organisms that produce plant growth stimulating compounds or are antagonistic to plant pathogenic fungi. This stage may last from several weeks to several months. Longer curing gives best quality compost.

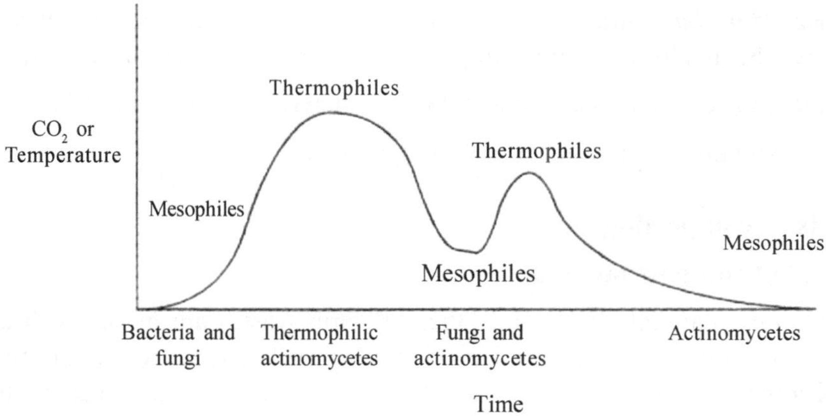


**Figure 1.** Different stages of composting

### Compost making includes three important principles

1. Narrowing down of C: N ratio to a satisfactory level (10:1 to 2:1).
2. Total destruction of harmful pathogens.
3. Destruction of weed seeds by high temperature (60-65o C).

### Controlling Pathogens

Pathogens are organisms-such as bacteria, fungi, and nematodes-that incite disease. There are pathogens that are dangerous to humans, animals, and plants. The heat produced during composting helps to control all of these.

### Organic Resources available for Composting

- Livestock and human wastes.
- Crop residues, tree litter and weeds.
- Green manure.
- Urban and rural wastes.
- Agro industrial by-products.
- Marine wastes.
- In India, it is estimated that the annual production of dung and urine of livestock and human beings amounts to about 2000 mt with the potential to supply 6.4, 2.0 and 2.7 mt of N, $P_2O$ and $K_2O$, respectively.

### Requirements of composting

- ***Bulky Organic Refuse:*** stubbles, cotton stalks, tur stalks, ground nut shell, weeds, leaves, dust bin refuse etc.
- ***Starter Material:*** Cattle dung, urine, night soil, sewage, urea, rockphosphate or any other readily available nitrogenous substance and microbial cultures.
- ***Water:*** to keep the moisture content of the material at a level of 50%.
- ***Air:*** especially in initial stages of decomposition.

### Methods of composting

#### 1. Activated compost method

The basic raw material (straw and other off-farm waste) are mixed with a mixture of cowdung and urine as decoction so as each and every part of raw material comes in contact with a mixture of cowdung and urine and fermentation takes place.

- The material is piled up in a heap of 1-1.5 m height. It is turned over from time to time.
- The material is kept moist with the mixture of cowdung and urine decoction.
- Very high temperature is attained, when temperature begins to drop then volume of material gets reduce.
- Then basic material is added again. About 25% new mixture of raw material mixed with cowdung and urine is added.
- If methodology is carried out properly, then the compost will be ready in 5-6 weeks.

## 2. Indore Method

Waste materials are chopped into 5-10 cm pieces and dried to 40-50 % moisture level. They are spread in layers of 10-15 cm thickness either in pits or in heaps of 1.0 m width, 4.0-5.0 m length and 1.0 m depth. The heap is properly moistened with dung using earth or night soil.

- Sufficient quantity of water is sprinkled over the heap to wet the composting materials to the level of 50 % moisture.
- Periodical turnings (15, 30 and 60 days) are given to aerate and material is covered with a thin layer of soil (2-3 thickness).
- The average composition of manure: 0.8% N, 0.3% $P_2O$ and 1.5 % $K_2O$, respectively.

## Disadvantages

- Requires labour for turning
- Not practicable if large quantity of material is there
- Site should be at elevated level near a cattle shed and water source.

## 3. Bangalore method of composting

Dr. C. N. Acharya (1949) developed this method for the utilization of town residues and night soil. It is hot fermentation method.

- The compost production depot is located on the city outskirts.
- First, refuse is filled in the trench to about 15 cm height.
- The night soil is spread over this to a layer of 5 cm. After filling the pit with refuse and night soil in alternate layers, the pit is filled to 15 cm above the ground level with a final layer of 15 cm on the top.
- This may be made dome-shaped and converted with thin layer of soil with red earth or mud to prevent moisture loss and breeding of flies.

- The materials are allowed to remain as such without any turning and pot watering for about three months.
- The compost obtained by this method contains 1.5% N, 1.0% $P_2O$ and 1.5 % $K_2O$, respectively.

### 4. Nadep compost

The Nadep method of making miracle compost was first invented by a farmer named N.D. Pandharipande (also popularly known as "Nadepkaka") living in Maharashtra (India).

- The method, which has become quite popular among the farmers in Western India, now bears his name.
- The Nadep method of making compost is unique not because it is successful in making good compost, which other methods can also lay claim to; its real secret lies in the large quantities of compost the process can deliver with a minimum of human effort within a specific period of time.
- The process basically involves placing select layers of different types of compostible materials in a simple, mud-sealed structure designed with brick and mud water.
- The system permits conversion of approximately 1 kg of animal dung into 40 kg of rich compost which can then be applied directly to the field.
- The multiplication factor is significant in view of the fact that in the tropics, there is rapid decomposition of organic materials in the soil.
- This organic matter must be replaced and replenished if agricultural fertility is not to go on declining.
- The problem is a scarcity of compostable materials, particularly animal dung, prevailing within the country. (A good proportion of animal dung is dried and used as fuel in many rural areas.) Thus, even if all available organic materials, including dung, were religiously and scrupulously collected, they would still not be sufficient to replace the organic constituents of the vast quantities of India's fast-degrading soils. Compost can be prepared from a wide range of organic materials including dead plant material such as crop residues, weeds, forest litter and kitchen waste.
- Compost making is an efficient way of converting all kinds biomass into high value fertilizer that serves as a good alternative to farmyard manure, especially for crop-growing households without livestock.
- The Nadep method of composting actually enables the farmer to get around the difficulty of generation of mass and to increase the quantity of compost

rapidly within a given frame of time and without any significant additional expense.

## Description

- This method of making compost involves the construction of a simple, rectangular brick tank with enough spaces maintained between the bricks for necessary aeration.
- The recommended size of the tank is 10 ft (length) x 5 ft (breadth) x 3 ft (height).
- All the four walls of NADEP tank are provided with 6// vents by removing every alternate brick after the height of 1ft. from bottom for aeration.
- Tank can be constructed in mud mortar or cement mortar.

## Establishment Activities

Raw materials required for filling NADEP tank

- Agricultural waste (Dry & green)-1350-1400 kgs.
- Cattle dung or biogas slurry-98-100 kgs.
- Fine sieved soil-1675 kgs.
- Water-1350-1400 litres.
- The important technique in the manufacture of Nadep compost is that the entire tank should be filled in one go, within 24 hours and should not go beyond 48 hours, as this would affect the quality of the compost.
- Thatched roof, Brick Wall Flooring, Air Vents, Green-farm technologies for small and marginal farms
- Before filling: the tank is plastered by dilute cattle dung slurry to facilitate bacterial activity from all four sides. It is also filled in definite layers each layer consisting of the following sub layers.

## Sub-layer-1

- 4 to 6 inch thick layer of fine sticks, stems, (To facilitate aeration) followed by 4 to 6 inch layer of dry and green biomass.

## Sub-layer-2

- 4 kgs. Cow dung is mixed with 100 litres of water and sprinkled thoroughly on the agricultural waste to facilitate microbial activity.

**Sub-layer-3**

- 60 kgs. of fine dry soil is spread uniformly over the soaked biomass for moisture retention and acts as a buffer during biodegradation. Thus the proportion of organic materials for each layer is 100 kgs.
- Organic biomass: 4 kgs. cowdung + 100 litres water + 60 kgs soil. In this way, approximately 10 -12 layers are filled in each tank. After filling the tank, biomass is covered with 3// thick layer of soil and sealed with cow dung + mud plaster.

**Maintainance**

- After 15-30 days of filling the organic biomass in the tank gets automatically pressed down to 2 ft.
- The tank is refilled by giving 2-3 layers over it and is resealed.
- After this filling the tank is not disturbed for 3 months except that it is moistened at intervals of every 6-15 days.
- The entire tank is covered with a thatched roof to prevent excessive evaporation of moisture.
- Cracks should not be allowed to develop. If they do, they should be promptly filled up with slurry.

**Benefits**

- Reduced cash expenses on chemical fertilizer, improved soil fertility, increased crop yield.
- Supports organic crop production, reduced dependence on outside inputs.
- From each NADEP tank approximately 2.5 tons of compost is prepared within 90-120 days.
- The use of compost reduced the need for mineral fertilizer thus reducing production costs and outside dependence.

**5. Coimbatore method**

- Composting is done in pits of different sizes depending on the waste material available. A layer of waste materials is first laid in the pit. It is moistened with a suspension of 5-10 kg cow dung in 2.5 to 5.0 Lt of water and 0.5 to 1.0 kg fine bone meal sprinkled over it uniformly. Similar layers are laid one over the other till the material rises 0.75 m above the ground level. It is finally plastered with wet mud and left undisturbed for 8 to 10 weeks. Plaster is then removed, material moistened with water, given a turning and made into a rectangular heap under a shade. It is left undisturbed till its use.

- In Coimbatore method, there is anaerobic decomposition to start with, following by aerobic fermentation. It is the reverse in Bangalore method. The Bangalore compost is not so thoroughly decomposed as the Indore compost or even as much as the Coimbatore compost, but it is bulkiest.

**6. Windrow composting**

Windrow composting consists of placing the mixture of raw materials in long narrow piles or windrows which are agitated or turned on a regular basis. The turning operation mixes the composting materials and enhances passive aeration. Typically the windrows are initially from 3 feet high for dense materials like manures to 12 feet high for fluffy materials like leaves. The width varies from 10 to 20 feet. The equipment used for turning determines the size, shape, and spacing of the windrows. Bucket loaders with a long reach can build high windrows. Turning machines produce low, wide windrows. Windrows aerate primarily by natural or passive air movement (convection and gaseous diffusion). The rate of air exchange depends on the porosity of the windrow. Therefore, the size of a windrow that can be effectively aerated is determined by its porosity. A light fluffy windrow of leaves can be much larger than a wet dense windrow containing manure. If the windrow is too large, anaerobic zones occur near its centre which release odours when the windrow is turned. On the other hand, small windrows lose heat quickly and may not achieve temperatures high enough to evaporate moisture and kill pathogens and weed seeds. For small to moderate scale operations, turning can be accomplished with a front end loader or a bucket loader on a tractor. The loader simply lifts the materials from the windrow and spills them down again, mixing the materials and reforming the mixture into a loose windrow. The loader can exchange material from the bottom of the windrow with material on the top by forming a new windrow next to the old one. This needs to be done without driving onto the windrow in order to minimize compaction. Windrows turned with a bucket loader are often constructed in closely spaced pairs and then combined after the windrows shrink in size. If additional mixing of the materials is desired, a loader can also be used in combination with a manure spreader. A number of specialized machines have been developed for turning windrows. These machines greatly reduce the time and labour involved, mix the materials thoroughly, and produce a more uniform compost. Some of these machines are designed to attach to farm tractors or front-end loaders; others are self-propelled. A few machines also have the capability of loading trucks or wagons from the windrow. It is very important to maintain a schedule of turning. The frequency of turning depends on the rate of decomposition, the moisture content and porosity of the materials, and the desired composting time. Because the decomposition rate is greatest at the start of the process, the frequency of turning decreases as the windrow ages. Easily degradable or high nitrogen mixes may require daily turnings at the start of the process. As the process continues, the turning frequency can be reduced to a single turning per week. By the end of the first week of composting, the windrow height diminishes appreciably and by the end of the second week it may be as low as 2 feet. It may be prudent to combine two windrows at this stage and continue the turning schedule as before. Consolidation of windrows is a good wintertime practice to retain the heat generated during composting. This is one of the advantages of windrow composting. It is a versatile system that can be adjusted to different conditions caused by seasonal changes. With the windrow method, the active composting stage generally lasts three to nine weeks depending upon the nature of the materials and the frequency of turning. Eight weeks is a common period for manure composting operations. If three weeks is the goal, the windrow requires turning once or twice per day during the first week and every three to five days thereafter.

### 7. Chinese rural composting

The pit method, composting is carried out in a corner of a field and in a circular or rectangular pit. Rice straw, animal dung (usually pig), aquatic weeds or green manure crops are used and often silt pumped from river beds is mixed with the crop residues. The pits are filled layer by layer, each layer being 15 cm thick. Usually, the first layer is of a green manure crop or water hyacinth, the second layer is a straw mixture and the third layer is of animal dung. These layers are alternated until the pit is full, when a top layer of mud is added; a water layer of about 4 cm depth is maintained on the surface to create anaerobic conditions which help to reduce losses of nitrogen. Approximate quantities of the different residues in tons per pit are: river silt 7.5, rice straw 0.15, animal dung 1.0, aquatic plants or green manure 0.75 and superphosphate 0.02. Three turnings are given in all, the first one month after filling the pit and, at this time, the superphosphate is added and thoroughly mixed in. Water is added as necessary. The second turning is done after another month and the third two weeks later. The material is allowed to decompose for three months and produces about eight tons of compost per pit.

### 8. Japanese method of composting (Bokashi composting)

Bokashi composting uses anaerobic bacteria (those able to thrive in low oxygen environments) and inoculated bran to ferment food waste. Popular among busy professionals who don't have time to sort kitchen scraps, this Japanese method accommodates foods typically banned from composting piles, such as meat, dairy and grains.

Begin by placing food scraps into a bucket with a tight seal. Most Bokashi buckets are small and portable, making them a practical composting system for urban dwellers. Once your refuse is in the bucket, pour about a handful of bran on top and close the lid. When the bucket is full, set it aside for another 10 to 14 days, draining the leachate every other day. (SCD Probiotics' Bokashi bucket comes with a spigot made for this purpose.) You'll eventually end up with what some experts call "pre-compost." The product at this stage can be very acidic, so you'll want to cure it by burying it underneath soil, away from plant roots. After another two weeks, this phase should be complete, allowing you to plant above the plot or transfer the finished compost elsewhere.

## Factors influencing the bio composting process

a) ***Particle Size:*** If particle is small, the space for the growth of microorganisms will be more which ultimately increases the microbial activity and fastened the composting process.

b) ***Nutrients:*** Any waste materials which do not contain heavy metals/toxic waste should be used. The list of nutrient value as in Table-1 should be taken in account to increase the nutrition value of the waste. Preferably, the local waste with some available agro waste is always ideal.

c) ***Moisture:*** The total net weight moisture should be maintained 50-60% which is optimum for microbial growth. Hence, this percentage of moisture should be maintained throughout process which will increase the process of decomposition. If the moisture is below 50%, the microbial activity is less, which results slow down the composting process. If the moisture is above 60%, the anaerobic condition (clogging) will take place which also slows down the activity of microorganisms.

d) ***Aeration and Agitation:*** For the growth and activity of microorganism, oxygen supply is most essential part. Any process to be adopted only to ensure the oxygen supply. The agitation, turning by any tools or by means of Aero tilling may be used.

## Advantages of composting

- Volume reduction of waste
- Final weight of compost is very less
- Composting temperature kill pathogen, weed seeds and seeds
- Matured compost comes into equilibrium with the soil.
- During composting number of wastes from several sources are blended together
- Excellent soil conditioner

- Saleable product
- Improves manure handling
- Reduces the risk of pollution
- Pathogen reduction
- Additional revenue.
- Suppress plant diseases and pests
- Reduce or eliminate the need for chemical fertilizers
- Promote higher yields of agricultural crops
- Facilitate reforestation, wetlands restoration, and habitat revitalization efforts by amending contaminated, compacted, and marginal soils
- Cost-effectively remediate soils contaminated by hazardous waste
- Remove solids, oil, grease, and heavy metals from storm water runoff
- Capture and destroy 99.6 percent of industrial volatile organic chemicals (VOCs) in contaminated air
- Provide cost savings of at least 50 percent over conventional soil, water, and air pollution remediation technologies, where applicable.

**Drawbacks of using composts**

Agricultural use of composts remains low for several reasons:

- The product is heavy and bulky, making it expensive to transport
- The nutrient value of compost is low compared with that of chemical fertilizers, and the rate of nutrient release is slow so that it cannot usually meet the nutrient requirement of crops in a short time, thus resulting in some nutrient deficiency
- The nutrient composition of compost is highly variable compared to chemical fertilizers
- Agricultural users might have concerns regarding potential levels of heavy metals and other possible contaminants in compost, particularly mixed municipal solid wastes. The potential for contamination becomes an important issue when compost is used on food crops
- Long-term and/or heavy application of composts to agricultural soils has been found to result in salt, nutrient, or heavy metal accumulation and may adversely affect plant growth, soil organisms, water quality, and animal and human health.

***Longterm benefits of composts in agriculture:*** Compost has been considered as a valuable soil amendment for centuries. Most people are aware that using composts is an effective way to increase healthy plant production, help save money, reduce the use of chemical fertilizers, and conserve natural resources. Compost provides a stable organic matter that improves the physical, chemical, and biological properties of soils, thereby enhancing soil quality and crop production. When correctly applied, compost has the following beneficial effects on soil properties, thus creating suitable conditions for root development and consequently promoting higher yield and higher quality of crops.

### i) Physical properties of soils

- Reduces the soil bulk density and improves the soil structure directly by loosening heavy soils with organic matter, and indirectly by means of aggregate-stabilizing humus contained in composts. Incorporating composts into compacted soils improves root penetration and turf establishment.
- Increases the water-holding capacity of the soil directly by binding water to organic matter, and indirectly by improving the soil structure, thus improving the absorption and movement of water into the soil. Therefore, water requirement and irrigation will be reduced.
- Protects the surface soil from water and wind erosion by reducing the soil-dispersion action of beating raindrops, increasing infiltration, reducing water runoff, and increasing surface wetness. Preventing erosion is essential for protecting waterways and maintaining the quality and productivity of the soil.
- Helps bind the soil particles into crumbs by the fungi or actinomycetes mycelia contained in the compost and stimulated in the soil by its application, generally increasing the stability of the soil against wind and water erosion.
- Improves soil aeration and thus supplies enough oxygen to the roots and escapes excess carbon dioxide from the root space.
- Increases the soil temperature directly by its dark color, which increases heat absorption by the soil, and indirectly by the improved soil structure.
- Helps moderate soil temperature and prevents rapid fluctuations of soil temperature, hence, providing a better environment for root growth. This is especially true of compost used as surface mulch.

### ii) Chemical properties of soils

- Enables soils to hold more plant nutrients and increases the cation exchange capacity, anion exchange capacity, and buffering capacity of soils for longer periods of time after composts are applied to soils. This is important mainly for soils containing little clay and organic matter.

- Composts contain the major nutrients required by all plants such as N,P,K, and secondary nutrients such as calcium (Ca), magnesium (Mg) and sulphur (S) plus essential micronutrients or trace elements, such as copper (Cu), zinc (Zn), iron (Fe), manganese (Mn), boron (B) and molybdenum (Mo).
- The nutrients from mature composts are released to the plants slowly and steadily. The benefits will last for more than one season.
- Stabilizes the volatile nitrogen of raw materials into large protein particles during composting, thereby reducing N losses.
- Provides active agents, such as growth substances, which may be beneficial mainly to germinating plants.
- Adds organic matter and humus to regenerate poor soils.
- Buffers the soil against rapid changes due to acidity, alkalinity, salinity, pesticides, and toxic heavy metals.

### iii) Biological properties of soils

- Supplies food and encourages the growth of beneficial microorganisms and earthworms.
- Helps suppress certain plant diseases, soil borne diseases, and parasites.
- Research has shown that composts can help control plant diseases (e.g. Pythium root rot, Rhizoctonia root rot, chili wilt, and parasitic nematode) and reduce crop losses. A major California fruit and vegetable grower was able to cut pesticide use by 80% after three years of compost applications as part of an organic matter management system. Research has also indicated that some composts, particularly those prepared from tree barks, release chemicals that inhibit some plant pathogens.

### iv) Control of diseases in soils

- Successful competition for nutrients by beneficial microorganisms;
- Antibiotic production by beneficial microorganisms;
- Successful predation against pathogens by beneficial microorganisms;
- Activation of disease-resistant genes in plants by composts; and
- High temperatures that result from composting kill pathogens.

### v) Control of weeds

Reduces and kills weed seeds by a combination of factors including the heat of the compost pile, rotting, and premature germination.

### vi) Buffering and exchange capacity

From 20 to 70 % of the exchange capacity of many soils is caused by colloidal humic substances. Total acidities of isolated fractions of humus range from 300 to 1400 meg/100g. As far as buffer action is concerned, humus exhibits buffering over a wide pH range.

### vii) Adsorption of pesticides and other organic chemicals

The immense importance of organic matter in influencing the behaviour of organic pesticides in soil, including effectiveness against the target species, phytotoxicity to subsequent crops. Organic matter content has been found to be directly related to the sorption of most herbicides.

### viii) Incidence of plant pathogens

A plentiful supply of organic matter may favor the growth of saprophytic organisms relative to parasitic ones and thereby reduce populations of the latter. Biologically active compounds in soil, such as antibiotics and certain phenolic acids, may enhance the ability of certain plants to resist attack by pathogens.

## Organic residues

Organic residues are the part of the substance that remain after its initial usage as in the case of nutrients (both organic and inorganic forms) as well as unutilized byproducts of crop cultivation with beneficial/harmful effects on subsequent cropping. Crop residues are defined as the un-economic plant parts that are left in the field after the harvest and remains that are generated from packing sheds or that are discarded during crop processing. The greatest potential as a biomass resource appears to be from the field residues of rice, corn, wheat, soybean and grain sorghum and the processing residues of sugarcane (bagasse) and cotton (gin waste).

Crop residues, however, require preparation usually consisting of size reduction and drying, before they are suitable for use a biomass feedstock for conversion of energy. Two principal conversion methods are used

- ***Thermochemical:*** including direct combustion, pyrolysis, liquefaction and gasification with air or oxygen.
- ***Biological:*** including digestion and hydrolysis followed by fermentation.

Processing of crop residues produces a directly usable form of energy as hot combustion gas or an indirect energy form, such as liquid, solid or gaseous fuel.

The estimated residue production from commonly cultivated crops is given as follow.

| Crop | Residue yield (t/ha) |
|---|---|
| Corn | 5.49 |
| Wheat | 2.04 |
| Soybean | 1.72 |
| Grain sorghum | 3.05 |
| Sugarcane | 85.07 |
| Cotton | 0.52 |

***Organic residue potential:*** A large number of crops are grown in India. After making use of their economic parts, the remaining portion is mostly wasted except for a few crops. The adoption of mechanized farming in many advanced regions in the country has resulted in leaving a sizable amount of crops/stalk in the field after harvesting. The potential of crop residues of major cereals, pulses, oilseeds and commercial crops for recycling of valuable plant nutrients for sustained crop production is enormous. On National basis, not more than one third of these residues is available for utilization and in general 50 % of the nutrients are mineralized in the soil on decomposition in a cropping season. The crop residue potential in India is given in Table 1.

**Table 1.** Residue availability, their nutrient content and nutrient potential.

| Crops | Residue yield ('000 t) | Nutrient content (%) | | | Nutrient potential ('000 ton) |
|---|---|---|---|---|---|
| | | N | P | K | |
| Rice | 80744 | 0.61 | 0.09 | 1.15 | 1493.8 |
| Wheat | 44987 | 0.48 | 0.07 | 0.98 | 688.3 |
| Sorghum | 11563 | 0.52 | 0.10 | 1.21 | 216.2 |
| Maize | 6219 | 0.58 | 0.09 | 1.25 | 119.4 |
| Pearlmillet | 8283 | 0.45 | 0.07 | 0.95 | 121.6 |
| Barley | 3180 | 0.52 | 0.08 | 1.25 | 88.8 |
| Sugarcane | 15645 | 0.45 | 0.08 | 1.20 | 270.7 |
| Potato | 5062 | 0.52 | 0.09 | 0.85 | 73.9 |
| Groundnut | 9580 | 1.65 | 0.12 | 1.23 | 277.3 |
| Total | 185263 | | | | 3320.0 |

*Source :* Bhardadwaj and Gaur (1985)

Among various crop residues available in country wheat straw is mainly used as feed for cattle. Though poor in quality as compared to green fodder, this is most widely used crop residue to meet the feed requirement, particularly fodder scarcity. Rice straw, though not good in feeding quality due to high silica content, however, basmati rice residues are used for feeding cattle. However, in regions where mechanical harvesting of wheat and rice is done, a sizeable quantity is left in the field. These are either burnt or incorporated into the soil while preparing the field for the next crop. Among different crop residues sugarcane trash is available in large quantity and its management assumes special significance because it is hard to decompose and farmers prefer to burn them in the sugarcane field. The nutrient content of common farm residues is presented in Table 2.

**Table 2.** Average nutrient content of common farm residues

| Residue | Nutrient content (%) | | |
|---|---|---|---|
| | N | P | K |
| FYM | 0.80 | 0.41 | 0.74 |
| Paddy straw | 1.59 | 1.34 | 3.37 |
| Wheat straw | 2.90 | 2.05 | 0.90 |
| Water hyacinth compost | 2.00 | 1.00 | 2.30 |
| Sugarcane trash | 2.73 | 1.81 | 1.31 |

*Source:* Rani Perumal (1994)

***Crop residue components:*** There are two major components of crop residue available

1. Harvest refuses that include straws, stubble, stover and haums of different crops.
2. Process waste like groundnut shell, oil cakes, rice husks, coir pith and cobs of maize, sorghum and bajra.

The residues can be directly applied to the field and ploughed in or used for fuel, animal feed, compost and other domestic purposes like fencing and roofing (Table 3). One way of profitably recycling the crop residues is their use as surface mulching materials. Mulches are thermo-insulators and barrier to vapour transfers and can also be used for moisture conservation.

**Table 3.** Crop residue uses

| Crop | Residue | Use |
|---|---|---|
| Cereals | Straw | Cattle feed, thatching of dwelling huts and cattle sheds |
| Pigeon pea | Stem | Fuel for cooking food, thatching of dwelling huts and cattle sheds |
| Cotton | Stem | Fuel for cooking food, thatching of dwelling huts and cattle sheds |
| Linseed | Stem | Fibre |
| Sugarcane | Tops Bagasse | Green fodder for cattleFuel in sugar factory; raw material for making hard board or paper |
| Peanuts | Greentops | Cattle feed |

*Source:* Rajendra Prasad and Power (1991)

***Effect on soil management:*** Soil degradation reduces its productivity and is serious problem on much of the land in semi-arid regions. To avert continued degradation, the soil productivity balance must be shifted from degrading processes to conservation practices. Crop residues management and conservation tillage are on the positive side of the balance. When adequate residues are available and conservation tillage is used, soil erosion is greatly reduced and water conservation is enhanced. Water conservation is important for improving crop yields in semi-arid regions, especially where irrigation is not used (Smith and Ellicot, 1990). A major constraint to residue management in many countries in low production and wide spread use of residues for other purposes. In such case, clean tillage and appropriate support practices such as contouring, furrow diking, strip cropping and terracing may provide adequate soil and water conservation benefits. Wherever residues are available, they can be recycled permitting more of them to be retained on land for soil and water conservation purpose. Some alternative practices include limited or selective residual removal, substituting high quality forages for residues as animal feed, alley, cropping, using wasteland areas more effectively, improving the balance between feed supplies and animal populations and using alternative fuel forces.

Crop residues have the potential for improving soil and water conservation and sustaining soil productivity and enhancing crop yield. However, in semi-arid regions without irrigation, the amount if residues produce is often limited and, in many cases, they are used for feed, fuel and shelter. The amount of surface residue needed to reduce erosion is relatively low. Unless practical alternatives are available, adequate residues should be retained on the land to control erosion. When residue exceeds the amount needed for the selected level of erosion control, the excess could be removed and used for other purpose. Certainly, the type of material must be considered. Also, the amounts show may not prevent erosion but erosion control and water conservation increase the amount of surface residues. Because of the competition for crop residues for many uses, management options, if implemented could reduce the demand of residues, yet provide some residues in some cases. These options include limited residual removal, selective residual removal and substituting high quality forages for residues.

Organic matter is regarded as a very important parameter of soil fertility and productivity. It has number of important roles to play in soil, both in their physical structure, sink for plant nutrients and medium for biological activities. Organic matter has greatest contribution to soil productivity. It provides nutrients to the soil, improves water holding capacity and helps the soil to maintain better aeration for seed germination and plant root development (Zia *et al*., 1993). Use of compost can be beneficial to improve organic matter status in soil because compost is rich source of nutrients with high organic matter content. Depletion of nutrients and poor organic matter contents of Pakistani soils can only be replenished by applying compost to these soils (Sarwar, *et al*., 2008).

Compost prepared from crop residues, leaves, grass clippings, plant stalks, wines, weeds, twigs and branches are very good alternative which proved useful in many countries of the world. Use of compost has not only been adopted to enhance soil organic matter and enrich it in different nutrients but also to control the environmental pollution from debris. Raw manure use has often been associated with imbalances in soil fertility (Kuepper, 2003) because it is often rich in specific nutrients like phosphate or potash. In contrast an effective composting process converts raw waste products into humus, which is relatively stable, and chemically active organic fraction found in fertile soils.

## USDA Target Ranges for Compost Piles

1. Temperature of 120 to 140°F
2. Moisture content of 50 to 60 percent.
3. pH of 6.5 to 8.5
4. Bulk density of less than 1,100 pounds per cubic yard (40 lb per cubic foot)

## Compost application on the organic farm

Calculating application rates for compost is similar to figuring the application rates for manure on cropland. First, estimate the nitrogen (N), phosphate (P) and potassium (K) requirement for the crop (based on realistic yield expectations for a particular soil or field). A chemical analysis will show the N-P-K content of the compost. Then calculate a compost rate based on N, P, or K as the critical or priority nutrient. The amount of compost to apply is calculated from the recommended rate of the priority nutrient and the plant-available nutrient content of the compost.

***Compost application rate formula:*** The amount of compost to apply equals the *Recommended Amount of Priority Nutrient* divided by the *Plant Available Priority Nutrient* (the total priority nutrient concentration of the compost times the availability coefficient). For example, if compost contains 25.0 kg total N/ton (wet weight basis) and the availability coefficient is 0.25 (for incorporated material), then the plant available N is 25 times 0.25,or 6.25 kg/ton. If the priority nutrient is N and the recommended amount of N for the crop is 125 kg/ha, then the appropriate application rate is 125 (recommended amount) divided by 6.25 (plant available N), or 20 tons of compost per hectare.

## Soil biota and deomposition of organic residues

***Soil biota:*** Soil biota, the biologically active powerhouse of soil, includes an incredible diversity of organisms. Tons of soil biota, including micro-organisms (bacteria, fungi, and algae) and soil "animals" (protozoa, nematodes, mites, springtails, spiders, insects, and earthworms), can live in an acre of soil and are more diverse than the community of plants and animals above ground. Soil biota are concentrated in plant litter, the upper few inches of soil, and along roots. Soil organisms interact with one another, with plant roots, and with their environment, forming the soil food web.

***Soil biota function:*** As soil organisms consume organic matter and each other, nutrients and energy are exchanged through the food web and are made available to plants. Each soil organism plays a role in the decomposition of plant residue, dead roots, and animal remains. The larger soil organisms, such as millipedes and earthworms, shred dead leaves and residue, mix them with the soil, and make organic material more accessible to immobile bacteria. Earthworms can completely mix the top 6 inches of a humid grassland soil in 10 to 20 years. Ants and termites mix and tunnel through soils in areas of arid and semiarid rangeland. Predators in the soil food web include scorpions, centipedes, spiders, mites, some ants, insects, and beetles. They control the population of soil biota. The smaller organisms, including mites, springtails, nematodes, and one-celled protozoa, graze on bacteria and fungi. Other organisms feed on dead roots, shredded residue, and the fecal by-products of the larger organisms. The smallest soil organisms, microscopic bacteria and fungi, make up the bulk of the biota in the soil. They finish the process of decomposition by breaking down the remaining material and storing its energy and nutrients in their cells. Algae and fungi are the first organisms to colonize rock and form "new soil" by releasing substances that disintegrate rock.

***Importance of soil biota:*** Through their interactions in the soil food web, the activities of soil biota link soil with the plants and animals above ground. Soil organisms perform essential functions that allow soil to resist degradation and provide benefits to all living things.

***Residue decomposition:*** Without the soil food web, the remains of dead plants and animals would accumulate on the earth's surface, making nutrients unavailable to plants. Soil biota decomposes these organic residues and some forms of organic matter in the soil. They convert these materials into new forms of organic matter and release carbon dioxide into the air. Many of the biota can break down pesticides and pollutants.

***Nutrient storage and release:*** Most of the annual nutrient needs are supplied through decomposition of organic matter in the soil. As soil organisms consume organic materials, they retain (immobilize) nutrients in their cells. This process prevents the loss of nutrients, such as nitrogen, from the root zone. When fungi and bacteria die or are eaten by other organisms, nutrients are mineralized, that is, slowly released to the soil in plant-available forms. Nutrient immobilization and mineralization occur continuously throughout the year. Some bacteria and fungi provide nutrients to plants in exchange for carbon. Special types of bacteria, called visible nodules.

The bacteria convert nitrogen from the air in the soil into a form that the plant host can use. When the leaves and roots die and decompose, nitrogen levels increase in the surrounding soil, improving the growth of other plants. Fungi produce hyphae that frequently look like fine white entangled threads in the soil. Some fungal hyphae (mycorrhizal fungi) attach to plant roots and act like an extended root system, providing nutrients and water to the plant.

***Water storage, infiltration, and resistance to erosion:*** Soil biota form water-stable aggregates that store water and are more resistant to water erosion and wind erosion than individual soil particles. Threads of fungal hyphae bind soil particles together. Bactenria and algae excrete material that "glues" soil into aggregates. As they tunnel through the soil, the larger soil biota form channels and large pores between aggregates, increasing the water infiltration rate and reducing the runoff.

## Earthworms and vermicompost

Microorganisms and earthworms are important biological organisms helping nature to maintain nutrient flows from one system to another and also minimize environmental degradation. The earthworm population is about 8–10 times higher in uncultivated area. This clearly indicates that earthworm population decreases with soil degradation & thus can be used as a sensitive indicator of soil degradation.

## What is vermicomposting?

Vermicomposting is a simple biotechnological process of composting, in which certain species of earthworms are used to enhance the process of waste conversion and produce a better end product. Vermicomposting differs from composting in several ways. It is a mesophilic process, utilizing microorganisms and earthworms that are active at 10–32°C (not ambient temperature but temperature within the pile of moist organic material). The process is faster than composting; because the material passes through the earthworm gut, the resulting earthworm castings (worm manure) are rich in microbial activity and plant growth regulators, and fortified with pest repellence attributes. In short, earthworms, through a type of biological alchemy, are capable of transforming garbage into 'gold'.

***Source of plant nutrients:*** Earthworms consume various organic wastes and reduce the volume by 40–60%. Each earthworm weighs about 0.5 to 0.6 g, eats waste equivalent to its body weight and produces cast equivalent to about 50% of the waste it consumes in a day. These worm castings have been analyzed for chemical and biological properties. The moisture content of castings ranges between 32 and 66% and the pH is around 7.0. The worm castings contain higher percentage (nearly twofold) of both macro and micronutrients than the garden compost (Table 4).

**Table 4.** Nutrient composition of vermicompost and garden compost

| Nutrient element | Vermicompost (%) | Garden compost (%) |
|---|---|---|
| Organic carbon | 9.8–13.4 | 12.2 |
| Nitrogen | 0.51–1.61 | 0.8 |
| Phosphorus | 0.19–1.02 | 0.35 |
| Potassium | 0.15–0.73 | 0.48 |
| Calcium | 1.18–7.61 | 2.27 |
| Magnesium | 0.093–0.568 | 0.57 |
| Sodium | 0.058–0.158 | <0.01 |
| Zinc | 0.0042–0.110 | 0.0012 |
| Copper | 0.0026–0.0048 | 0.0017 |
| Iron | 0.2050–1.3313 | 1.1690 |
| Manganese | 0.0105–0.2038 | 0.0414 |

The vermicompost provides all nutrients in readily available form and also enhances uptake of nutrients by plants.

***Improved crop growth and yield:*** Vermicompost plays a major role in improving growth and yield of different field crops, vegetables, flower and fruit crops. The application of vermicompost gave higher germination (93%) of mung bean (*Vigna radiata*) compared to the control (84%). Further, the growth and yield of mung bean was also significantly higher with vermicompost application. Likewise, in another pot experiment, the fresh and dry matter yields of cowpea (*Vigna unguiculata*) were higher when soil was amended with vermicompost than with biodigested slurry (Karmegam and Daniel 2000).

The efficiency of vermicompost was evaluated in a field study by Desai et al. (1999). They stated that the application of vermicompost along with fertilizer N gave higher dry matter (16.2 g plant$^{-1}$) and grain yield (3.6 t ha$^{-1}$) of wheat (*Triticum aestivum*) and higher dry matter yield (0.66 g plant$^{-1}$) of the following coriander (*Coriandrum sativum*) crop in sequential cropping system.

***Reduction in soil C:N ratio:*** Vermicomposting converts household waste into compost within 30 days, reduces the C:N ratio and retains more N than the traditional methods of preparing composts (Gandhi et al. 1997). The C:N ratio of the unprocessed olive cake, vermicomposted olive cake and manure were 42, 29 and 11, respectively. Both the unprocessed olive cake and vermicomposted olive cake immobilized soil N throughout the study duration of 91 days.

***Role in nitrogen cycle :*** Earthworms play an important role in the recycling of N in different agro-ecosystems, especially under *jhum* (shifting cultivation) where the use of agrochemicals is minimal. Bhadauria and Ramakrishnan (1996) reported that during the fallow period intervening between two crops at the same site in 5- to 15-year *jhum* system, earthworms participated in N cycle through cast-egestion, mucus production and dead tissue decomposition. Soil N losses were more pronounced over a period of 15-year *jhum* system. The total soil N made available for plant uptake was higher than the total input of N to the soil through the addition of slashed vegetation, inorganic and organic manure, recycled crop residues and weeds.

***Improved soil physical, chemical and biological properties:*** The Vermicompost application help to improve physic-chemical and biological properties of soil. Studies on vermicompost indicated that it increases macropore space ranging from 50 to 500 ìm, resulting in improved air-water relationship in the soil which favorably affect plant growth (Marinari et al. 2000). The application of organic matter including vermicompost favorably affects soil pH, microbial population and soil enzyme activities (Maheswarappa et al. 1999). It also reduces the proportion of water-soluble chemical species, which cause possible environmental contamination (Mitchell and Edwards 1997).

***Types of earthworms:*** Earthworms are invertebrates. There are nearly 3600 types of earthworms in the world and they are mainly divided into two types:

1. Burrowing; and
2. Non-burrowing.

The burrowing types *Pertima elongata* and *Pertima asiatica* live deep in the soil. On the other hand, the non-burrowing types *Eisenia fetida* and *Eudrilus eugenae* live in the upper layer of soil surface. The burrowing types are pale, 20 to 30 cm long and live for 15 years. The non-burrowing types are red or purple and 10 to 15 cm long but their life span is only 28 months. The non-burrowing earthworms eat 10% soil and 90% organic waste materials; these convert the organic waste into vermicompost faster than the burrowing earthworms. They can tolerate temperatures ranging from 0 to 40°C but the regeneration capacity is more at 25 to 30°C and 40–45% moisture level in the pile. The burrowing type of earthworms come onto the soil surface only at night. These make holes in the soil up to a depth of 3.5 m and produce 5.6 kg casts by ingesting 90% soil and 10% organic waste.

***Earthworm multiplication:*** Numerous organic materials have been evaluated for growth and reproduction of earthworms as these materials directly affect the efficacy of vermicompost. Zajonc and Sidor (1990) evaluated and compared various non-standard materials for the preparation of vermicompost. A mixture of cotton waste with cattle manure in the ratio of 1:5 was found to be the best. The use of grape cake alone increased earthworm weight slightly. Tobacco (*Nicotiana tabacum*) waste, used as substrate, increased earthworm weight but the earthworms failed to reproduce. A mixture of tobacco waste with rabbit manure in the ratio of 1:5 was found to be lethal to the earthworms.

A multiplication with three kinds of earthworm cultures (*Eisenia fetida, Eudrilus eugenae* & *Perionyx excavatus*) using wheat straw, chickpea straw, tree leaves (*Peltophorum* sp) and *Parthenium* mixed with cow dung as feed materials showed an increase in earthworm population and size during incubation for 90 days. The three types of earthworms multiplied 12 to 18 times when grown individually using legume tree leaves and cow dung mixture as raw material. However, mixed culture (of all three species) showed higher multiplication rate (27 times) than the individual species. The earthworm population decreased when grown in mixture of *Gliricidia* stems and cattle manure. So, *Gliricidia* loppings could not be used for multiplication of earthworms as *Gliricidia* bark is known to possess toxic properties as it is used as rat poisoning bait.

The *Gliricidia* and tobacco leaves are not suitable for multiplication of earthworms. Perhaps the alkaloids and other principal compounds present in these leaves may effect the survival of earthworms. Also, soil and rabbit manure should not be mixed with earthworm feed material.

## Temperature changes during the process

Change in temperature was observed during the process of vermicomposting (from 5 to 65 days) with different farm residues (*Parthenium* and grass). In the beginning of the process, ie, up to 15 days, the temperature was high (32 to 33°C) in both *Parthenium* and grass substrates when compared to outside temperature (26 to 30°C). Later, there was a gradual decrease in temperature, which reached a minimum of about 24°C. However, higher temperature was recorded in *Parthenium* compost (decline from 32.8 to 27.5°C) than in grass compost (decline from 31.5 to 26.8°C) during the whole period of digestion process. Generally more heat was evolved from control treatment (without earthworms) than the vermicompost treatments (with earthworms). From these studies, it was suggested that the most suitable period for releasing the earthworms into organic residues would be between 15 and 20 days after heaping of the organic residues when the temperature is about 25°C (Nagavallemma et al. 2004).

## Methods of vermicomposting

***Pits below the ground:*** Pits made for vermicomposting are 1 m deep and 1.5 m wide. The length varies as required.

***Heaping above the ground:*** The waste material is spread on a polythene sheet placed on the ground and then covered with cattle dung. Sunitha et al. (1997) compared the efficacy of pit and heap methods of preparing vermicompost under field conditions. Considering the biodegradation of wastes as the criterion, the heap method of preparing vermicompost was better than the pit method. Earthworm population was high in the heap method, with a 21-fold increase in *Eudrilus eugenae* as compared to 17-fold increase in the pit method. Biomass production was also higher in the heap method (46-fold increase) than in the pit method (31-fold). Consequent production of vermicompost was also higher in the heap method (51 kg) than in the pit method (40 kg).

***Tanks above the ground:*** Tanks made up of different materials such as normal bricks, hollow bricks, stones, asbestos sheets and locally available rocks can be utilized for vermicompost preparation. Tanks can be constructed with the dimensions suitable for operations depending upon space available. The tanks dimensions with 1.5 m (5 feet) width, 4.5 m (15 feet) length and 0.9 m (3 feet) height provides manageable vermibed. The commercial biodigester contains a partition wall with small holes to facilitate easy movement of earthworms from one tank to the other.

***Cement rings:*** Vermicompost can also be prepared above the ground by using cement rings. The size of the cement ring should be 90 cm in diameter and 30 cm in height.

***Commercial model:*** The commercial model for vermicomposting developed by ICRISAT consists of four chambers enclosed by a wall (1.5 m width, 4.5 m length and 0.9 m height). The walls are made up of different materials such as normal bricks, hollow bricks, stones, asbestos sheets and locally available rocks. This model contains partition walls with small holes to facilitate easy movement of earthworms from one chamber to another. Providing an outlet at one corner of each chamber with a slight slope facilitates collection of excess water, which is reused later or used as earthworm leachate on crop. The four components of a tank are filled with plant residues one after another. The first chamber is filled layer by layer along with cow dung and then earthworms are released. Then the second chamber is filled layer by layer. Once the contents in the first chamber are processed the earthworms move to chamber 2, which is already filled and ready for earthworms. This facilitates harvesting of decomposed material from the first chamber and also saves labor for harvesting and introducing earthworms. This technology reduces labor cost and saves water as well as time.

***Materials required for vermicomposting:*** A range of agricultural residues, all dry wastes, for example, sorghum straw and rice straw (after feeding cattle), dry leaves of crops and trees, pigeonpea (*Cajanus cajan*) stalks, groundnut (*Arachis hypogaea*) husk, soybean residues, vegetable wastes, weed (*Parthenium*) plants before flowering, fiber from coconut (*Cocos nucifera*) trees and sugarcane (*Saccharum officinarum*) trash can be converted into vermicompost. In addition, animal manures, dairy and poultry wastes, food industry wastes, municipal solid wastes, biogas sludge and bagasse from sugarcane factories also serve as good raw materials for vermicomposting. The quantity of raw materials required using a cement ring of 90 cm in diameter and 30 cm in height or a pit or tank measuring 1.5 m × 1 m × 1 m is given below:

| | |
|---|---|
| Dry organic wastes (DOW) | 50 kg |
| Dung slurry (DS) | 15 kg |
| Rock phosphate (RP) | 2 kg |
| Earthworms (EW) | 500–700 |
| Water (W) | 5 L every three days |

The various ingredients are used in the ratio of 5:1.5:0.2:50–75:0.5 of DOW:DS:RP:EW:W. In the tank or pit system 100 kg of raw material and 15–20 kg of cow dung are needed for each cubic meter of the bed.

## Vermicompost preparation

***Steps in the process:*** Vermicomposting involves the following steps

- Cover the bottom of the cement ring with a layer of tiles or coconut husk or polythene sheet.
- Spread 15–20 cm layer of organic waste material on the polythene sheet.
- Sprinkle rock phosphate powder if available (it helps in improving nutritional quality of compost) on the waste material and then sprinkle cow dung slurry. Fill the ring completely in layers as described. Paste the top of the ring with soil or cow dung. Allow the material to decompose for 15 to 20 days.
- When the heat evolved during the decomposition of the materials has subsided (15–20 days after heaping), release selected earthworms (500 to 700) through the cracks developed.
- Cover the ring with wire mesh or gunny bag to prevent birds from picking the earthworms. Sprinkle water every three days to maintain adequate moisture and body temperature of the earthworms.
- The vermicompost is ready in about 2 months if agricultural waste is used and about 4 weeks if sericulture waste is used as substrate.

- The processed vermicompost is black, light in weight and free from bad odor.
- When the compost is ready, do not water for 2–3 days to make compost easy for shifting. Pile the compost in small heaps and leave under ambient conditions for a couple of hours when all the worms move down the heap in the bed. Separate upper portion of the manure and sieve the lower portion to separate the earthworms from the manure. The culture in the bed contains different stages of the earthworm's life cycle, namely, cocoons, juveniles and adults. Transfer this culture to fresh half decomposed feed material. The excess as well as big earthworms can be used for feeding fish or poultry. Pack the compost in bags and store the bags in a cool place.
- Prepare another pile about 20 days before removing the compost and repeat the process by following the same procedure as described above.

### Precautions during the process

The following precautions should be taken during vermicomposting:

- The African species of earthworms, *Eisenia fetida* and *Eudrilus eugenae* are ideal for the preparation of vermicompost. Most Indian species are not suitable for the purpose.
- Only plant-based materials such as grass, leaves or vegetable peelings should be utilized in preparing vermicompost.
- Materials of animal origin such as eggshells, meat, bone, chicken droppings, etc are not suitable for preparing vermicompost.
- *Gliricidia* loppings and tobacco leaves are not suitable for rearing earthworms.
- The earthworms should be protected against birds, termites, ants and rats.
- Adequate moisture should be maintained during the process. Either stagnant water or lack of moisture could kill the earthworms.
- After completion of the process, the vermicompost should be removed from the bed at regular intervals and replaced by fresh waste materials.

### How to use vermicompost?

- ***Vermicompost can be used for all crops:*** agricultural, horticultural, ornamental and vegetables at any stage of the crop.
- ***For general field crops:*** Around 2–3 t/ha vermicompost is used by mixing with seed at the time of sowing or by row application when the seedlings are 12–15 cm in height. Normal irrigation is followed.

- ***For fruit trees:*** The amount of vermicompost ranges from 5 to 10 kg per tree depending on the age of the plant. For efficient application, a ring (15–18 cm deep) is made around the plant. A thin layer of dry cow dung and bone meal is spread along with 2–5 kg of vermicompost and water is sprayed on the surface after covering with soil.
- ***For vegetables:*** For raising seedlings to be transplanted, vermicompost at 1 t/ha is applied in the nursery bed. This results in healthy and vigorous seedlings. But for transplants, vermicompost at the rate of 400–500 g per plant is applied initially at the time of planting and 45 days after planting (before irrigation).
- ***For flowers:*** Vermicompost is applied at 750–1000 kg/ha.
- For vegetable and flower crops vermicompost is applied around the base of the plant. It is then covered with soil and watered regularly.

**Benefits to farmers**

- Less reliance on purchased inputs of nutrients leading to lower cost of production
- Increased soil productivity through improved soil quality
- Better quantity and quality of crops
- For landless people provides additional source of income generation

**To industries**

- Cost-effective pollution abatement technology

**Benefits to environment**

- Wastes create no pollution, as they become valuable raw materials for enhancing soil fertility

**To national economy**

- Boost to rural economy
- Savings in purchased inputs
- Less wasteland formation

**Advantages over chemical fertilizers**

- It restores microbial population which includes nitrogen fixers, phosphate solubilizers etc.

- Provides major and micro nutrients to the plants.
- Improves soil texture and water holding capacity of the soil.
- Provides good aeration to the soil, improving root growth and proliferation of beneficial soil micro-organisms.
- Improves structural stability of the soil, prevents soil erosion.
- Enhances quality of grains/fruits due to increased sugar content.

## Vermiwash production

Vermiwash is the coelomic fluid of earth worms. It contains anti bacterial, anti fungal and growth promoting substances. Vermiwash is a liquid that is collected after the passage of water through a column of worm action and is used as a foliar spray.

## Method

i) A plastic barrel with 200 litres capacity is used for making vermiwash. The tap should be attached at bottom of the burrel to collect the vermiwash.

ii) A sand layer of two inch thickness is placed at the bottom. Afterwards, a 2 inch thick small gravel layer is spread followed by two inch thick medium sized and large sized gravel layer is placed.

iii) Half feet thickness of one month old cowdung is spread over the gravel layer. And above this layer two feet thick slow degradable material is spread. Over this layer, pre decomposed waste material is spread upto 80 per cent volume of the barrel.

iv) One inch thick soil layer is laid over the layers of waste material. This soil layer is covered with half feet thick layer of old cowdung.

v) Earth worms are released.

vi) Facilities should be created for water to trickle down the barrel.

vii) Vermiwash is collected in the bottle from the tap.

The unit starts yielding good quality vermiwash after one week.

The vermiwash is coppery brown in colour and it should be stored in the cool dry place under shade.

## Uses

### Root dip/stem dip

The seedlings before transplanting are dipped in the vermiwash solution which is diluted five times with water for 15-20 minutes and afterwards these are transplanted. Similarly, the cuttings can also be dipped in the solution.

### Foliar spray

The diluted vermiwash five times with water is used for foliar spray on crop foliage. It provides the plants with vital nutrients and at the same time helps to control plant diseases.

### Soil drench

Vermiwash is diluted ten times with water and the soil is drenched with the solution to prevent some of the soil borne pathogens.

### Enemies of vermicomposting

Termites, ants, rats, centipedes, frog, birds, pig, etc. Termites and ants are main enemies.

### Control

- Maintain 40-50% moisture in the pit (for controlling termites, ants and rats).
- Plant turmeric plants around the pit.

### Liquid manure

Liquid manure is the preparations obtained by active fermentation of plant or animal residues (Walia 2021). The liquid organic manure is by using cow dung and cow urine alongwith locally available green leaves, tender stems of the plants.

i) Green leaves: Sunhemp, Dhaincha, Cowpeas, Moongbean and other leguminous plants.

ii) Weeds: Weeds before flowering stage

***Materials required:*** Cowdung- 3 kg; cow urine-3 litres; green leaves/weeds-3 kg and 20 litres capacity bucket.

### Methodology

1. The green leaves/weeds are cut into small pieces and these are put into 20 litres bucket.
2. The three litres cow urine and three kg cow dung is added into the bucket.
3. The bucket is filled with water then covered with a gunny bag.
4. The contents are stirred every day for first week then once a week for succeeding three weeks.
5. The biomass gets fully decomposed and the liquid organic manures becomes ready for use in one month.

### Use

The liquid organic manure diluted ten times with water can be used for foliar application on foliage.

### Advantages

1. It acts as a growth stimulant to the plants.
2. Depending upon raw material, it acts as insecticide/fungicide.
3. It improves nutrient assimilator.

### Precautions

1. Stirring of liquid manure is required for rapid fermentation.
2. The concentrated liquid manure should not be sprayed on foliage as it causes scorching of foliage.
3. The liquid organic manure should be properly sieved to avoid chocking of spray pump.
4. The liquid organic manure should be sprayed in morning or evening or cool sunshine hours.

### Dung brew

It is used to provide nutrition to the crop plants by spraying fermented solution of dung brew.

**Materials required:** Cattle dung-5kg; cow urine-15 litres and plastic bucket of 20 litres.

### Methodology

1. Add 5 kg cowdung and 15 litres cow urine in the bucket.
2. Stir the mixture well.
3. Cover the bucket with gunny bag
4. Stir the materials on alternate day for 15 days.

### Use

The cow dung brew solution is diluted ten times before spray. The sieved solution is used for spray on crop foliage.

### Cow urine spray

The cow urine contains more than 24 nutrients & acts as a growth promoter for all crops. However, the cow urine is not utilized by the farmers and goes waste.

***Materials required:*** Cow urine and water

### Methodology

The diluted cow urine by ten times with water can be used as a foliar spray. However, the concentrated cow urine may result in scorching.

### Stem paste

Stem paste provides nutrition through stem (cambium) in orchard plants. It also prevents the attack of stem boring insects, mealy bubs and scales.

***Materials required:*** Cow dung-50 kg; cow urine -25 litres; sieved sand-10 kg; soil from field-10 kg

### Methodology

- Mix cow dung, sand and field soil with cow urine to form a paste. Add sufficient cow urine till consistency of paste is reached.
- The paste is plastered over clean stem and branches of fruit tree.
- The paste can also be used for healing the cut of tree after pruning.

### Panchgavya

Panchgavya is an ancient Indian preparation consists of five products of cow namely dung, urine, milk, curd (youhurt), ghee (clarified butter) and it acts as tonic to soil and crops.

***Materials required:*** Fresh cow dung-5 kg; cow urine-3 litres; cow milk-2 litres, cow ghee-500 grams, curd-2 kg and a mud pot.

### Methodology

1. Mix thoroughly cow dung and ghee in a container for 15 minutes three times a day for initial three days.
2. Rest of ingredients are added on fourth day and stir the contents twice a days for two weeks.
3. The panchgavya preparation becomes ready after 18 days and can be stored for six months.

### Usage

#### As foliar spray

Three per cent panchgavya spray is ideal for foliar spray. It should be filtered before spraying to avoid clogging of spray pump nozzle.

### Through irrigation

The panchgavya solution can be mixed with irrigation water at the rate of 50 litres/hectare by surface irrigation or drip irrigation systems.

### Seed/seedling treatment

Three per cent panchgavya solution can be used for drenching in the nursery bed, soaking the seeds or dipping the nursery seedlings. The rhizomes of turmeric, colocasia and ginger etc can be soaked in 3 per cent panchgavya for 20-30 minutes.

***Amrithakaraisal:*** Take ten kg fresh cow dung, ten litre cow urine and one kg jaggery and mix all the ingredients in fifty litres capacity plastic drum and stir the material manually until the whole solution becomes homogenous. The mixture is allowed to settle down for 24 hours. The solution is used for seed priming/ treatment and fertigation etc.

***Pitcher khad (manure):*** Take 15 kg cow dung, 15 litre cow urine and 15 litre water and 250 gm jaggery and mix all the ingredients in a container and cover the container with gunny bag or cloth. The material is allowed to ferment for 4-5 days and the fermented mixture is mixed with 200 litres of water and spray material is ready for spraying over one acre crop area. Two to three sprays are done for short duration crops.

***Harit pani:*** Take 25 kg chopped weeds in a 200 litres container. Add 500 g jaggery, 500 g tamarind and 250 g salt to the container having finely chopped weeds. Add 100 litre of water into the container so that the added water covers the ingredients and stir the mixture daily. The mixture decomposes with 2-3 weeks. Use 10 per cent solution for foliar application or as soil drench.

***Sanjivak:*** Take 100 kg cow dung, 100 litre cow urine and 500 g jaggery in a 500 litres drum. Allow the mixture to ferment for 10 days. Dilute the mixture twenty times with water and spray over 0.4 ha area by sprinkling or with irrigation water.

***Jeev amrit:*** Mix 10 kg cow dung, 10 litres cow urine, 2 kg jaggery, 2 kg pulse grain flour and one kg virgin soil in 200 litre water. Ferment the material for 5-7 days. Stir the material/solution regurarly three times a day. Use the fermented material with irrigation water in 0.4 ha area.

***Amrit pani:*** Mix 10 kg cow dung with 500 grams honey and mix it to form a creamy paste. Then add 250 grams cow desi ghee(clarified butter) and mix at high speed. Add 200 litres of water and then sprinkle the suspension over 0.4 ha field area or with irrigation water. Apply second dose after one month in between crop rows or with irrigation water.

## Inclusion of legumes in cropping system

The contribution of legume N to succeeding crops in rotation can be stated as follows:

1. legume N can contribute substantially to N needs of the succeeding crops grown in a rotation, but amounts of nitrogen fixed will depend on the time during which legumes are grown, amount of biomass incorporated and species of succeeding crop grown.
2. Increased yields of crop following legumes may result from increased N availability, therefore shows residual effect.

## Green manuring and bio-fertilizers

It is very difficult to estimate the total size of green inputs market in India because of its diversity in terms of products and also due to the nature of it being unorganized market. Green inputs into agriculture include bio-fertilizers, bio-pesticides, compost, Farm Yard Manure (FYM), green manure etc. As most of this inputs are either not traded and even if they are traded, it is only at informal levels available information regarding production capacity, demand and sales is at best sketchy estimation and hence inadequate. The limited number well-established firms that have their presence in this market today as a result it this is predominately controlled by the small and local producers of bio-fertilizer, vermicompost and other input producers who are in large numbers.

Relatively speaking there is more rigour in estimation of bio-fertilizer market in India, because of the presence of some large producers in production of bio-fertilizers and comparatively this is more organized than other green inputs market. Based on the gross cropped area in India (190 million hectare) and recommended doses of bio-fertilizers, potential demand is estimated to be about 6,27,000 MT. This demand can further be segregated into different categories of bio-fertilizer, such as *Rhizobium, Azotobactors, Azospirillium*, BGA, and Phosphate Solubilizer etc. whose demand differ widely (Table 5).

**Table 5.** Estimated Total Potential Demand for Bio-fertilizers in India

| Category of Bio- Fertilizer | Amount in Million Tonnes |
|---|---|
| Rhizobium | 35,730Mt |
| Azotobacter | 162,610Mt |
| Azospirillum | 77,160Mt. |
| BGA | 267,510Mt |
| Phosphate solubilizer | 275,510Mt |
| TOTAL | 818,730Mt |

***Green Manuring:*** Ghai and Thomas (1989) have defined green manuring as "a farming practice where a leguminous plant which has derived enough benefits from its association with appropriate species of Rhizobium is ploughed into the soil and then a non legume is grown and allowed to take the benefits of the already fixed nitrogen". The practice of green manuring was started from several century B.C. in India and China. During the course of time availability of chemical fertilizers decreased the significance of green manuring. In recent years, due to hike in price of chemical fertilizers, the practice of green manuring is reemphasized.

In agriculture, a green manure is a type of cover crop grown primarily to add nutrients and organic matter to the soil. Typically, a green manure crop is grown for a specific period, and then plowed under and incorporated into the soil while green or shortly after flowering. Green manure crops are commonly associated with organic agriculture, and are considered essential for annual cropping systems that wish to be sustainable. Historically, the practice of green manuring can be traced back to the fallow cycle of crop rotation, which was used to allow soils to recover.

## Functions

1. Green manure crops may include legumes such as cowpeas, soybeans, annual sweet clover, vetch, sesbania, sannhemp, guara, moongbean, and velvet beans, as well as non-leguminous crops such as sudangrass, millet, sorghum, and buckwheat.
2. Leumes are often used as green manure crops for their nitrogen fixing abilities, while non-leguminous crops are used primarily for weed suppression and addition of biomass to the soil. Green manures usually perform multiple functions, that include soil improvement and soil protection.
3. Leguminous green manures such as clover and vetch contain nitrogen-fixing symbiotic bacteria in root nodules that fix atmospheric nitrogen in a form that plants can use.
4. Green manures increase the percentage of organic matter (biomass) in the soil, thereby improving water retention, aeration, and other soil characteristics.
5. The root systems of some varieties of green manure grow deep in the soil and bring up nutrient resources unavailable to shallower-rooted crops.
6. Common cover crop functions of weed suppression and prevention of soil erosion and compaction are often also taken into account when selecting and using green manures.

   Some green manure crops, when allowed to flower, provide forage for pollinating insects.

7. Incorporation of cover crops into the soil is immediately followed by an increase in abundance of soil microorganisms that aid in the decomposition of this fresh material.
8. The degradation of plant material allows the nutrients held within the green manure to be released and made available to the succeeding crop.
9. This additional decomposition also allows for the re-incorporation of nutrients that are found in the soil on a particular farm such as nitrogen (N), potassium (K), phosphorus (P), calcium (Ca), magnesium (Mg), and sulfur (S). Microbial activity in the soil also leads to the formation of mycelium and viscous materials which benefit the health of the soil by increasing its soil structure (i.e. by aggregation).
10. Soil that is well-aggregated has increased aeration and water infiltration rates, and is more easily turned or tilled than non-aggregated soil. Further aeration of the soil results from the ability of the root systems of many green manure crops to efficiently penetrate compact soils. The amount of humus found in the soil also increases with higher rates of decomposition, which is beneficial for the growth of the crop succeeding the green manure crop.
11. Green manure crops are also useful for weed control, erosion prevention, and reduction of insect pests and diseases. The deep rooting properties of many green manure crops make them efficient at suppressing weeds.
12. Green manure crops often provide habitat for many native pollinators as well as predatory beneficial insects, which allow for a reduction in the input of insecticides where cover crops are planted.
13. Some green manures are also successful at suppressing plant diseases, especially Verticillium wilt in potato.
14. Incorporation of green manures into a farming system can drastically reduce, if not eliminate, the need for additional products such as supplemental fertilizers and pesticides.

***Green manure crops:*** Another important contribution of green manure to an agricultural field is the nitrogen fixing ability and consequent nitrogen accumulation in the soil, particularly of those leguminous crops used. Depending on the species of cover crop grown, the amount of nitrogen released into the soil lies between 45 and 225 kg per hectare. With green manure use, the amount of nitrogen that is available to the succeeding crop is usually in the range of 40-60% of the total amount of nitrogen that is contained within the green manure crop.

## Criteria for the selection of green manure crops

a) Fast growing

b) Produces abundant and succulent shoots (foliages),

c) Well adapted to the local condition,

d) Have a high rate of nitrogen fixation

e) Have high water-use efficiency when used in drier regions. The legume should use as little water as possible while still producing substantial quantities of top-growth, Ghai and Thomas (1989) have given a list of various leguminous plants to be used as green manure (Table 6). Some of them are: cultivated annual legumes (e.g. *Crotalaria juncea, C. striata, Cassia mimosoides, Cyamopsis pamas, Glycine wightii, Indigofera linifolia, Sesbania rostrata, Vigna radiata*), perennial legumes (e.g. *Acacia nilotica, Cassia hirsuta, Sesbania aegyptica, Leucaena Jeucocephala*), and wild annual legumes (e.g. *Cassia cobanensis, Lathyrus sativus, Mimosa invisa, Mucuna bracteata*).

**Table 6:** Contribution of nitrogen by nodulated legumes used as green manure/cover crop

| $N_2$ fixing system | Amount of N contributed (kg/ha) |
|---|---|
| **1. Green manure legumes :** | |
| *Sesbania aculeata* – Rhizobium | 70-120 |
| *Leucaena leucocephala* - Rhizobium | 500-600 |
| Beans (broad beans / lupines /soybean / lentil, etc.) – Rhizobium | 60-210 |
| Fodders (*Trifolium / Medicagol Melilotus*, etc. ) – Rhizobium | 100-300 |
| **2. Cover crop legumes** | |
| *Lablab purpureus* – Rhizobium | 240 |
| *Glycine jawanica* – Rhizobium | 210 |
| **3. Non-legumes :** | |
| *Casuarina equisitifolia* - Frankia | 100 |
| *Alnus* – Frankia | 30-300 |
| **4. Others :** | |
| Azolla – *Anabaena* | 25-190 |
| Grasses – *Azospirillium* | 15-100 |

*Source :* Ghai and Thomas (1989)

***Sesbania:*** Many legumes are grown and then turned into the soil while they are still green to enrich soil nitrogen. Sesbania is a green manure plant which forms both root and stem nodules in association with rhizobium and thereby fixes more atmospheric nitrogen. These legumes produce ten times more nodules than other legumes and have a very high capacity to fix atmospheric nitrogen. Apart from enrichment of soil nitrogen, green manuring enriches the phosphorus, calcium, sulphur and other micronutrient of the soil.

***Use in organic farming:*** Organic farming relies on soil health and cycling of nutrients through the soil using natural processes. Green manures perform the vital function of fertilization, in concert with the addition of animal manures if those are used. While there are a variety of beneficial aspects of the use of green manure, there are also limitations to consider. Time, energy, and resources (monetary and natural) are required to successfully grow and utilize these cover crops. Consequently, it is important to choose green manure crops based on the growing region and annual precipitation amounts to ensure efficient growth and use of the cover crop(s).

## Essential precautions

a) Broadcast or sow the seeds after the harvest of main crop, preferably in the water sign/ zodiac of the ascending period of the month,

b) Slash the green manure crops shoots and incorporate them into the soil,

c) Slashing and incorporation of green manure crop should be done while the soil has adequate moisture. If there is no moisture an irrigation is essential. Without moisture incorporation is difficult and if done, it will not be effective. Meanwhile moisture is a critical factor for the microbes to act over the incorporate material.

d) Sow or transplant the next crops 1-2 weeks after the incorporation of the green manure, since in the initial stage of decomposition there will be a release heat of decomposition that is very Unfavourable for germination and the growth of young plants

e) Use disc harrow to incorporate the green manure crop. Disc harrow have become the implement of choice for incorporating all the heaviest perennial stands. Disc harrow implements adequately mix the legume residues into the soil, promoting even decomposition and nutrient release near to the soil surface.

## There are different green leaf manure crops that can be cultivated and they are

1. ***Cowpea:*** Cowpea is one of the important leguminous green leaf manure crops. The cowpeas plant decomposes easily and very well suited for green manure purpose. June-July months are best suited for sowing of this manure. Even though it is being cultivated in summer months (March to April). The inoculation with effective Rhizobium bacteria increases the nitrogen fixation up to 40 kg/ha.

2. ***Dhaincha (Sesbania Aculeata) :*** Dhaincha/Jantar is suitable for its sowing in loamy and clayey soils. It is fairly resistant to drought as well as stagnation of water. It grows well even in alkaline soils and corrects alkalinity if grown repeatedly for 4-5 years. The roots have plenty of nodules. It yields about 15-20 tonnes of fresh green maatter per ha and requires a seed rate of 50 kg/ha. Use of effective Rhizobium strain with seeds fixes the nitrogen 1 kg / day.

3. ***Sesbania Speciosa :*** It is a valuable green manure for wetlands and can be grown in a wide range of soils. Seed production is prolific however, pods are frequently attacked by insects. This green manure can be raised on the field 15 borders. Sesbania seedling (21days) can be planted in a single line at 5-10 cm apart in the borders of the fields. In about 90 days it produces about 2-4 tonnes of green manure per ha. It does not affect the rice yield by shading or root effect. If second rice crop is planted immediately after the first crop, the manure can be incorporated into the field. About 300-400g of seeds are sufficient to raise nursery and plant the seedlings around the boundary of one hectare. To control insects *Verticillium lacanii* (Liquid ) fungi is useful.

4. ***Sunnhemp (Crotalaria Juncea) :*** It is a quick growing green manure crop and gets ready for incorporation in about 45 days after sowing. It does not withstand heavy irrigation leading to flooding. The crop is at times subject to complete damage by leaf eating caterpillars. The crop can produce about 8-12 tonnes of green biomass per ha. The seed requirement is 30 kg/ha.

5. ***Sesbania Rostrata :*** One of the important features of this green manure is that in addition to the root nodules, it produces nodules in the stem. The stem nodulation is an adaptation for waterlogged situation since flooding limits growth of green manures and may reduce root nodulation. Under normal condition, both root and stem nodules are effective in N fixation. It has higher N content of 3.56% on dry weight basis. Biomass production is higher during summer (April-June) than in winter (Dec.-Jan.) season. This green manure can also be produced by raising seedlings (30 days old) and planted in the paddy field along the bunds or as intercrop with rice. Use of Rhizobium bacteria increase the nitrogen fixation about 60-100 kg/ha/year.

6. ***Wild Indigo (Tephrosia Purpurea) :*** This is a slow growing green manure crop and cattle do not prefer to graze it. The green manure is suitable for light textured soils, particularly in single crop wetlands. It establishes itself as a self sown crop and the seeds remain viable till the harvest of rice. On an average about 3-4 tonnes of green manure is obtained in one ha. The seed rate is 30 kg/ha. The seeds have a waxy impermeable seed coat and

hence scarification is required to induce germination. Soaking seeds in boiling water for 2-3 minutes is also equally effective in promoting germination.

7. ***Indigo (Indigofera Tinctoria) :*** It resembles wild indigo and is along duration crop with more leafy growth. It comes up well in clayey soils with one or two irrigations.

8. ***Pillipesara (Pheseolus Trilobus) :*** This is a dual purpose crop yielding good fodder for the cattle and green manure. Pillipesara comes up well in hot season with sufficient soil moisture. Loamy or clayey soils are best suited. After taking one or two cuttings for fodder or light grazing by animals, the crop can be incorporated into the soil. About 5-8 tonnes of manure can be obtained from one ha.

9. ***Glyricidia (Glyricidia Maculeata) :*** This is a shrubby plant that comes up well in moist situations. Under favourable conditions, it grows well like a tree. It can be easily grown in waste lands, farm road sides, field bunds, etc. The crop can be established by stem cuttings or seedlings planted in the field borders. It can be pruned for its tender loppings and compound leaves for green leaf manuring at the time of puddling rice. On an average, a well-established plant yields 12-15 kg green matter. About 400 plants on the peripheral bunds yields 5-6 tonnes green manure/ha.

10. ***Karanj (Pongamia Glabra) :*** It is a leguminous tree grown in wastelands. On an average, a tree can yield 100-120kg of green matter. The leaves contain about 3.7% N (on dry weight basis).

11. ***Calatropis (Calotropis Gigantca) :*** On roadsides and fallow lands, the plant grows wild under different soil and climatic conditions. The leaves are more succulent and a plant can produce about 4-5 kg of green matter. Besides it also helps in controlling soil born pests like termite.

## Biofertilizer

Biofertilizer is a large population of a specific or a group of beneficial microorganisms for enhancing the productivity of soil either by fixing atmospheric nitrogen or by solubilising soil phosphorus or by stimulating plant growth through synthesis of growth promoting substances. Bio-fertilizers based on renewable energy source are cost effective supplement to chemical fertilizers, eco-friendly and can help to economise on the high investment needed for chemical fertilizer use as far as nitrogen and phosphorus are concerned.

**Source:** Biofertilizer is a 100% natural and organic fertilizer that helps to provide and keep in the soil all the nutrients and microorganisms required for the benefits of the plants. There are millions of Microscopic Organisms near the plants that conform a in a micro environment that provides nutrients to the plants also helps to keep the water and retain the nutrients in the soil easy available to the plants.

***Need for biofertilizers:*** Biofertilizers have definite advantage over chemical fertilizers. Chemical fertilizers supply over nitrogen whereas biofertilisers provide in addition to nitrogen certain growth promoting substances like hormones, vitamins, amino acids, etc., crops have to be provided with chemical fertilizers repeatedly to replenish the loss of nitrogen utilised for crop growth. On the other hand biofertilizers supply the nitrogen continuously throughout the entire period of crop growth in the field under favourable conditions. Continuous use of chemical fertilisers adversely affect the soil structure whereas biofertilizers when applied to soil improve the soil structure. The deleterous effects of chemical fertilizers are that they are toxic at higher doses. Biofertilizers, however, have no toxic effects.

The soil microorganisims used in biofertilizers are: Phosphate Solubilizing microbes, *Mycorrhizae, Azospirillum, Azotobacter, Rhizobium, Sesbania, Blue Green Algae,* and *Azolla*. Let's go through these groups in a little detail in order to understand their role in bio-fertilizers, which can be used to make rich, living soil that is suitable for coffee plants.

***When you fill them with chemicals:*** Most of them die forever, losing the capacity of the soil to be sustainable at long term. The end product of the various phases of Biodigestion, is a liquid which is dark in colour due to the presence of Humus in it. We call this Pure Biofertilizer, and it can be used on soil as a high quality organic fertilizer, or even as a corrector of pH, bacterial life and texture. The Biofertilizer has a relative high nutrient concentration, and even so, it can be used directly over soil before planting. Once diluted, it constitutes a high quality foliar fertilizer, and in this form, it is known as Diluted Biofertilizer.

***Diluted Biofertilizer:*** Also has all the needed conditions to be used as a complete Nutrient Solution in Organic Hydroponics.

***Bacteria:*** Many free-living and symbiotic bacteria are discussed in Biological Nitrogen Fixation which fixes atmospheric $N_2$. Therefore, certain measures are adopted to increases number of such bacteria in soil which may increase the gross yield of nitrogen. The two methods, bacterization and green-manuring are the most widely used techniques.

## Method of application

***Seed treatment:*** The carrier based inoculants is mixed with little quantity of water and then thoroughly mix it with seed on clean pucca floor. Let it dry in shade and sow the seed immediately.

Mix half kg packet of Azospirillum biofertilizer with 100 litre of water. Dip the root of rice nursery seedlings for one acre in this solution for 45 minutes and transplant immediately.

***Soil treatment:*** The inoculation material; mixed with soil is spread over field. For example, application of Azotobacter/Consortium biofertilizer @ 10 kg/ha in the furrows at the time of planting would be helpful in increasing the cane yield.

***Soil application:*** If Rhizobium/Consortium inoculum is not available then two qt of surface soil (2-10 cm) is carried from the field where particular crop was sown during previous year, broadcast the soil in the field where the crop is to be sown during current year. eg, soybean, moong etc.

## Mode of application

Inoculation with biofertilizer should be done after treating the seed with fungicide and insecticide, preferably in the evening because the direct sun light kills the bacteria.

Insecticide ____________ Fungicide ___________ Biofertilizer inoculation

**There are variety of nitrogen fixing micro-organisms present in the nature. These are broadly divided into three categories, viz.,**

- Symbiotic micro-organism e.g. Legume -*Rhizobium*, Symbiosis
- Asymbiotic or free living e.g. *Azotobacter*, blue green algae.
- Associative Symbiosis, e.g. *Azospirillum*

These micro-organisms supply in addition to nitrogen, considerable amount of organic matter enriching structure of soil. Inoculants of these micro-organisms have proved their technical feasibility, economic viability and social acceptability. They are therefore called as 'biofertilizer'. Hence the term 'biofertilizer' or microbial inoculants may be defined as preparations containing living or latent cells of efficient strains of nitrogen fixing, phosphorous solubilizing or cellulolytice micro-organism.

***Bacterization:*** Bacterization is a technique of seed-dressing with bacteria (as water suspension) for example, *Azotobacter*, *Bacillus, Rhizobium* etc. It has been proved that bacteria can successfully be established in root region of plants which in turn improve the growth of hosts. Bacterial fertilizers named *'azotobakterin'* (containing cells of *Azotobacter chroococcum*) and *'phosphobacterin'* (containing cells of *Baccilus megatehum var. phosphaticum*) have been used in erstwhile U.S.S.R. and East European countries, respectively. These increased the crop yield about 10-20 per cent. Subsequently bacterization of seeds in Russia, Czechoslovakia, Rumania, Poland, Bulgaria, Hungary, England and India has clearly demonstrated the increase in crop yield such as wheat, barley, maize, sugarbeet, carrot, cabbage and potato. In rhizosphere, bacteria secrete growth substances and antibiotic secondary metabolites which contribute to seed germination and plant growth.

***Mass cultivation of microbial inoculants:*** Bacteria to be inoculated in soil as biofertilizer need to be multiplied on artificial media to harvest on a large scale so that it can be supplied to farmers. Mass cultivation of *rhizobia, Azotobacter* and *Azospirillum* is discussed below:

**Rhizobium:** Rhizobium bacteria, basically form root nodules in leguminous plants and fix atmospheric nitrogen in a symbiotic association. The Rhizobium bacteria gives nitrogen to the plant and the plant gives protection to the bacteria from oxygen damage by harbouring it inside the root nodule. Nitrogen fixing ability varies from 40 kg to 200 kg/hectare/season. The crops like groundnut could fix 50-60, lentil 90-100, red gram 168-200, soybean 60-88, cover crops like alfalfa 100-300, black gram/green gram 50-55, cluster bean 37-196, chick pea 85-100, cowpea 80-85kg/ha/season.

Species of *Rhizobium* are kept in different specialization groups. Inoculum of different strains are prepared separately and cultivated on large scale, as required. Strains of *Rhizobium sp.* are grown in Yeast Extract Mannitol (YEM) broth in a small or large container as needed. Chemical composition of YEM broth is as below:

| S.No. | Material | Quantity | S.No. | Material | Quantity |
|---|---|---|---|---|---|
| 1. | Yeast extract | 1g | 5. | NaCl | 0.1 g |
| 2. | Mannitol | 10 g | 6. | Distilled Water | 1000 ml |
| 3. | $K_2HPO_4$ | 0.5 g | 7. | pH | 6.5-7.0 |
| 4. | $Mg\ SO_4\ 7H_2O$ | 0.2 g | | | |

## Methods of cultivation

Following are the steps of mass cultivation of *Rhizobium*.

a) Sterilize the growth medium and inoculate with broth of mother culture prepared in advance, incubate for 3-4 days at 30 - 32°C

b) Test the cultures for its purity and transfer to a large fermenter, wait for 4-9 days for bacterial growth (for good bacterial growth make the device for its aeration),

c) Allow to grow the bacteria either in a large fermenter containing broth or in small flasks as per demand, check the quality of broth,

d) Blend the broth with sterile carrier e.g. peat, lignite, farmyard manure and charcoal powder.

e) Pack the culture in polyethylene bags and keep at 25°C.

f) Check the quality of carrier culture, store at 4°C in a controlled-temperature room, and supply to farmers.

During the blending of broth a variety of carriers are used, for example, peat, lignite, farmyard manure, charcoal powder, etc. In India powdered farmyard manure and charcoal powder are good carrier and an alternative to peat and lignite. Good quality of carrier culture is that which contains sufficient amount of rhizobial cells i.e. $1000 \times 10^6$ to $4000 \times 10^6$ rhizobia/g carrier. Seed inoculation with aqueous suspension of carrier culture during sowing has revealed the luxuriant nodulation and good yield of crops.

***Methods of seed inoculation with rhizobial culture:*** The steps of seed inoculation with rhizobial culture are given in the following Figure. Dissolve 10 per cent sugar or gur (jaggery) in water by boiling it for some time. Leave the content to cool down. Gum arabic solution (10%) may also be added to the solution. This serves as sticker for Rhizobium cells to seeds. Mix this carrier based culture of Rhizobium to form the inoculum slurry. For one hectare, 400 g charcoal based culture would be sufficient for mixing the seeds. Transfer the inoculum slurry on seeds and mix properly. The number of rhizobial cells/ seed should be between $10^5$ to $10^6$. Spread the seeds in shade for drying on cement floor or plastic sheets. While using rhizobial cultures, certain precautions are taken into account. For example, use of culture before expiry date, use of small amount of pesticides when required, immediate sowing of seeds after mixing, etc. Seeds must be stored at 4°C when not used immediately to protect the rhizobial cells.

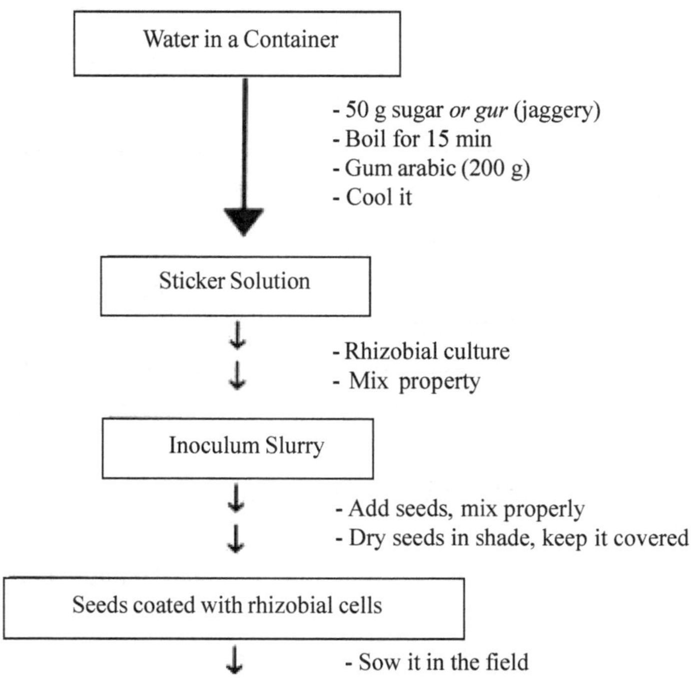

***Pelleting:*** When soil has the adverse conditions such as dryness, acidity, excess fertilizers and pesticides, etc., the rhizobial cells are protected by adopting special method of inoculation. One of these methods is pelleting i.e. preparation of pelleted seeds. High amount of gum arabic (40%) or carboxymethylcellulose (20%) is added to the inoculum slurry before mixing with seeds. Finally, pelleting agent is mixed when inoclated seeds are moist (before seed drying ) to get the seeds evenly coated. The commonly used pelleting agent are rock phoshate, charcoal powder and gypsum.

***The Blue-Green Algae (Cyanobacteria) Blue:*** Blue Green Algae (BGA) or Cyanobacteria have the ability to carry out both photosynthesis as well as nitrogen fixation. They belong to the order Nostocales and Stigonematales. Algal flakes are grown and then broadcasted.

Role of the blue-green algae (e.g. *Aulosira, Anabaena, Cylindrospermum, Nostoc, Plectonema, Tolypothrix*) in the paddy fields was realized. In water-logging condition, the cyanobacteria multiply, fix atmospheric $N_2$ and release it into the surroundings in the form of amino acids, proteins and other growth promoting substances.

***Algalization:*** It were Japanese workers (Watanabe and coworkers) who developed techniques for mass cultivation of blue-green algae to be used as biofertilizer in paddy fields. Venkataraman (1961) coined the term 'algalization' to denote the process of application of blue-green algal culture in field as biofertilizer. He initiated algalization technology in India and demonstrated the way how this technology could be transferred to farmer level who hold small lands (Vekataraman, 1972).

***Mass cultivation of cyanobacterial biofertilizers:*** For outdoor mass cultivation of cyanobacterial biofertilizers, the regional specific strains should be used. However, many germplasm collection laboratories have been established by the D.B.T. in different parts of the country for the development of starter inoculum. Mixture of 5 or 6 regional acclimatized strains of cyanobacteria, e.g. species of *Anabaena, Aulosira, Cylindrospermum, Gloeotrichia, Nostoc, Plectonema, Tolypothrix* are generally used for starter inoculum. The following four methods are used for mass cultivation:

i) Cemented tank method.

ii) Shallow metal troughs method,

iii) Polythene lined pit method, and

iv) Field method.

The polythene lined pit method is most suitable for small and marginal farmers to prepared algal biofertilizer. In this method, small pits are prepared in field and lined with thick polythene sheets. Mass cultivation of cyanobacteria is done by using any of the four methods under the following steps:

i) Prepare the cemented tanks, shallow trays of iron sheets or polythene lined pits in an open area. Width of tanks or pits should not be more than 1.5 m. This will facilitate the proper handling of culture.

ii) Transfer 2 -3 Kg soil (collected from open place for l m 2 area of the tank) and add 100 g of superphosphate. Water the pit to about 10 cm height. Mix lime to adjust the pH 7. Mix well and allow to settle down soil particles.

iii) When water becomes clear, sprinkle 100 g of starter inoculum on the surface of water.

iv) When temperature remains between 35-40° during summer, optimum growth of cyanobacteria is achieved. Always maintain the water level to about 10 cm during this period,

v) After drying, the algal mat will get separated from the soil and forms flakes. During summer about 1 kg pure algal mat per $m^2$ area is produced. These are collected, powdered, kept in sealed polythene bags and supplied to the farmers.

vi) The algal flakes can be used as starter inoculum if the same process is repeated.

The 127 strains of blue green algae fix atmospheric nitrogeneither symbiotically or free living. In India, mixed cultures of different high $N_2$ fixing blue green algae like Aulosira, Nostoc, Anabaena, Calothrix, Westlopsis, Gloeotrichia, Wollea, Cylindrospermum, Aphanthece are used for increasing rice production. Blue green algae on an average fix nitrogen about 20-30 kg/ha/season besides excrete large quantity of polysacchrites and other organic acids which assist to reclaim soil.

## Azolla

Azolla is an aquatic heterosporous fern which contains an endophytic cyanobacterium, Anabaena azollae, in its leaf cavity. It fixes nitrogen symbiotically with blue green algae, Anabeana, azollae. The significance of Azolla as biofertilizer in rice field was realized in Vietnam. Recently, it has become very popular in China, Indonesia, Philippines, India and Bangladesh.

Azolla is a floating fern which harbours a blue green algae in its leaf cavities. The fern multiplies very fast with the symbiotic association of the algae and this rapid multiplication creates a huge amount of biomass on the surface of the water. It is then harvested, dried and used as biofertilizer. Azolla could double its biomass in about 3 to 4 days. Azolla could contribute nitrogen about 60-80 kg/ha in addition to other nutrients (Table 7).

**Table 7.** Chemical composition of azolla

| **Constitutes** | **% on dry weight basis** |
|---|---|
| Ash | 10.5 |
| Crude fat | 3.0-3.6 |
| Crude protein | 23-30 |
| Nitrogen | 2.4-3.5 |
| Phosphorus | 0.5-0.9 |
| Calcium | 0.1-1.0 |
| Potassium | 2.0-4.5 |
| Magnesium | 0.5-0.65 |
| Iron | 0.06-0.26 |
| Chlorophyll | 0.34-0.55 |
| Starch | 6.5-6.54 |
| Soluble sugars | 3.4-3.45 |

A total of six species of Azolla are known so far viz., *A. caroliniana, A. filiculoides, A. mexicana, A. microphylla, A. nilotica, A. pinnata and A. rubra.* Out of these *A. pinnate* is commonly found in India. The global collections of several species of Azolla are maintained at CRRI (Cuttack). Within the leaf cavity filaments of *Anabaena azollae* are present.

***Mass cultivation of Azolla:*** Microplots ($20m^2$) are prepared in nurseries in which sufficient water (5-10 cm) is added. For good growth of Azolla, 4-20 Kg $P_2O_5$/ha is also amended. Optimum pH (8.0) and temperature (14-30°C) should be maintained. Finally, microplots are inoculated with fresh Azolla (0.5 to 0.4 Kg/ $m^2$). After three, week of growth mat formed by Azolla is harvested and the same microplot is inoculated with fresh Azolla to repeat the cultivation, azolla mat is harvested and dried to use as green manure. There are two methods for its application in field:

a) Incorporation of Azolla in soil prior to rice plantation, and

b) Transplantation of rice followed by water draining and incorporation of Azolla. However, reports from the IRRI (Philippines) reveal that growing of Azolla in rice field before rice transplantation increased the yield equivalent to that obtained from 30Kg N/ha as urea or ammonium phosphate.

Moreover, Azolla shows tolerance against heavy metals viz. As, Hg, Pb, Cu, Cd, Cr, etc. It tolerates low concentration but at high levels a setback in biochemical pathways is caused. *A. pinnata* absorbs heavy metals into cell walls and vacuoles through evolution of specific metal resistant enzymes. Therefore, heavy metal resistant species such as A. pinnata can also be incorporated as green manure in rice field near the polluted areas where heavy metal concentration is between 0.01 and 1.5 mg/liter. Due to development of chemical industries and discharge of effluents into water bodies, heavy metal concentration is gradually increasing day by day. Industries where work of electroplanting, fertilizers, tanning, etc. are done, they act as a chief source for soil and water pollution. For example, disturbed vegetation in aquatic system around Damodar river valley in India has received a great attention.

***Phosphate solubilizing microbes:*** Phosphorus is an important nutrient for plants. There are several microorganisims which can solubilize the cheaper sources of phosphorus, such as rock phosphate. Bacteria like *Pseudomonas striata*, and *Bacillus megaterium* are also important phosphorus solubilizing soil microorganisims. Many fungi like *Aspergillus* and *Penicillium* are potential solubilizers of bound phosphates. They solubilise the bound phosphorus and make it available to the plant, resulting in improved growth and yield of crops. Soil phosphates are rendered available to plants by soil microorganisims through secretion of organic acids. Therefore, phosphate dissolving soil microorganisims play some part in correcting phosphorus deficiency in plantation soils. They may also release soluble inorganic phosphate into soil through decomposition of phosphate rich organic compounds. These microbial inoculants can substitute almost 20 to 25% of the phosphorus requirement of plants. Phosphate solubilising microbes can also be inoculated to coffee husk along with rock phosphate while preparing compost to enrich the compost with available phosphorus.

***Azospirillum*:** Azospirillum are nitrogen-fixing bacteria that lives in a symbiotic relationship in the root cortex of several tropical crops. They stimulate plant growth through nitrogen fixation and production of growth substances like auxins, gibberellins and cytokinin. It is estimated that almost 10 to 15% of the required nitrogen (15-25 kg N/ha/season) can be met by Azospirillum biofertilizer. Azospirillum biofertilizers are largely used in India, in rice, wheat, sugarcane, maize, fodder crops, fruir crops, etc. Azospirillum remains endophytically inside the roots of sugarcane.

***Azobacter:*** Azotobacter are free-living, nitrogen-fixing bacteria and are known to produce several plant growth promoting substances. In addition to nitrogen fixation by these bacteria, they are also known to protect plants against pathogenic microorganisims either by discouraging their growth or by destroying them. These inoculants need more attention in view of their triple action of nitrogen fixation, bio-control, and production of plant growth regulators. Azotobacter biofertilizers has been used in India for cereals, millets, vegetables, cotton, mulberry, sunflower, fruit crops, plantation crops and flower crops. Azotobacter fix about 10-25 kg N/ha/season, under normal conditions and much higher under favourable conditions besides also act as a crop protectant, since it releases anti-metabolites, which prevents spread of pathogens.

**Table 8.** Biofertilizers and their generalized beneficiary crops

| Name of biofertilizer | Beneficiary crops |
|---|---|
| Rhizobium | Crop specific biofertilizers for legume like Groundnut, Soybean, Red gram, Green gram, Black gram, Lentil, Cowpeas, Bengal gram. |
| Azotobector | Cotton, Vegetables, Mulberry, Plantation Crops, Rice, Wheat, Barley, Ragi, Jowar, Mustard, Safflower, Niger, Sunflower, Tobacoo, Fruit species, Condiments, Ornamental flowers. |
| Azospirillum | Sugarcane, Vegetables, Maize, Pearlmillet, Rice, Wheat, Fodders, Oilseeds, Fruits and Flowers. |
| Acetobector | Sugarcane |
| Blue green algae | Rice |
| Azolla | Rice |
| Phosphate solubilizing micro-orgnisms | All crops (non-specific) |
| Micrrrhizae | All crops (non-specific) |

## Mycorrhiza as biofertilizers

**Mycorrhiza** (fungus roots) is a distinct morphological structure which develops as a result of mutualistic symbiosis between some specific root - inhabiting fungi and plant roots. Plants which suffer from nutrient scarcity, especially P and N, develop mycorrhiza i.e. the plants belong to all groups e.g. herbs, shrubs, trees, aquatic, xerophytes, epiphytes, hydrophytes or terrestrial ones. In most of the cases plant seedling fails to grow if the soil does not contain inoculum of mycorrhizal fungi.

The term “mycorrhizae” refers to fungus associated with plant roots. These fertilizers are divided into ectotrophic and endotrophic or the vesicular arbuscular mycorrhiza (VAM) categories. Most plants depend on their mycorrhizal association for adequate uptake of nutrients (especially the immobile ions such as phosphate, zinc and micronutrients) and survival in natural ecosystems. Mycorrhixal association stimulates branching of the root and increases the absorption surface of the root. Other benefits include tolerance to drought, high soil temperature, soil toxins, and extreme Ph levels, as well as protection against root pathogens. This is why, When trees are introduced to new regions, inoculation of soil with mycorrhizal fungi is a necessary prerequisite for the establishment of the trees.

In recent years, use of artificially produced inoculum of mycorrhizal fungi has increased its significance due to its multifarous role in plant growth and yield, and resistance against climatic and edaphic stresses, pathogens and pests. Microrrhizae in association with host plant contribute to mobilize phosphates from the soil which are being used conventionally in large number of crops like wheat, maize, millets, potato, soybean, tomatoes, grapes, apples, banana, tea, coffee, cocoa, rubber etc. Micorrhizae could supply 15-30 kg phosphate/ha/ season besides mobilize micronutrients from soil. So, in organic farming mycorrhizae play an important role in contributing Phosphate.

***Mechanism of symbiosis:*** The mechanism of symbiosis is not fully understood. Biorkman (1949) postulated the carbohydrate theory and explained the development of mycorrhizas in soils deficient in available P and N, and high light intensity. Slankis (1961) found that at high light intensity, surplus carbohydrates are formed which are exuded from roots. This in turn induces the mycorrhizal fungi of soil to infect the roots. At low light intensity, carbohydrates are not produced in surplus, therefore, plant roots fail to develop mycorrhizas.

***Types of mycorrhizas:*** By earlier mycologists the mycorrhizas were divided into the following three groups

i) *Ectomycorrhiza.* It is found among gymnosperms and angiosperms. In short roots of higher plants generally root hairs are absent. Therefore, the roots are infected by mycorrhizal fungi which, in turn, replace the root hairs (if present) and form a mantle. The hyphae grow intercellularly and develop Hartig net in cortex. Thus, a bridge is established between the soil and root through the mycelia.

ii) *Endomycorrhiza.* The morphology of endomycorrhizal roots, after infection and establishment, remain unchanged. Root hairs develop in a normal way. The fungi are present on root surface individually. They also penetrate the cortical cells and get established intracellulary by secreting extracellular enzymes. Endomycorrhizas are found in all groups of plant kingdom.

iii) *Ectendomycorrhiza.* In the roots of some of the gymnosperms and angiosperms, ectotrophic fungal infection occur. Hyphae are established intracellularly in cortical cells. Thus, symbiotic relation develops similar to ecto- and endo-mycorrhizas.

**Marks (2004) classified the mycorrhizas into seven types on the basis of types of relationships with the hosts**

i) Vesicular-arbuscular (VA) mycorrhizas (coiled, intracellular hyphae, vesicle and arbuscules present)

ii) Ectomycprrhizas (sheath and inter-cellular hyphae present),

iii) Ectendomycorrhizas (sheath optional, inter and intra-cellular hyphae present),

iv) Arbutoid mycorrhizas (seath, inter-and coiled intracellular hyphae present),

v) Monotropoid mycorrhizas (sheath, inter- and intra- cellular hyphae and peg like haustoria present)

vi) Ericoid mycorrhizas (only coiled intracellular hyphae, long coiled hyphae present), and

vii) Orchidaceous mycorrhizas (only coiled intracellujlar hyphae present).

Type (i) is present in all groups of plant kingdom; Types (ii) and (iii) are found in gymnosperms and angiosperms. Types (iv), (v) and (vi) are restricted to Ericales, Monotropaceae and Ericales respectively. Types (vii) is restricted to Orchidaceous only. Types (iv) and (v) were previously grouped under ericoid mycorrhizaes.

## Methods of Inoculum Production and Inoculation

Methods of inoculum production of VAM fungi differ; however, some of these two are briefly described here.

a) ***Ectomycorrhizal fungi:*** The basidiospores, chopped sporocarp, sclerotia, pure mycelia culture, fragmented mycorrhizal roots or soil from mycorrhizosphere region can be used as inoculum. The inoculum is mixed with nursery soils and seeds are sown.

b) ***VA mycorrhizal fungi :*** VA mycorrhizas can be produced on a large scale by pot culture technique. This requires the host plants, mycorrhizal fungi and natural soil. The host plants which support large scale production of inoculum are sudan grass, strawberry, sorghum, maize, onion, citrus, etc.

The starter inoculum (spores) of VA mycorrhizal fungi can be isolated from soil by wet sieving and decantation technique. VA mycorrhizal spores are surface sterilized and brought to the pot culture. Commonly used pot substrates are sand: soil (1:1, w/w) with a little amount of moisture. An out line for inoculum production.

## There are two methods of using the inoculum

i) Using a dried spore-root- soil to plants by placing the inoculum several centimeters below the seeds or seedlings,

ii) Using a mixture of soil-roots, and spores in soil pellets and spores adhered to seeds with adhesives.

Commercially available pot culture of VA mycorrhizal hosts grown under aseptic conditions can provide effective inoculum.

In India, Forest Research Institute, Dehra Dun has established mycorrhizal bank in different states of the country. Inocula of these can be procured as needed and used in horticulture and forestry programmes.

### Benefits from mycorrhizas to plants

i) They increase the longevity of feeder roots, surface area of roots by forming mantle and spreading mycelia into soil and, in turn, the rate of absorption of major and minor nutrients from soil resulting in enhanced plant growth.

ii) They play a key role for selective absorption of immobile (P, Zn and Cu) and mobile (S, Ca, K, Fe, Mn, Cl, Br, and N) elements to plants. These are available to plants in less amount.

iii) Some of the trees like pines cannot grow in new areas unless soil has mycorrhizal inocula because of limited or coarse root hairs.

iv) VA mycorrhizal fungi enhance water uptake in plants,

iv) VA mycorrhizal fungi reduce plant response to soil stress such as high salt levels, toxicity associated with heavy metals, mine spoils, drought and minor element (e.g. Mn) imbalance.

v) VA mycorrhizal fungi decrease transplant socks to seedlings. They produce organic 'glues' which bind soil particles into semistable in aggregates. Thus, they play a significant role in augmenting soil fertility and plant nutrition.

vi) Some of them produce metabolites which change the ability of plants to induce roots from woody plant cuttings and increase root development during vegetative propagation.

vii) They increase resistance in plants and with their presence reduce the effects of pathogens and pests on plant heal.

### Plant Growth Promoting Rhizobacteria (PGRP)

Currently PGPR are being used alegely for deriving better crop yield through controlling the harmful micro-organisms of iron in their energy metabolism via production and extraction of iron, thereby allowing the plant to achieve increased growth. The bacteris used are Agrobacterium, Arthorobacter, Azotobacter, Bacillus pseudomonas, Cellulomonas, Enterobacter, Erwina, Florabacteria and Rhizobium.

### Importance of biofertilizers

- The increase in the productivity during the green revolution period is accompanied by an exponential increase in consumption of non-renewable sources of energy.
- In view of the fast diminishing energy sources comined with their escaling cast.

### Benefits of biofertilizers

- Germination increase up to 20 percent.
- Improved seedling emergence and growth.
- Increase yield from 10 to 40 percent.
- Improve the quality of fruit and keeping quality.
- Saving of 25 to 35 percent inorganic fertilizers.
- Increase the availability and up take of N and P in plants.
- Improve the status of soil fertility maintain good soil health and crop productivity.

### Methods of compost enrichment

Farm compost is poor in P content as it contains 0.4-0.8 percent available P. Addition of P makes the compost more balanced and supplies nutrient to micro-organisms for their multiplication and faster decomposition. The addition of P also reduces N losses and compost can be enriched by following methods:

- ***Application of superphosphate, bone meal or phosphate rock**:* 1 kg of superphosphate or bone meal is applied over each layer of animal dung. Low-grade phosphate rock can also be used for this purpose.
- ***Use of animal bones**:* these can be broken into small pieces, boiled with wood ash leachate or lime water and drained, and the residue applied to the pits. This procedure of boiling bones facilitates their disintegration. Even the addition of raw bones, broken into small pieces and added to the pit, improves the nutrient value of compost significantly.
- ***Wood ash waste:*** can also be added to increase the K content of compost.
- ***Addition of N-fixing and P-solubilizing cultures**:* The quality of compost can be further improved by the secondary inoculation of *Azotobacter*, *Azospirillum lipoferum*, and *Azospirillum brasilence* (N-fixers); and *Bacillus megaterium* or *Pseudomonas* sp. (P solubilizers). These organisms, in the form of culture broth or water suspension of biofertilizer products, can be sprinkled when the decomposing material is turned after one month. By this time, the temperature of the compost has also stabilized at about 35 °C. As a result of this inoculation, the N content of straw compost can be increased by up to 2 percent. In addition to improving N content and the availability of other plant nutrients, these additions help to reduce the composting time considerably.
- ***Accelerator technology:*** Through microbial enrichment or addition of rock phosphate (phospho-compost technique) or of pyrites (sulpho-compost) or

addition of both rock phosphate and pyrites (phospho-sulpho compost) to convert into valuable organic fertilizer

i) Efficient cellulytic cultures (Aspergillus, Tricoderma and Penicillium) accelerate recycling process and reduced the composting period from 6 months to 3-4 months.

ii) If cultures are not available, fresh cow dung can be used as base carrier of cellulytic organisms

The nutrient content is more in enriched compost

## Conclusion

Ill effects of the conventional farming system are felt in India in terms of the unsustainablity of agricultural production, environmental degradation, health and sanitation problems, etc. Organic agriculture is gaining momentum as an alternative method to the modern system. Many countries have been able to convert 2-10 per cent of their cultivated areas into organic farming. The demand for organic products is growing fast (at the rate of 20 per cent per annum in the major developed countries).

It appears that India is lagging far behind in the adoption of organic farming. So far, the only achievement seems to be the laying down of the National Standards for Organic Production (NSOP) and the approval of 4 accreditation agencies (all government bodies) whose expertise is limited to a few crops. The following are some of the issues, which require attention at the government policy making levels if we want to lay the spadework for the spread of organic agriculture in the country.

1. Substantial financial support by governments (Central, state and lower level bodies) is absolutely necessary to promote organic farming. A major factor behind the progress made by the major organic countries has been the very liberal subsidies provided by the governments. In India, organic farmers do not receive the benefits of government subsidies as they are targeted at the conventional cultivation. Given the low risk bearing capacity, the need to make the organic farming an attractive proposition at least during the initial period, the likely prospect of loss of productivity for some time, and the non-existence of marketing channels for organic produces the financial support must be adequate.

   The suggestion to begin the introduction of organic farming in the north-east region and in the dry farming areas of India is a well thought out proposition. However, it should be remembered that these regions are inhabited by the poorest and least advantaged groups whose dependence on agriculture for a livelihood is total. A programme for organic agriculture in these areas must be fully supported by the full compensation both in cash and kind to the farmers in the event of the loss of production they would suffer till it reaches the levels of the days prior to the adoption of organic farming. A fair, quick and efficient delivery system for such assistance, perhaps by keeping the government bureaucracy at a distance should also be in place beforehand.

2. Market development for the organic products is a crucial factor to promote domestic sales. Supplies do not match the demand for organic products in the country and the absence of proper links between the two has been pointed out for the tardy growth of organic farming in the country. An important role of the government in this direction is giving various supports to the producer and consumer associations to market the products.
3. The producer organizations must be encouraged to get accredited for inspection and certification in accordance with the NSOP. They can also have own standards and even symbols. This may also reduce the costs of certification besides the simplification of the process.
4. A vigorous campaign to highlight the benefits of organic farming against the conventional system is essential to increase the awareness of the farmers and consumers.
5. Identification of crops for cultivation on the organic farms is important. The examples of soyabean in Madhya Pradesh and cotton in the rainfed areas could be kept in view in the process.

# 5

# Organic Weed Management

**Weeds:** Weeds are unwanted and undesirable plants which interfere with the utilization of land and water resources and thus adversely affect human welfare. They can also be referred to as plants out of place. Usually, this means that weeds grow where we either want other plants to grow or where we want no plants at all. In cropland and forests, weeds compete with the beneficial and desired vegetation, reducing the yield and quality of produce. Undesirable vegetation also flourishes in aquatic systems, forestry and non-cropped areas such as industrial sites, roadsides, railway lines, airfields, landscape plantings, water tanks, water ways, etc. Thus, all plants may become weeds in particular situations.

Weeds are an important factor in the management of all land and water resources, but their effective impact is greatest on agriculture. There is no reliable study of world-wide damage due to weeds. However, it is widely known that losses caused by weeds exceed the losses from any category of agricultural pests, such as insects, nematodes, diseases, rodents, etc. Of the total annual loss of agricultural produce from various pests, weeds account for 45%, insects 30%, diseases 20% and other pests 5%.

## Harmful Effects Caused by Weeds

### Reduction in crop yield

Weeds compete with crop plants for nutrients, soil moisture and sunlight. The intensity of weed competition depends upon: a) type of weed species, b) severity of weed infestation, c) duration of weed infestation, d) competing ability of crop plants and e) climatic conditions which affect weed and crop growth.

Reduction in crop yield has a direct correlation with weed competition. Generally, an increase in one kilogram of weed growth corresponds to a reduction in one kilogram of crop growth. Weeds remove plant nutrients more efficiently than crop plants. In a drought situation, they thrive better than crop plants. When left undisturbed, some weeds can grow faster and taller than crop plants and inhibit tillering and branching. They can curtail sunlight and adversely affect photosynthesis and plant productivity.

Depending on the degree of competition, weeds reduce crop yields by 10% to 25%. In other words, if weed competition were completely eliminated from the entire cropland, the world's farm production could be increased by 10% to 25%. At a crop yield loss of 15 in a developing country like India, total weed control in the entire farmland would add US $5 billion to its economy at the current price level. This figure would be far greater if the direct and indirect impact of weeds on aquatic systems, forestry and industrial sites are also considered. This loss, if allowed to continue, amounts to a big drain in the nation's economy.

The economic impact of weeds on the US economy equals or exceeds US $20 billion, with the agricultural sector alone accounting for US $15 billion. Parker and Fryer estimated that the world was losing annually 11.5% of the total food production. That is, if all the weeds in food crops were controlled, the current world's food production would be higher by 11.5% or 450 million tons. This is too staggering a figure to be ignored by the world leaders and weed scientists.

Crop yield loss from weeds is highest in the tropics. For example, in rice, the mainstay of Asia's economy, proper weed control increases gain yields by 20% to 75%. In extreme weed situations, weed management could triple the yield of rice.

### Reduction in land value

Heavy infestation of perennial weeds can make the land unsuitable or less suitable for cultivation, resulting in loss of its monetary value. Millions of hectares of cultivable area in the rice-growing regions of Asia have been abandoned or are not being regularly cultivated due to severe infestation of Cyperus rotundas (nutgrass or purple nutsedge), Cynodon dactylon (bermudagrass), Imperata cylindrica (thatchgrass, cogongrass, or alang-alang) and other perennial weeds.

### Limited choice of crops

Crops differ in their ability to compete with weeds. In many instances the presence of particular weeds in fields limits the choice of crops to be grown. Very heavy weed infestations render some economically important crops, particularly pulses, vegetables, cotton, jute and forage crops unsuitable or less suitable for cultivation.

### Loss of quality

The quality of leafy and other vegetable crops suffers in the presence of weeds. In addition, Allium canadense (wild onion), Avena fatua (wild oats) and wild rice may impair the quality of onion, wheat and rice produce respectively. The leaves of Mikamia mlcrantha may contaminate tea leaves during plucking and reduce the value of tea. Contamination of other noxious weed seeds greatly reduces the value of crop seed and grain and sometimes even renders them unsaleable. The presence of weeds and weed debris in crop grain and other farm products reduces their market value and causes spoilage during storage.

## Reduced human efficiency

Weeds reduce human efficiency through physical discomfort caused by allergy and poisoning. Weeds such as Parthenium hysterophorus (parthenium), Rhus radicans (poison ivy) and Ambrosia spp. (ragweed) that cause itching, hay fever and other debilitating allergies contribute markedly to chronic human illness and suffering. The presence of spiny and thorny weeds such as Amaranthus spinosus, Argernone mexicana, Polygonum perfoliatum, etc. restricts movement of farm workers in carrying out, efficiently, farm practices such as fertilizer application (top-dressing), insect and disease control measures, harvesting, etc.

## Increased costs of insect and disease control

Weeds serve as alternative hosts to several crop insects, nematodes and pathogens. Insects such as aphids, thrips, weevils and stem flies survive on Brassica kaber (wild mustard), Daucus carota (wild carrot), Ambrosia spp. (ragweed), Arnaranthus retroflexus (redroot pigweed), etc. Weeds such as Avena fatua (wild oats) and some perennial grasses harbour pathogens of black stem rust of wheat. By harbouring these insects and pathogens, their attack on crop plants is increased. This results in increased cost of their control.

## Problems caused by aquatic weeds

Aquatic weeds reduce markedly the flow of water in irrigation and drainage canals, channels and streams. Obstruction to water flow raises water levels in canals and streams, resulting in: a) flooding, b) seepage into adjoining areas, c) breaks in canal banks and d) inadequate delivery of irrigation water to farms located at a distance from the main water source. In addition, reduced velocity of flow causes increased siltation and reduced carrying capacity, necessitating more frequent mechanical cleaning. Weeds cause enormous loss of water through transpiration. Aquatic weeds also obstruct weirs, gates and other water-regulating structures. Algae and fragments of plant material clog irrigation equipment. Aquatic weeds form breeding grounds for obnoxious insects like mosquitoes. They reduce recreational values by interfering with fishing, swimming, boating, hunting and navigation on streams and canals. Weeds on watershed areas and floodplains utilize large amounts of water and thereby deny its use for the production of food, feed and fibre. Some of the prominent aquatic weeds include Eichhornia crassipes, Typha angustifolia, Hydrilla verticillata, Potamogeton spp., Salvinia molesta, Ipomoea aquatica, Nymphaea spp., etc.

## Methods of Weed Propagation (Reproduction)

***Sexual reproduction:*** requires pollination of a flower, leading to seed production. Weeds are very prolific seed producers (common annual weeds may produce from 5,000 to over 500,000 seed per plant). This contributes to "weediness" of species and difficulty of control. Annual and biennial weeds and simple perennials reproduce sexually. Creeping perennials may reproduce sexually or asexually.

***Asexual reproduction:*** production of new plants from vegetative reproductive structures. Does not require flowers and pollination, seed are not produced. Process confined to creeping perennials. It also called vegetative reproduction. Creeping perennial have various vegetative organs for reproduction. See definitions below.

***Rhizome:*** A horizontal, underground stem which can produce adventitious roots and shoots (i.e., new plants) at the nodes. A rhizome can be distinguished from a root because rhizomes have nodes, internodes, and scale leaves (rudimentary leaves). Roots do not have nodes or leaves. Johnsongrass is example.

***Stolon:*** An above-ground stem that grows flat on the ground and can produce adventitious roots and shoots (i.e., new plants) at the nodes. Bermudagrass is example.

***Tuber:*** Enlarged terminal portions of rhizomes. Possess extensive storage tissue and axillary buds. Yellow nutsedge is example.

***Bulb:*** Specialized underground storage organ consisting of fleshy leaves with a short stem at the base. Food storage in the leaves. Wild garlic is example.

***Creeping roots:*** Horizontal roots modified for food storage and vegetative reproduction (can give rise to shoots). Often deep in the soil. Carolina horsenettle is example.

## Annuals and biennials

Annual and biennial weeds depend on seed production as the sole means of propagation and survival while perennial weeds are less dependent on this mechanism. Production of abundant and small seeds is a common adaptation that ensures a high probability of dispersal and re-infestation. A single plant of an annual weed can produce enough seeds in one season to cover an entire area of one acre with this weed species next year. For example, one plant of Sisymbrium altissinum (tumble mustard) produces over a half million (511,208) seeds which are enough, if evenly scattered, to sow at least 11 seeds ft' in an acre of land or 3200 seeds on every acre of a 160-acre farm. Similarly, Amaranthus retroflexus (redroot pigweed), Portulaca oleracea (common purslane), and Solanum nigrum (black nightshade) produce 196,405, 193,213 and 178,000 seeds plant respectively, while Brassica nigra (black mustard) produces 58,363 seeds per plant.

Some weed species have the ability to produce seeds between intervals of normal disturbance associated with a crop situation. Avena fatua (wild oats) germinate at the same time the wheat crop is sown and shatter their mature seeds before crop harvest. Many weeds can produce a large number of viable seeds even after having been cut off soon after flowering. A few weed species produce seeds through apomixis, i.e. without fertilization. Weeds such as ferns reproduce by spores rather than seeds.

Weed species differ in their regenerative strategies. Seeds of some species germinate soon after they are shed. Seeds of these species have a short life in the soil and their persistence is dependent on annual seed production and dispersal. In other species, seeds remain in the soil for long periods with intermittent germination of a part of the population. Some of the weed seeds are very long-lived, but these represent a small portion of the total seedbank. In agronomic situations, the majority of the seeds that germinate during the first two years represent a major threat for crop yield loss and control costs.

## Perennials

Vegetative reproduction is a common trait of perennial weeds. As discussed above, vegetative propagation includes rhizomes, stolons, tubers, bulbs, corms, roots and stems. Some perennial weeds propagate in one or more ways. Aside from their remarkable capacity for vegetative reproduction, most perennial weeds are also prolific seed producers. Their ability to produce both vegetative propagules and seeds make them very competitive and aggressive in cropping systems. Even if the aerial parts of perennial weeds die, the vegetative propagules continue to live and send up new growth, leading to flowering of shoots and production of a seed crop year after year.

A seedling of a perennial weed species growing from seeds is not a perennial when it first emerges from the soil. For example, Convolvulus arvensis (field bindweed) becomes a perennial when it has about 20 true leaves and Cyperus esculentus (yellow nutsedge) only 4 to 6 wk after it emerges from seed. Similarly, Elytrigia repens (quackgrass) becomes a perennial within 6 to 8 wk of emergence from seeds and Sorghum halepense (Johnsongrass) after only 3 to 6 wk of emergence. The population of certain perennial weeds and relative importance of various propagules are presented in Table 1.

**Table 1.** Propagation of some important perennial weeds

| Weed species | Common name | Propagules (in order of importance) |
|---|---|---|
| *Achillea* spp. | Yarrow | stolons, terminal rosettes of leaves, seeds |
| *Agrotis gigantea* | Redtop | rhizome (with dormant underground buds) |
| *Allium canadense* | Wild onion | bulbs and bulbils |
| *Asclepias syriaca* | Common milkweed | root stocks |
| *Cirsium arvense* | Canada thistle | creeping roots |
| *Convolvulus arvensis* | Field bindweed | creeping roots, seeds |
| *Cynodon dactylon* | Bermudagrass | rhizomes, nodes, seeds |
| *Cuperus rotundus* | Purple nutsedge | tubers, bulbs (nuts), seeds |
| *Elytrigia repens* | Quackgrass | rhizomes, seeds |
| *Equisetum arvense* | Field horsetail | rhizomes, spores |
| *Euphorbia esula* | Leafy spurge | creeping roots, seeds |
| *Imperata cylindrica* | thatchgrass, cogongrass | rhizomes, seeds |
| *Oxalis* spp. | Wood sorrel | bulbils, rhizomes, taproots |
| *Panicum repens* | Torpedograss | rhizomes, seeds |
| *Pteridium aquilinu* | Braken fern | rhizomes, spores |
| *Rumex crispus* | Curly dock | taproots, rosettes of leaves, seeds |
| *Ranunculus repens* | Creeping buttercup | procumbent stems, seeds |
| *Sonchus arvensis* | Perennial sowthistle | creeping roots, seeds |
| *Taraxacum officinale* | Dandelion | taproots, seeds |
| *Veronica filiformis* | Slender speedwell | creeping stems |

## Methods of weed control

**Eradication:** Complete kill or removal of a weed species (including un-germinated seed and vegetative propagules) from a defined geographical area. Insure that the weed will not reappear unless introduced into the area. Eradication is the ideal of weed control but is rarely achieved.

**Weed management:** Limiting weed population to a level that will not cause economic damage.

**Weed control *versus* weed management:** Terms are used interchangeably. However, weed management implies more knowledge of the problem; an attempt to determine how much control is justified, and more planning in deciding how to best handle the problem.

**Integrated weed management:** The use of all weed control methods (or those which are applicable) in a planned, coordinated management scheme.

**Prevention:** Steps taken to prevent introduction, establishment, and/or spread of a specified weed species into an area not currently infested with that species.

**Critical period of weed control:** This period has been defined as an interval in the life cycle of the crop when a must be kept weed free to prevent yield loss.

If weeds have been controlled throughout the critical period, the weeds that emerge later will not affect the economic yield and can be controlled prior to harvest or with a harvest and desiccate the crop.

## Strategies for weed management

I. Know the weeds Obtain correct identification of the major weeds present on the farm. Monitor fields regularly throughout the season. Keep records on what weeds emerge at different seasons, and on the efficacy of any preventive and control measures taken. Learn each weed's life cycle, growth habit, seasonal pattern of development and flowering, modes of reproduction and dispersal, seed dormancy and germination, and how the weed affects crop production.

II. Design the cropping system to minimize niches for weed growth In planning the crop rotation; avoid creating open niches in time or space. Plan tight rotations that follow one crop harvest promptly with the next planting. Open niches in space between crop rows can be reduced by using a narrower row spacing, intercropping, relay cropping, over seeding cover crops into established vegetables, or no-till management of cover crops prior to transplanting vegetables.

III. Keep the weeds guessing with crop rotations Plan and implement diversified crop rotations that vary timing, depth, frequency, and methods of tillage; timing and methods of planting, cultivation, and harvest; as well as crop plant family. Alternate warm- and cool-season vegetables. Rotate vegetable fields into perennial cover for two or three years to interrupt life cycles of annual weeds adapted to frequent tillage. Schedule tillage and cultivation operations when they will do the most damage to the major weed species.

IV. Design the cropping system and select tools for effective weed control Develop control strategies to address anticipated weed pressures in each of the farm's major crops. Choose the best cultivation implements and other tools for cost-effective preplant, between-row, and within-row weed removal. Plan bed layout, as well as row-and plant spacing, to facilitate precision cultivation.

## Preventive methods

These are methods and practices that check the spread of weeds. This can be achieved by careful management of agricultural machinery, using weed free seed and proper irrigation methods. Prevention focuses on keeping new weeds out and preventing the further spread of weed seed or perennial plant parts. Stopping the addition or introduction of weed seed to the soil can be particularly critical for successful weed management.

Understanding weed biology is an important component in developing a preventive approach. Weed species have strengths and weaknesses that make them vulnerable or resilient at different stages in their life cycle. Therefore, proper identification and knowledge of weed life cycle and reproduction and spread are important factors for developing management strategies. For example, disking or field cultivating a creeping perennial such as quackgrass or hedge bindweed in the spring may make the problem worse by spreading underground rhizomes or other vegetative structures.

**Some preventive tactics can be classified as sanitation**: removing or destroying weeds in fields or near fields before they flower and release weed seed (Table 2). Weed seeds can live for a number of years, depending on the species and whether the seed is exposed or buried beneath the soil surface. If necessary, weeds may need to be removed from the field by hand before they produce seed. Weeds can also be introduced into fields through manure, compost, hay, straw, animal feed, contaminated crop seed, or other materials. Whenever you plant, apply, or drive something in a field, make an effort to learn whether weed seeds are present and weigh the benefits against potential risks.

**Table 2.** Cultural practices for weed management

| Cultural practice | Prevailing effect | Example |
|---|---|---|
| Crop rotation | Reduction of weed emergence | Alternation between winter spring-summer crops |
| Cover crops (used as green manures or dead mulches) | | Cover crop grown in-between two cash crops |
| Primary tillage | | Deep ploughing, alternation between ploughing and reduced tillage |
| Seed bed preparation | | False (stale)-seed bed technique |
| Soil solarization | | Use of black or transparent |
| Irrigation and drainage system | | Irrigation placement (microtrickle-irrigation), clearance of vegetation growing along ditches |
| Crop residue management | | Stubble cultivation |

## Preventive measures of weed control

The following different measures are adopted for preventing further introduction and spread of weeds for min missing weed population.

1. Use clean seed that is free from weed seeds for sowing purpose.
2. Use well decomposed compost and F.Y.M. During decomposition much heat is generated and most of the weed seeds lose their viability. Avoid adding mature weeds to manure pits.

3. Destroy the weeds before formation of flowers for avoiding seedling and further spread of weeds.
4. Remove weed growth: The cropped area, bunds irrigation channels, waste lands, public places etc. should be kept clean or free of weeds.
5. Do not permit the livestock to move from weed infested area to clean area.
6. Avoid feeding animals with fodder and other material containing weed seeds.
7. Farm implements and machinery should be properly cleaned before moving from weed infested area to clean area.
8. Avoid shifting of soil, gravel sand etc from weed infested area to clean area.
9. Keep threshing yard, compost and F.Y.M. Pits free from weeds.
10. Inspect seed, necessary stocks for the presence of weed seeds, tubers, rhizomes, weed seedlings etc.
11. Use vigilance: Inspect your farm frequency for any strange looking weed. Destroy such patches of weeds by digging deep or by using suitable herbicides.
12. Follow the legal and quarantine measures while importing crop seeds, food grains, seedlings etc.

### Cultural strategies

Preventing weeds from emerging and spreading stage through various approaches are employed for creating favourable condition for the crop. These practices if used properly, helps in controlling weeds.

1. **Field preparation**: The field has to be kept weed free. The Flowering of weeds should not be allowed. This helps in the prevention of buildup of weed seed population.
2. **Pre-germination of weeds**: In pre-germination irrigation or rainfall germinates weed seeds just before the cash crop is planted. The newly germinated weeds can be killed by light cultivation or flaming.
3. **Stale seed bed/dab system:** This method is very effective in reducing weeds population but its only drawback is that sometimes sowing is delayed. Stale seed bed is one where initial 1 to 2 flushes of weeds are destroyed by harrowing before planting or sowing of the crop. This is achieved by soaking a well prepared field with either irrigation or after receiving rain and allowing the weeds to germinate. These newly emerged weeds are destroyed by harrow with spike tooth or blade harrow. This should be followed immediately by sowing the crop.

4. **Blind tillage**: Tillage of the soil after sowing a crop before the crop plants emerge.
5. **Variety selection**: Careful selection of crop varieties is essential to limit weeds and pathogen problems and to satisfy market needs. A quick growing and early canopy cover providing cultivar competes better with weeds. The traditional tall growing varieties perform better than modern dwarf varieties under unweeded situations thaths why most of organic growers grow tall varieties at their farm. Short statured erect leave varieties permit more light compared to tall and leafy traditional varieties. So, vigorous and fast growing crop varieties are better competitors with weeds.
6. **Crop rotation:** Crop rotation involves alternating different crops in a systematic sequence on the same land. It is an important strategy for developing a sound long term weed control program. Weeds tend to thrive with crops of similar growth requirements as their own and cultural practices designed to contribute to the crop may also benefit the growth and development of weeds. Monoculture, that is growing the same crop in the same field year after year, results in a build-up of weed species that are adapted to the growing conditions of the crop. When diverse crops are used in a rotation, weed germination and growth cycles are disrupted by variations in cultural practices associated with each crop (tillage, planting dates, crop competition, etc). Many weed problems exist with us because of practicing monoculture i.e growing of same crop year after year in the same field. Monoculture leads to increase the infestation of certain crop associated weeds. For example *Echinochloa* spp in Rice; *Phalaris minor* and *Chenopodium album* in Wheat; Striga spp. in Sorghum, Maize, Pearl millet and Sugarcane; *Orobanche* spp. in Tobacco.

Wild oats can be eliminated by growing rice. *Phalaris minor* population can be eliminated or lowered by growing wheat with other rabi crops such as berseem, potato, raya or gobhi sarson, winter maize, oats (fodder), sugarcane, etc. The replacement of rice-wheat with sugarcane-wheat almost eliminates the *Phalaris minor* infestation.

Cotton and Sunflower stimulate germination of *Striga* spp but being non-host of this weed cause its complete suppression. The proper crop rotations or inclusion of pulse crops in rotation is helpful for control of crop associated and parasitic weeds. E.g Striga on Jawar or Orobanche crops like groundnut, soybean, cotton etc in the crop rotation.

So, alternate different crops in a systematic sequence on the same land for developing a sound long term weed control programme.

7. **Growing of intercrops:** Intercropping involves growing a smother crop between rows of the main crop. Intercrops are able to suppress weeds.

However, the use of intercropping as a strategy for seed control should be approached carefully. The intercrops can greatly reduce the yields of the main crop if competition for water or nutrients occurs. Many a time intercropping with crops like soybean , cowpea, red gram, green gram , groundnut etc. helps in suppression of secondary growth of weeds e.g Intercropping of soybean or cow pea in sorghum or pearl millet is helpful to suppress secondary growth of weeds as the inter row space is covered by intercrop at latter stage of crop growth. In this method at initial stage of crop growth it is necessary to control the weeds by mechanical methods. Intercropping of cowpeas in maize and harvesting it at 40-45 days after sowing before twining, apart from fodder provides effective control of weeds.

8. **Relay cropping:** A pulse crop is usually broadcast as relay crop into standing rice crop 10-15 days before harvest. As soon as wheat/oilseed crop is harvested, the pulse crop cover the field in dry season and suppress the weed growth.

9. **Cover crops**: Rapid development and dense ground covering by the crop will suppress weeds. Cover crops such as rye, mustard, cowpea, groundnut or fast growing crops like wheat, barley or dense forage crops in the cropping system can suppress weed growth.

10. **Sowing time**: Manipulation of sowing time is an important agronomic practice to manage the weeds. The sowing of crop is planned in a period which is unfavourable for germination of weed seeds but is not so far for crop performance. *Phalaris minor* infestation is less in October sown wheat. Though, late sown wheat (December) results in less drymatter accumulation by *Phalaris minor* but crop yields are also affected adversely.

11. **Proper crop stand and narrow spacing:** Weed crop competition is more in gappy and lower plant population. Therefore, practice like use of good quality seed, proper method of sowing, optimum seed rate, spacing and time of sowing and plant protection measures against insect pests and diseases are very important to obtain proper and uniform stand of crop, capable of offering initial competition to young weeds. Higher plant density and narrow spacing is useful to minimize weed crop competition than the lower plant density and wider spacing. Lack of adequate plant population is prone to heavy weed infestation, which becomes, difficult to control later.

12. **Sowing method:** Weed management is easy and better in bed planted and transplanted crops. In bed planted wheat the weeds are less as the weeds emerged can be removed mechanically while reshaping the beds with bed planter. It also facilitates manual removing of weeds from wheat fields near maturity of crop and helps preventing the weed seed dispersal in the field.

Transplanted rice provides better control of weeds as compared to direct seeded rice as in direct seeded rice both rice and other rainy season crop weeds emerged.

13. **Seed rate/planting density:** Crop population, spatial arrangement, and the choice of cultivar (variety) can affect weed growth. The basic idea of increasing seed rate or planting density is to attain more number of cop plants per unit area in order to boost initial smothering potential of crop plants on weeds. High seed rate in wheat (125 kg/ha) in wheat significantly reduces weed population and dry matter accumulation by weeds. Higher plant density in rice 44 plants $m^{-2}$ reduces weed biomass. The weed population also reduces in closer row spacing of wheat (15-20 cm) over 22.5 cm row spacing in timely sown wheat and 15 cm in late sown wheat and consequently increases wheat grain yield. For example, the studies have shown that narrow row widths and a higher seeding density will reduce the biomass of later-emerging weeds by reducing the amount of light available for weeds located below the crop canopy. Similarly, fast growing cultivars can have a competitive edge over the weeds.
14. **Inter-cultivation**: It is very effective and the cheapest method of weed control in line sown crop. Most of the inter-cultivation implements have a blade which cuts the weeds just below the soil surface and thus kill weeds.
15. **Lowering area under bunds:** Reduction in area under bund and water channels of crop fields is essential to minimize weed infestation as weeds on bunds and channels are often ignored by the farmers since they do not complete directly with crops.
16. **Irrigation and drainage:** Frequent irrigation or rain during initial stage of crop growth induced several flushes of weed (Lowland rice, where standing water is present most of the time, germination of weed is less). There are a number of ways that careful irrigation management can help you reduce weed pressure on crops.
17. **Farmyard manure:** Farm yard manure should be well demomposed before its field application as the undecomposed manure serves as a seed bank of weeds.
18. **Buried drip irrigation**: Drip pipes buried below the surface of the planting bed can provide moisture to the crop and minimize the amount of moisture that is available to weeds closer to the surface. The Punjab Agricultural University, Ludhiana had recommended direct seeded rice/maize-wheat cropping system under sub-surface drip irrigation method.
19. **Flooding:** It helps in controlling weeds like kans. (*Saccharum spontaneum*) which grows luxuriantly in heavy ill drained soils during rainy seasons. The

fields are embanked with tall bunds to impound water. The weeds get submerged under water and are smothered.

20. **Smothering:** A quick growing dense crop can successfully complete with weeds. Past growing and fast shading crop is called smother crop. Sweet potatoes, Lucerne, soybean, sun hemp, etc are good smother crops. These crops cover the field within short period due to higher plant density, fast growth and canopy coverage suppress weeds by fast shading effect and also make the under ground parts of weed feeble. The smother crops are sown at close spacing with high plant population to suppress weeds.

21. **Solarisation:** In this method the soil temperature is further increased by 5 to 10 °C by covering a pre-soaking fallow field with transparent plastic sheet. In some regions the technique raised the soil temperature to as high as 56 °C which was sufficient to kill vegetative propagation. This techniques followed for four weeks or longer period which reduced weed seed viability by 80 % or more. Israel has so far made maximum use of this technique as a part of integrated weed management. However, this is costly technique.

22. **Sanitation**: It is possible to prevent many new weeds from being introduced onto the farm and to prevent existing weeds from producing large quantities of seed. Use of clean seed, mowing of weeds around the edges of fields, prevention of weed seed mixing with crop seed, application of well rotten FYM and thoroughly composting manure before application can greatly reduce the introduction of weed seeds and problematic weed species and also thorough cleaning of any machinery is required before using in crop fields.

23. **Burning:** This method is adopted to destroy weeds in non-cropped areas like waste lands road sides, railway lines, bunds etc. The flame throwers and steam boxes are used for burning weeds in advanced countries. The flame should be so adjusted that is causes only wilting but not charring of weeds. Heat kills the living cells by coagulating the protoplasm and inactivating enzymes. The thermal death point (TDP) of most plant cells lies between 45 to 55° c (113 to 131° F). Rabbing in Konkan area is widely practiced to kill weeds seeds particularly from paddy nursery areas. It is not a good method, since the useful vegetation and organic matter is also destroyed along with weeds.

24. **Use of fertilizers or selective crop stimulation:** Nitrogen fertilizer can affect the competition between crops and weeds and in the subsequent crops. For example, nitrate is known to promote seed germination and seed production in some weed species. Nitrogen fertilization may result in increased weed growth instead of increased crop yield. Selective placement of nitrogen

in a band can favour the crop over the weed. The band application ( near to crop roots) of nitrogen for cereals, sugarcane and Sugarbeet etc is said to result in their vigorous growth that carries them beyond weed competition.

25. **Feed the crop, not the weeds**: Avoiding pre-plant broadcasting of soluble nutrients which might be utilized by fast-growing weeds than slow-growing crops, and may even stimulate weed germination. Expensive bagged organic fertilizer may be applied at low rates at planting time or sidedress near rows as they are likely to be captured by the crop, relying on mid-season release of nutrients from compost and / or green manures for primary fertility.

**Table 3.** Important cultural practices for weed management

| Cultural practice | Prevailing effect | Example |
|---|---|---|
| Sowing/planting time and crop row | Improvement of crop competitive ability spatial | Use of transplants, higher seeding rate, lower inter-distance, anticipation or delay of sowing/transplant date |
| Crop genotype choice | | Use of varieties characterized by quick emergence, high growth and soil cover rates in the early stages |
| Cover crops (used cultural method as living mulches) | | Legume cover crop sow in the inter-row of a row crop |
| Intercropping | | Intercropped cash crops |
| Fertilization | | Use of slow nutrient releasing organic fertilizers and amendments, fertilizer placement, anticipation or delay of pre-sowing or topdressing N fertilization |

## Mechanical practices

1. **Mechanical weeding**: Row cultivators are used two to three times to control weeds between rows (Table 4).

**Table 4:** Important mechanical practices for weed management

| Implements/Tools | Purpose |
|---|---|
| CRIDA Manual weeder | For weeding in upland crops |
| Self propelled power weeder | For weeding and intercultural operation in upland crops *viz.*, groundnut, soybean, maize and pigeonpea etc. |
| Dryland weeder | For weeding in row crops for removing shallow rooted weeds. It has been designed ergonomically for easy operation. Useful in dryland and garden land crops and is ideal at a soil moisture content of 8 to 10%. |
| Power rotary weeder | For weeding and intercultural operation in between row crops |

| | |
|---|---|
| Tractor-drawn Weeding-cum-earthing up equipment | For weeding and intercultural operation in between row crops like sugarcane, cotton, maize, etc. |
| Tractor-operated multi-row rotary weeder | For weeding between rows of paddy crop |
| Cono-weeder | For weeding between rows of paddy crop |
| Two-row finger type paddy rotary weeder | For weeding in paddy row crops |
| Battery operated portable wetland weeder | For weeding in SRI field |

2. **Mechanical or Physical Methods of Weed Control:** This includes the use of hand tools, implements and machinery operated either with the help of man power, or machine power for control of weeds. These are the costly and time consuming methods. However, these methods cause minimum damage to the environment.

3. **Hand Pulling or Hand Weeding:** Pulling the weeds by hand or hand weeding with the help of weeding hook is the oldest and most effective methods for control of weeds. Weeds can be easily uprooted after a good soaking irrigation or rain. Weeding should be done before flowering stage of weeds for avoiding seedling and further spread of weeds. Because as per the old saying. One year's seeding is seven years weeding. This is costly and time consuming method.

4. **Tillage System**: Tillage systems alter the soil seed bank dynamics and depth of burial of weed seeds. Studies have found that almost 75% of the seedbank was concentrated in the upper 5 cm of soil in no-till fields. In the moldboard plough system however, the seedbank is more uniformly distributed over depth. Other conservation tillage systems are intermediate to these two systems. Sometimes ploughings help in controlling *Cyperus rotundus*. Puddling of rice reduces incidence of weeds. Night tillage under organic agriculture is also one of the strategies for weed management.

   It is one of the practical methods of destroying weeds of all categories.

5. **Deep Ploughing**: Weeds are buried deep in the soil and also exposed to heat of the sun by deep ploughing. Many annual and perennial weeds can be controlled by deep ploughing and exposing the field to heat of the sun.

6. **Discing**: Help in cutting and burying of weeds. Harrowing with blade harrow is very effective for destroying newly germinated weeds before sowing of the crop.

   Inter-row cultivations: Inter-culturing with different types of hoes or mechanical weeders is useful to control weeds in between rows of the crop. Inter-row cultivations are normally used to control weeds between the crop rows. It cut or uproots weeds, and then either completely buries them or leaves them on the soil surface for desiccation.

7. **Mowing and Sickling:** This method is used lands, pastures, gardens and roadsides. The implement mower is use for cutting weeds. It does not destroy the weeds completely but prevents seed production by cutting growing parts. Cutting above ground parts of weed with sickle is called sickling and it prevents in seed formation.

8. **Summer Tillage or Summer Fallow:** Deep ploughing after harvest of the rabi crop and exposing underground part of weeds to strong sunlight during summer months is helpful for destroying many annual and perennial weeds or left field without any crop in summer season (summer fallow). This method is widely practised in dry farming areas.

9. **Digging:** This method is useful for controlling perennial weeds like nut grass, hariali etc. Digging is very useful for removing the under ground propagation parts of weeds form the deeper layer of soil digging , hand collection and destruction of under ground parts of weeds is adopted when noxious weeds are observed in the patches in the field. It is not economical on large areas as it is costly and time consuming method.

10. **Dredging and Chaining:** These methods are useful for controlling aquatic weeds. Removing of weeds along with their roots and rhizomes from the water with the help of mechanical force is called dredging. The floating aquatic weeds are removed by changing. A heavy chain in pulled over the water bodies to collect the weeds.

## Mulching

The principle aim of this method is to cut off light and avoid all top growth of weeds. The organic mulches such as plant residues i.e Stubbles, wheat straw , rice hulls, sugarcane trash etc are spread in between crop rows @ 5 tonnes/ha and thickness about 10 to 15 cm. The inorganic mulches such as black or white polythene sheet is spread in between crop rows. However, it is expensive and used for high value crops only. The black polythene is more effective for controlling weeds. Allelopathic chemicals in the mulch also can physically suppress seedling emergence. There are many forms of mulches available.

1. **Living mulch**: Living mulch is usually a plant species that grows densely and low to the ground such as clover. Living mulches can be planted before or after a crop is established. eg.., cowpeas in maize crop(1:1).

2. **Organic mulches**: Such materials as straw, bark, and composted material can provide effective weed control. Producing the material on the farm is

recommended since the cost of purchased mulches can be prohibitive, depending on the amount needed to suppress weed emergence. eg., 10 t/ha rice residue in organic turmeric and soybean.

## Thermal weed control

1. **Flamers**: Flamers are useful for weed control. Thermal weed control involves the use of flaming equipment to crate direct contact between the flame and the plant. This technique works by rupturing plant cells when the sap rapidly expands in the cells. Sometimes thermal control involves the outright burning down of the weeds. Flaming can be used either before crop emergence to give he crop a competitive advantage or after the crop has emerged.
2. **Soil solarization**: During summer and fall, organic farmers sterilize their soil through solarization. In this process, a clear plastic film is placed over an area after it has been tilled and tighly sealed at the edges. Solarization works when the heat crated under the plastic film becomes intense enough to kill weed seeds.
3. **Infrared weeders**: Infrared weeders are a further development of flame weeding in which the burners heat ceramic or metal surfaces to generate the infrared radiation directed at the target weeds. Some weeders use a combination of infrared and direct flaming to kill the weeds.
4. **Freezing**: Freezing would be advantageous only where there is an obvious fire risk from flaming. Liquid nitrogen and solid carbondioxide (dry ice) can be used for freezing weeds.

**Table 5**. Important practices for weed management

| Practice | Effect |
|---|---|
| Tillage | Kills growing weeds; damages perennial roots & rhizomes; buries seeds too deeply to emerge; brings weed seeds to the surface. |
| Post-planting cultivation | Removes weeds from the crop. |
| Stale seedbed | Flushes weeds from the soil before planting. |
| Organic fertility sources | Favor crops over faster-growing weeds due to slow release of nutrients. |
| Drip irrigation | Directs water to the crops rather than weeds. |
| Mulch | Smothers weeds: delays emergence of weeds |
| Using transplants | Competitive advantage to crop. |
| Competitive cultivars | Improve competitive ability of crop against weeds. |
| Increase plant density | Suppresses weeds by shading. |
| Rapid cleanup after harvest | Prevents seed set by residual weeds. |
| Cover crops | Suppress weeds, improves soil health. |

## Chemical weed management

Chemical weed control is generally not allowed in organic crop production systems. Only a limited number of herbicides are organically acceptable, and these include contact materials such as acetic acid, citric acid and solutions of sodium nitrate as well as pre emergent material, corn gluten. Herbicides can be used for selective weed control by manipulating the timing of application or placement of material or by exploiting differences in the chemical tolerances of the crop targeted weed. Weeds that emerge before the crop can be killed with contact herbicides (acetic acid etc). These herbicides kill plants that have emerged later. Corn gluten is a pre emergent material that is applied to the soil to suppress weeds as they germinate. Currently, the efficiency of these organically acceptable herbicides is marginal at best.

**Table 6.** Herbicides listed by the Organic Materials Review Institute (OMRI) for use in organic production.

| Product | Active Ingredients | Manufacturer |
|---|---|---|
| Alldown | Citric acid plus garlic plus acetic acid | Summerset Products 952-820-0363 |
| Corn gluten meal | Corn gluten meal | Numerous-for more information. |
| Ground force | Citric acid plus garlic plus acetic acid | Abby Laboratories |
| Herbicidal soap | Various salts of fatty acids | Several brands-may be synthetic and used for nonfood crops only |
| Matran II | Clove oil | EcoSmart Technologies, Inc. 888-326-7233 |
| Xpress | Thyme and clove oil | Bio Huma Netics 800-961-1220 |

The combination of mechanical and cultural methods is effective for minimizing the weed population.

## Biological weed control

The use of living organism (Bio-agents) such as insects, disease organisms, herbivorous fish, snails and even competitive plants to limit weed infestation is called biological control of weeds. It is also known as the classical biological weed control.

The philosophy behind the biological weed control is that in nature the growth of many weeds remains limited due to biotic stresses from their natural enemies, largely the insect pests and diseases. Therefore, the bio-agent which is specific to particular host wed can be used for control of the weed.

## Criteria for selection of bio-agent or qualities of good bio-agent for weed control

1. Host Specificity

   a) The bio-agent recommended for weed control should be host specific i.e. the bio-agent must feed or kill only the host plant (Weed).

   b) It should not harm or feed the crops and other useful (economical) plants. Starvation test is conducted for this purpose before using the bio-agent for weed control.

2. Bio-agent Hardiness

   a) The bio-agent should be hardy and must adopt or withstand the new environment in which it is to be used.

   b) It should be free from its natural enemies like predators and parasites.

   c) It should be capable of surviving in a situation of food shortage for longer period.

3. Feeding Habit

   a) It must be able to kill the host or at least prevent seed production or reproduction of host plant.

   b) The bio-agent should be capable of seeking the host.

   c) It should be a heavy feeder and consume or cause the death of large population of host rapidly.

4. Base of Multiplication

The bio-agent should have high rate and ease of natural reproduction to control the host plant in short time. If the natural rate of reproduction is less than more times required for control of host.

## Different bio-agent used for controlling weeds

The long term experiments with great care are conducted for searching a host specific bio agent. The outstanding and feasible examples of biological control of weeds are given below:

### Use of insects

a) Lantana camara (Ghaneri)-was controlled by using Crocidosema lantana Busck, a moth as most promising bio-agent in Haweli- This bio-agent feeds on flowers and seeds.

b) Lantana bug (*Teleonemia scruplosa* stall) was used in Australia for controlling Ghaneri.

c) Hispine beetles (*Octotoma scabripennis*) and *Uroplate girandi* were also used in Australia for control of Ghaneri.

d) Prickly pear/cactus (*Opuntia spp*) by using cochineal scale insects. Dactylopis indicus and Dactylopius tomentosus in south India and in Australia by using Cactoblastis cactorum.

e) Alligator weed (*Altermathere philoxoroides* ) in U.S.A has been effectively controlled with flea beetle (*Agasicles hygrophyla*)

f) Fern (*Salvinia molesta*) in kerela (India) by using a beetle called *Cytrobagous saviniae*.

g) *Cypenis rotundus* (nustsedge) in India U.S.A and Pakistan by *Bacteria spp*. a shoot borer moth.

**Use of carp fish**

Common carp (*Cyperus carpio*) and Chinese carp (White amur or grass carp) called Clenopharyngodon idella for control of aquatic weeds.

**Snails**

Snail (*Marisa cornuarietis*) and certain fresh water snails are useful to control aqutic weeds like water hyacinth, salvinia spp. and water lettuce.

**Use of mites**

A spider mite (*Tetranychus desertorum*) was used in Australia foe control of prickly pear (Cactus).

**Use of fungi**

a) *Acacia glauca* was controlled in Hawali by injecting suspension of spare of *Cephalosporium spp*.

b) *Rhizoctonia blight* for control of water hyacinth.

c) Use of Mycoherbicides/ Bioherbicides.

**Biological weed management**

Biological control would appear to be the natural solution for weed control in organic agriculture.

**Allelopathy as weed management**

Allelopathy is the direct or indirect chemical effect of one plant on the germination, growth or development of neighboring plants. It is now commonly regarded ascomponent of biological control. Species of both crops and weeds exhibit this ability. Allelopathic crops include barley, rye, annual rye grass, buckwheat, oats, sorghum, sudan sorghum hybrids, alfalfa, wheat, red clover, and sunflower. Vegetables, such as horseradish, carrot and radish, release, particularly powerful allelopathic chemicals from their roots. Suggestions have been made that allele chemicals and other natural products or their derivatives could form the basis of bioherbicides. However, it is unclear whether the application of natural weed killing chemicals would be acceptable to the organic standard authorities. There are six classes of allele chemicals namely alkaloids, benzoxazinones, cinnamic acid derivatives, cyanogenic compounds, ethylene and other seed germination stimulants, and flavonoids which have been isolated from over 30 families of terrestrial and aquatic plants. All these chemicals possess actual or potential phytotoxicity.

Incorporated plant residues can become toxic to weeds by the release of allelopathic chemicals. There are numerous reports of allelopathy and of the isolation of allelopathic compounds from plants. Examples of successful weed control leading to an increase in crop yield following incorporated cover crop residue include incorporation of *Sorghum bicolor* stalks before *Triticumaestivum*, incorporation of *Brassica napus* before *Solanum tuberosum*, and incorporation of *Trifoliumincarnatum* before *Zea mays* rice (*Oryza sativa*) residues against *Echinochloacrusgalli* var. *Oryzicola* (barnyard grass), an associated weed of rice.

## Beneficial organisms

Little research has been conducted on using predatory parasitic microorganisms or insects to manage weed populations. However, this may prove to be a useful management tool in the future. Natural enemies that have so far been successful include a weevil for the aquatic weed salvinia, a rust for skeleton weed and probably the most famous, a caterpillar (*Cactoblastis* sp.) to control prickly pear. There is also considerable research effort aimed at genetically engineering fungi (myco-herbicides) and bacteria so that they are more effective at controlling specific weeds. Myco-herbicides are a preparation containing pathogenic spores applied as a spray with standard herbicide application equipment.

Weeds are subject to disease and insect attacks just as crop are. Most biological control of weeds occurs in range or non-crop areas. As a result, biological control has little relevance for vegetable growers. Geese have been used for weed control in trees, vine, and certain row crops. Most types of geese will graze weeds, but Chinese weeder geese are considered the best for row crops. Chinestweeder geese are smaller than other types and tend to walk around delicate crop plants rather than over them. Geese prefer grass species and rarely eat crops. If confined, geese will even dig up and eat Johnson grass and Bermuda grass rhizomes. Care must be taken to avoid placing geese near any grass crops such as corn, sorghum, or small grains, as this is their preferred food. Fruiting vegetables, such as tomatoes when they begin to color, might also be vulnerable, so geese would have to be removed from tomato fields at certain times. Geese require drinking water, shade during hot weather, and protection from dogs and other predators.

**Table 7.** Use of biocontrol agents for weed control

| Name of the weed | Bioagent |
|---|---|
| Cyperusrotundus | Bactra verutana |
| *Ludwigiaparviflora* | *Halticacynea* (Steel blue beetle) |
| Parthenismhysterophorus | Zygrogrammabicolarata |
| Lantana camara | Crocidosema lantana, Teleonnemiascrupulosa |
| *Opuntiadilleni* | *Dactylopiustomentosus, D. Indicus* (cochineal scale insect) |
| *Eichhorneacrassipes* | *Neochetinaeichhornea, N. Bruchi* (Hyachinth weevil) *Sameodes alliguttalis* (hyancinth moth) |
| *Salviniamolesta* | *Crytobagussingularis* (weevil) *Paulinia acuminate* (grass hopper), *Sameamutiplicalis* |
| *Alternantheraphiloxaroides* | *Agasideshygrophilla* (flea beetle) *Amynothripsandersoni* |
| *Tribulusterrestris* | *Microlarinuslypriformis, M. lareynii* |
| Solanum elaegnifolium | Frumentanephalomicta |

**Table 8.** Use of fish for weed control

| Name of the weeds | Fish |
|---|---|
| Lemma, Hydrilla, Potamogeton | Grass carp or white amur |
| Algae | Silver carp, common carp |

**Table 9.** Use of competitive plants for weed control

| Name of the weeds | Competitive plants |
|---|---|
| Partheniumhysterophorus | Cassia sericea |
| *Typha* sp. | *Brachiaria mutica* |

## Biological weed control

**Allelopathy**: Allelopathy is the direct or indirect chemical effect of one plant on the germination, growth or development of neighboring plants. It is now commonly regarded as component of biological control. Species of both crops and weeds exhibit this ability. Allelopathic crops include barley, rye, annual ryegrass, buckwheat, oats, sorghum, sudan sorghum hybrids, alfalfa, wheat, red clover, and sunflower.

**Beneficial organisms:** Natural enemies that have so far been successful include a weevil for the aquatic weed salvinia, a rust for skeleton weed and probably the most famous, a caterpillar (*Cactoblastis* sp.) to control prickly pear. There is also considerable research effort aimed at genetically engineering fungi (myco-herbicides) and bacteria so that they are more effective at controlling specific weeds (Table 10-12). Myco-herbicides are a preparation containing pathogenic spores applied as a spray with standard herbicide application equipment.

**Table 10.** Bioagents for weed management

| Name of the weed | Bioagent |
|---|---|
| Cyperusrotundus | Bactra verutana |
| *Ludwigiaparviflora* | *Halticacynea* (Steel blue beetle) |
| Parthenismhysterophorus | Zygrogramma bicolarata |
| Lantana camara | Crocidosema lantana, Teleonnemiascrupulosa |
| *Opuntiadilleni* | *Dactylopiustomentosus, D. Indicus* (cochineal scale insect) |
| *Eichhorneacrassipes* | *Neochetinaeichhornea, N. Bruchi* (Hyachinth weevil) *Sameodes-alliguttalis* (hyancinth moth) |
| *Salviniamolesta* | *Crytobagussingularis* (weevil) *Paulinia acuminate* (grass hopper), *Sameamutiplicalis* |
| *Alternantheraphiloxaroides* | *Agasideshygrophilla* (flea beetle) *Amynothripsandersoni* |
| *Tribulusterrestris* | *Microlarinuslypriformis, M. lareynii* |
| Solanumelaegnifolium | Frumentanephalomicta |

**Table 11.** Commercial mycoherbicides for weed management

| Trade name | Pathogen | Target weed |
|---|---|---|
| Devine | *Phyophthorapalmivora* | *Morreriaodorata* (Strangler vine) in citrus |
| Collego | *Colletotrichumgleos-poroidesf.sp. aeschynomene* | *Aeschynomenevirginica* (northen joint vetch) in rice and soyabean |
| Biopolaris | *Biopolarissorghicola* | *Sorghum halepense* (Johnson grass) |
| Biolophos | *Streptomyces hygroscopius* | General vegetation(non-specific) |
| LUBAO 11 | *Colletotrichumgleosporoides* f.sp. *Cuscuttae* | Cuscutta sp. (Dodder) |
| 1 | *Alternariacassiae* | *Cassia abtusifolia* |
| ABG 5003 | *Cercosporarodmanii* | *Eichhorneacrassipes* (water hyancinth) |

Bioherbicides approach involves spray of specific fungal spores or its fermentation products, against the target weed. These preparations are called mycoherbicides (Bioherbicides). Mycoherbicides can be specific (Selective) or non-specific (Non selective) and are used just like chemical herbicides in developed countries. Control of weeds by using the mycoherbicides is called bio-herbicidal control of weeds.

**Table 12.** List of commercial mycoherbicides in use abroad

| S.No. | Product | Content | Weed controlled |
|---|---|---|---|
| 1. | De Vine | A liquid suspension of fungal spores of Phytophthora palmivora. | Strangle Vine (Morrenia odorata) in Citrus. |
| 2. | Collego | Wettable powder contaning fungal spores of collectotrichum spp. | Joint-Ventch (Aeschymone spp) |
| 3. | Biolaris | A suspension of fungal spores of Bipolaris sorghicola. | Johnson grass (Sorghum helepense) |
| 4. | Biolophos | A microbial toxin produced as fermentation product of Streptomyces hygroscopicus | Non selective for killing generated vegetation. |

## Use of competitive plants

a) Use of Rajagira (Amaranthus panjculatus) for control of nutsedge (nutgrass) and other weeds as it is the best smother crop and also produces allelopathic effect.

b) Use of smother crop like soybean, cowpea, greengram, Sanhemp, Sweet potato, etc. in crop rotation for control of weeds.

c) Proper crop rotation is helpful for management of crop associated and parasitic weeds.

## Biological weed control v/s other methods of weed control

a) The initial cost of biological weed control appears to be high, but when compared to the cost of developing new herbicides, it is quite reasonable.

b) This method is so far adopted for control of weeds in non cropped areas and not useful for controlling different weed species in cropped lands.

c) Minimum damage to the environment by biological method of weed control as compared to other methods of weed control.

d) This method is not solution for all weeds but used for larger weeds or particular weed species.

## Constraints/limitation for adoption of organic approach of plant protection

1. The major constraints of plant protection in organic farming like high cost of organic pesticides inputs, no market for organic pesticides product, unavailability of organic pesticides inputs, less yield and no price advantage for organic product are found to be the major constraints according to their ranking as first, second, third, fourth and fifth.

2. The next important constraints are found to be no consumers demand for organic product.
3. Natural insecticides are generally less stable than synthetic materials and degrade quickly in the environment, meaning that they are also less potent and have shorter residual periods than their synthetic counterparts. Therefore, satisfactory arthropod pest management can only be achieved when insecticide use is integrated with other strategies.
4. If the quality and efficacy of natural products like teas, extractsand fermentation products could be enhanced by commercialresearch and development programs, better solutions for sometypical problems of plant protection in organic farming could befound.
5. The other constraints in order to importance are lack of awareness, low employment potentiality and lack of experience of organic pesticides.
6. In regards to the relative importance of different constraints it is found that socio-economic constraints is the main hurdle followed by infrastructural, technological and situational in the process of adoption of organic plant protection.

## Indigenous practices

1. *Calotropis* (*Akk*): The practice of keeping cut branches of *calotropis (Akk)* at the entarance of irrigation channel in the field minimizes the striga population (Patel, 1992).
2. Cotton shells: *Cyperus rotundus* a noxious weed, can be controlled by using empty shells of cotton balls. The nutshells are broadcasted in 3 inch thick layer in the field before the rainy season. This burns/kills the root/ sedge of the weed. The field remains weed free for about two-three years. It also act as organic manure upon decomposition and helps in increasing crop yield.
3. Coriander: Coriander (*Coriandrum sativum*) is used to control *Striga asiatica* in Sorghum. Coriander seeds are planted between the sorghum rows. About 200 g of coriander seeds are sown for every three kg of sorghum seeds. Once the coriander plants are big enough the strands twist themselves around striga weed and suppress its growth.

# 6

# Biological Insect Management

The use of living organisms to suppress the population of a specific pest organism, making it less abundant or less damaging than it would otherwise be.

**Definitions of 'pest' (noun)**

1. An annoying person or thing; a nuisance.
2. An injurious plant or animal, especially one harmful to humans.
3. A deadly epidemic disease; a pestilence.

**What are bio-pesticides?**

Bio pesticides include naturally occurring substance that control pest (biochemical pesticides), microorganism that control pest (microbial pesticide) and pesticidal substances produced by plants containing added genetic materials (Plant-in corporate protectants).

Biopesticides are biochemical pesticides that are naturally occurring substance that control pest by non toxic mechanism.

Biopesticides are natural plant products belonging to the secondary metabolites include alkaloids, terpenoids, phenolics etc.

**Why bio-pesticides?**

1. Sustainable food chain: economic, environmental, social
2. Integrated Pest Management (IPM)
3. Pesticide resistance
4. Zero detectable residues
5. Pesticide product withdrawals
6. Pesticide resurgence
7. Toxic food
8. Relatively specific

## List of bio-pesticides in use

| No. | Products | Formulation |
|---|---|---|
| **Bacteria** | | |
| 1. | Bacillus thuringiensis var. isaelensis | 5.0%WP, 5.0%AS |
| 2. | Bacillus thuringiensis var. kurstaki | 5.0%WP, 7.5%WP |
| 3. | Pseudomonas fluorescens | 0.5%, 1.0%WP |
| **Fungi** | | |
| 4. | Ampelomyces quisqualis | 2.0% WP |
| 5. | Beauveria bassiana | 1.0%, 1.15% or 2.15%WP, 10.0%SC |
| 6. | Metarhizium anisopliae | 1.0, 1.5% WP |
| 7. | Paecilomyces lilacinus | 1.00% |
| 8. | Trichoderma harzianum | 0.5%, 1.0%, 2.0%WP |
| 9. | Trichoderma viride | 1.0%WP |
| 10. | Verticillium chlamydosporium | 1.00% |
| 11. | Verticillium lecanii | 1.15%WP |
| **Virus** | | |
| 12. | NPV of Helicoverpa armigera | 0.43%, 0.5%, 0.64%, 2.0% |
| 13. | NPV of spodoptera litura | 0.5%, 2.0% |
| **Botanicals** | | |
| 14. | Neem – Azadiractin | 0.3%, 0.15%, 1.0% |
| 15. | Vanis (Citronella) | |

*Source:* CIB & RC website, Minutes of the Registration Committee meetings, June 2003 to Dec 2007.

**Biological control** or **biocontrol** is a method of controlling/managing pests such as insects, mites, weeds and plant diseases by using other organisms. The biocontrol pest management relies on predation, parasitism, herbivory, or other natural mechanisms. There are three basic strategies for biological pest control

i) Classical (importation): A natural enemy of a pest is introduced in the hope of achieving control;

ii) Inductive (augmentation): A large population of natural enemies are administered for quick pest control;

iii) Inoculative (conservation): Measures are taken to maintain natural enemies through regular reestablishment.

**Biological control agents**: Natural enemies of insect pests are also known as biological control agents. The natural enemies are living organisms that kill pests, decrease pest reproductive potential, and compete with pest organisms for use of your plants. These include predators, parasitoids, pathogens, and competitors. Biological control can have side-effects on biodiversity through attacks on non-target species by any of the same mechanisms, especially when a species is introduced without thorough understanding of the possible consequences.

## History

*Cactoblastis cactorum*larvae feeding on *Opuntia*prickly pear cacti

The term "biological control" was first used by Harry Scott Smith at the 1919 meeting of the Pacific Slope Branch of the American Association of Economic Entomologists, in Riverside, California.

An Entomologist Paul H. DeBach (1914–1993) brought the term biological conrol into more widespread use. He worked on citrus crop pests.

The first international shipment of an insect as biological control agent was made by Charles V. Riley in 1873, shipping to France the predatory mites Tyroglyphus phylloxera to help fight against the grapevine phylloxera (Daktulosphaira vitifoliae) that was destroying grapevines in France.

The United States Department of Agriculture (USDA) initiated research in classical biological control following the establishment of the Division of Entomology in 1881, with C. V. Riley as Chief.

The first importation of a parasitoidal wasp into the United States was that of the braconid *Cotesia glomerata* in 1883–1884, imported from Europe to control the invasive cabbage white butterfly, *Pieris rapae*.

In 1888–1889 the vedalia beetle, *Rodolia cardinalis*, a lady beetle, was introduced from Australia to California to control the cottony cushion scale, *Icerya purchasi*.

Nine parasitoids (solitary wasps) of gypsy moth, seven of brown-tail moth, and two predators of both moths became established in the US.

Prickly pear cacti were introduced into Queensland, Australia as ornamental plants, starting in 1788. They quickly spread to cover over 25 million hectares of Australia by 1920, increasing by 1 million hectares per year. Digging, burning and crushing all proved ineffective. Two control agents were introduced to control the spread of the plant, the cactus moth *Cactoblastis cactorum*, and the scale insect *Dactylopius*. Between 1926 and 1931, tens of millions of cactus moth eggs were distributed around Queensland with great success, and by 1932, most areas of prickly pear was destroyed.

### Status of biopesticides in India and at global level

- Biopesticides represent only 2.89% (as on 2005) of the overall pesticide market in India.

- Annual growth rate of about 2.3% in the coming years (Thakore, 2006).
- In India 15 types of biopesticides have been registered under the Insecticide Act, 1968.
- Consumption of biopesticides has increased from 219 metric tons in 1996-97 to 683 metric tons in 2006-07.
- About 85% of the biopesticides used are neem based products.
- Global biopesticide consumption is estimated at about 24 million kg. North America consumes the largest share (40%) of the global biopesticide production followed by Europe and Oceanic countries accounting for each.

The global market for biopesticides will increase from US$ 672 million in 2005 to over US$ 1 billion in 2010. Consumption level of biopesticides is approximately 1 kg/ha at global level.

## Types of biopesticides

Bio-pesticides fall into three major classes:

A. Microbial pesticides

B. Plant pesticides

C. Biochemical pesticides

A. **Microbial pesticides**: These consist of a microorganism (e.g., a bacterium, fungus, virus, or protozoan) as the active ingredient. Microbial pesticides can control many different kinds of pests, although each separate active ingredient is relatively specific for its target pest(s). For example, there are fungi that control certain weeds, and other fungi that kill specific insects.

   The most widely used microbial pesticides are subspecies and strains of *Bacillus thuringiensis*, or Bt. Each strain of this bacterium produces a different mix of proteins, and specifically kills one or a few related species of insect larvae. While some Bt's control moth larvae found on plants, other Bt's are specific for larvae of flies and mosquitoes. The target insect species are determined by whether the particular Bt produces a protein that can bind to a larval gut receptor, thereby causing the insect larvae to starve.

B. **Plant pesticides**: These are pesticidal substances that plants produce from genetic material that has been added to the plant. For example, scientists can take the gene for the Bt pesticidal protein and introduce the gene into the plant's own genetic material. Then the plant, instead of the Bt bacterium, manufactures the substance that destroys the pest. Both the protein and its genetic material are regulated by EPA; the plant itself is not regulated.

C. **Biochemical pesticides** are naturally occurring substances that control pests by non-toxic mechanisms. Conventional pesticides, by contrast, are generally synthetic materials that directly kill or inactivate the pest. Biochemical pesticides include substances, such as insect sex pheromones, that interfere with mating, as well as various scented plant extracts that attract insect pests to traps. Because it is sometimes difficult to determine whether a substance meets the criteria for classification as a biochemical pesticide, EPA has established a special committee to make such decisions.

## Types of biological pest control

There are three basic biological pest control strategies: importation (classical biological control), augmentation & conservation

**Importation:** Importation or classical biological control involves the introduction of a pest's natural enemies to a new locale where they do not occur naturally.

A biological control agent requires a colonizing ability which allows it to keep pace with changes to the habitat in space and time then bio-control agent will be most effective in controlling a pest. The control is greatest if the agent has temporal persistence, so that it can maintain its population level even in the temporary absence of the target species, and if it is an opportunistic forager, enabling it to rapidly exploit a pest population.

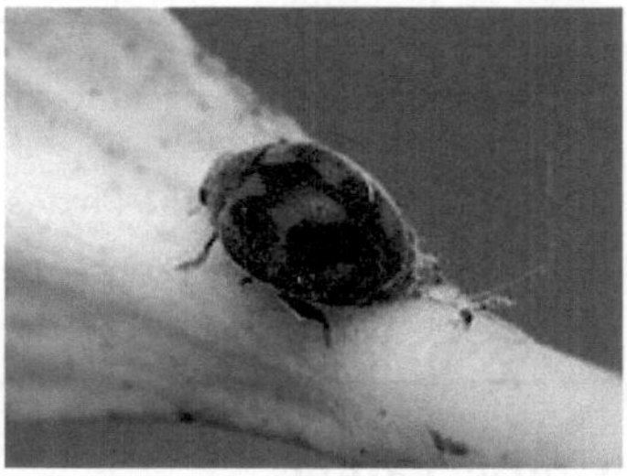

*Rodolia cardinalis*, the vedalia beetle, was imported from Australia to California in the 19th century, successfully controlling cottony cushion scale.

**Table 1.** Importation or classical biological control examples

| S.No. | Host plant | Country | Bio-control agent |
|---|---|---|---|
| 1. | Icerya purchasi (cottony cushion scale) | Australia | Predatory insect Rodolia cardinalis (the vedalia beetle) |
| 2. | Antonina graminis | Texas | Neodusmetia sangwani |
| 3. | Alternanthera philoxeroides (alligator weed) | Florida (U.S.) | Alligator weed flea beetle |
| 4. | The giant Salvinia (Salvinia molesta), a serious pest, covering waterways | Warm climates and in Zimbabwe | Salvinia weevil (*Cyrtobagous salviniae*) and the salvinia stem-borer moth (*Samea multiplicalis)* |

**Augmentation:** Augmentation involves the supplemental release of natural enemies that occur in a particular area, boosting the naturally occurring populations there. In inoculative release, small numbers of the control agents are released at intervals to allow them to reproduce, in the hope of setting up longer-term control, and thus keeping the pest down to a low level, constituting prevention rather than cure. In inundative release, in contrast, large numbers are released in the hope of rapidly reducing a damaging pest population, correcting a problem that has already arisen. Augmentation can be effective, but is not guaranteed to work, and depends on the precise details of the interactions between each pest and control agent.

*Hippodamia convergens*, the convergent lady beetle, is commonly sold for biological control of aphids.

**Table 2.** Augmentation biological control examples

| S.No. | Biocontrol agent | Host |
|---|---|---|
| 1. | Parasitoidal wasp, Encarsia formosa, | Greenhouse whitefly |
| 2. | Predatory mite Phytoseiulus persimilis | Two-spotted spider mite |
| 3. | Egg parasite Trichogramma | Moths |
| 4. | Nematodes | Soil-dwelling insect pests |

## Conservation

The conservation of existing natural enemies in an environment is the third method of biological pest control. Natural enemies are already adapted to the habitat and to the target pest, and their conservation can be simple and cost-effective, as when nectar-producing crop plants are grown in the borders of rice fields and these provide nectar to support parasitoids and predators of plant hopper pest. Predators of aphids were found to be present in tussock grasses growing as field boundary hedges in England, but they spread too slowly to reach the centres of fields. Control was improved by planting a metre-wide strip of tussock grasses in field centres, enabling aphid predators to overwinter there.

An inverted flowerpot filled with straw to attract earwigs

## Biological control agents

### Predators

Predators are mainly free-living species that directly consume a large number of prey during their whole lifetime. Given that many major crop pests are insects, many of the predators used in biological control are insectivorous species. Lady beetles, and in particular their larvae which are active between May and July in the northern hemisphere, are voracious predators of aphids, and also consume mites, scale insects and small caterpillars. The spotted lady beetle (*Coleomegilla maculata*) is also able to feed on the eggs and larvae of the Colorado potato beetle (*Leptinotarsa decemlineata*).

The larvae of many hoverfly species principally feed upon aphids, one larva devouring up to 400 in its lifetime.

Predatory lacewings are available from biocontrol dealers

Predatory *Polistes* wasp searching for bollworms or other caterpillars on a cotton plant

The parasitoid wasp *Aleiodes indiscretus* parasitizing a gypsy mothcaterpillar, a serious pest of forestry

## Parasitoids

Parasitoids lay their eggs on or in the body of an insect host, which is then used as a food for developing larvae. The host is ultimately killed. Most insect parasitoids are wasps or flies, and many have a very narrow host range.

*Encarsia formosa*, widely used in greenhouse horticulture, was one of the first biological control agents developed.

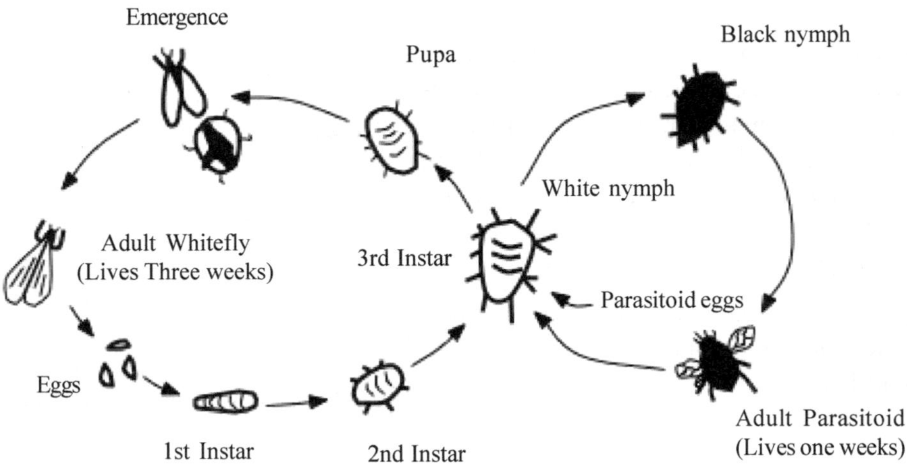


Life cycles of greenhouse whitefly and its parasitoid wasp Encarsia formosa

*Encarsia formosa* is a small predatory chalcid wasp which is a parasitoid of whitefly, a sap-feeding insect which can cause wilting and black sooty moulds in glasshouse vegetable and ornamental crops. It is most effective when dealing with low level infestations, giving protection over a long period of time.

## Pathogens

Pathogenic micro-organisms include bacteria, fungi, and viruses. They kill or debilitate their host and are relatively host-specific. Various microbial insect diseases occur naturally, but may also be used as biological pesticides.

## Bacteria

Bacteria used for biological control infect insects via their digestive tracts, so they offer only limited options for controlling insects with sucking mouth parts such as aphids and scale insects. *Bacillus thuringiensis*, a soil-dwelling bacterium, is the most widely applied species of bacteria used for biological control, with at least four sub-species used against Lepidopteran (moth, butterfly), Coleopteran (beetle) and Dipteran (true fly) insect pests. The bacterium is available to organic farmers in sachets of dried spores which are mixed with water and sprayed on vulnerable plants such as brassicas and fruit trees.

Genes from *B. thuringiensis* have also been incorporated into transgenic crops, making the plants to express some of the bacterium's toxins, which are proteins.

These confer resistance to insect pests and thus reduce the necessity for pesticide use if pests develop resistance to the toxins in these crops, *B. thuringiensis* will become useless in organic farming also. The bacterium *Paenibacillus popilliae* which causes milky spore disease has been found useful in the control of Japanese beetle, killing the larvae. It is very specific to its host species and is harmless to vertebrates and other invertebrates.

### Fungi

Green peach aphid, a pest in its own right and a vector of plant viruses, killed by the fungus *Pandora neoaphidis* (Zygomycota: Entomophthorales)

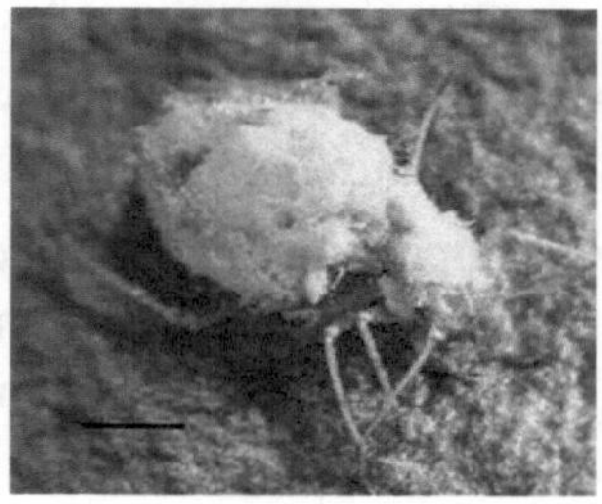

Entomopathogenic fungi, which cause disease in insects, include at least 14 species that attack aphids. *Beauveria bassiana* is mass-produced and used to manage a wide variety of insect pests including whiteflies, thrips, aphids and weevils. *Entomophaga* is effective against pests such as the green peach aphid.

### Viruses

Baculoviruses are specific to individual insect host species and have been shown to be useful in biological pest control. For example, the Lymantria dispar multicapsid nuclear polyhedrosis virus has been used to spray large areas of forest in North America where larvae of the gypsy moth are causing serious defoliation. The moth larvae are killed by the virus they have eaten and die, the disintegrating cadavers leaving virus particles on the foliage to infect other larvae.

### Competitors

The legume vine *Mucuna pruriens* is used in the countries of Benin and Vietnam as a biological control for problematic *Imperata cylindrica* grass. The legume vine *Mucuna pruriens* is extremely vigorous and suppresses neighbouring plants by out-competing them for space, moisture, nutrients and light.

### Advantages of biological control

- Absence of toxic effects
- No development of resistance by pests.
- No residue of poison in the soils and rivers.
- No build up of toxins in the food chain.
- No killing of pollinators or development of secondary pests throgh the destruction of their environment.

**Disadvantages**

- Control is slow
- It is often unpredictable.
- It is difficult and expensive to develop and apply.
- It requires expert supervision.
- Many of the most important pests are exotic, invasive species that severely impact agriculture, horticulture, forestry and urban environments.
- They tend to arrive without their co-evolved parasites, pathogens and predators, and by escaping from these, populations may soar.
- Importing the natural enemies of these pests may seem a logical move but this may have unintended consequences; regulations may be ineffective and there may be unanticipated effects on biodiversity.
- The adoption of the techniques may prove challenging because of a lack of knowledge among farmers and growers.
- Biological control can affect biodiversity through predation, parasitism, pathogenicity, competition, or other attacks on non-target species.

**Bio-pesticides**: Natural occurrence of diseases caused by micro-organisms is common in both insects and weeds and is a major natural mortality factor in most situations. Use of micro-organisms for pest control involves their culture in artificial media and later introduction of larger amounts of inoculums in to the field at appropriate time. Many fungi and bacteria can be handled in this way but insect viruses have the limitation that they have to be raised in living insects. As the biocontrol agents (microbial pathogens) are applied on the targeted pests in much the same way as chemical pesticides, they are often termed as bio-pesticides or natural pesticides.

- *Bacillus thuringensis*, a bacterial pathogen infesting a wide range of insect pests, is the most common microbial insecticide in use today. It is used against caterpillars that attack a wide range of crop. Unlike most other chemical insecticides, it can be used on edible products up to the time of harvest. It is selective in action and does not harm parasites, predators or pests. The bacteria come in several commericla formulations such as Dipel, Delfin, Halt, Spicturin, Biolep, BioAsp etc. Another bacterium *B popilaleis* also commonly available against white grub *Popillae japonica* and *Hototricha* sp. Amongst insect pathogenic fungi, commercial preparations of *Verticillium lecanii* are available for the control of aphids, thrips and white fly under glass house conditions.

**Botanicals**: Some weeds like Lantana, notchi, tulsi, adathoda etc act as natural repellant to many pests. Trees like pungam, wood apple, anona and their byproducts have excellent insecticidal value in controlling diamond back moth, heliothis, white flies, leafhoppers and aphid infestation.

- Most commonly used botanicals are neem (*Azadirachta indica*), pungamia (*Pungamia glabra*) and mahua (*Madhuca indica*). Neem seed kernel extract (2 to 5%) has been found effective against several pests including rice cutworm,
- Diamond back moth, rice BPH, rice GLH, tobacco caterpillar, aphids and mites. The pesticidal ingredients of neem formulations belonging to general class of natural products called triterpenes, more specifically, limonides. They act as repellents and also disrupt growth and reproduction in insects. Commonly known limonoids are azadirachtin, meliantriol, salannin, nimbin and nimbidin. The efficiency of vegetable oils in preventing infestation of stored product pests such as bruchids, rice and maize weevils has been well documented. Root extracts of asparagus work as a nematicide for plant parasitic nematodes. Similarly leaf extracts of many plants can inhibit a number of fungal pathogens.

## Bio-pesticides

The bio-pesticides are used in different field crops for insect-pest and disease management and are given in the following table:

**Table 3.** Bio-pesticides used for organic insect managemen

| Crop | Insect-pest and disease | Bio-pesticide/bio-product |
|---|---|---|
| Cotton | Bollworms | Traps, Lures, NPV, BT, Trichogramma |
| | Whitefly, Jassid, thrip | Neem based formulations |
| | Mites | *Chrysoperia, Verticillium, Baeluveria* |
| Chickpea, Pigeon pea, Green gram, Black gram | Pod borer/cut worm | Traps, Lures, BT, NPV, *Trichogramma* |
| Rice | Yellow stem borer, leaf folder | Traps, Lures, BT, *Trichogramma* |
| | Hoppers | *Baeluveria,* Neem based formulations |
| Sugarcane, Basmati rice, Maize | Borer | *Trichogramma* |
| Soybean | Catterpiller | Traps, Lures, BT, NPV, *Trichogramma* |
| | Whitefly, Pod borer | Neem based formulation |
| Mustard | Aphid | *Chrysoperia,* Neem based formulation |
| Potato | Cutworms | Traps, Lures, BT |

**Microbial organisms**: These are those micro-organims which are capable of causing diseases in insects as a result they lose their appetite, subjected to several physiological disturbances, leading to the ultimate death of the insect. The microorganisms exploited in biological control of insect-pests are broadly classified into a) Insect viruses b) bacteria c) Entomo Pathogenic Fungi d) Entomo Pathogenic Nematodes and other organisms like protozoans and Rickettesia etc. while several antagonistic fungi and bacteria are being successfully used in minimizing the plant disease incidence. Nematode pest management by using biotic agents is also one of the most promising areas and gaining much desired importance in the current scenario of organic farming.

**Pathogens causing diseases in insects and destroying them: these include**

- Viruses: Nuclear polyhedroses viruses (NPV): They cause typical "tree top disease" in insects. The infected insect loose appetite becomes restless and reaches the apical portion of the plant due to O2 depletion inside the insect system. Later the infected insect will be dying by hanging itself in an inverted 'V' shape on the apical portion of the plant. Effective against lepidopteran insects in different crops. *Ha* NPV is used for the management of *Helicoverpa armigera* while *Sl* NPV is meant for *Spodoptera litura*. Similarly, castor semilooper is managed by *Ach* NPV and red hairy caterpillar by *Am* NPV.
- Grunulosis viruses (GV) and Cytoplasmic viruses (CPV): Extensively used against insect-pests of sugarcane.
- Entomo: Pathogenic viruses are highly specific to host insects which make them exceptionally safe to non target organisms and nature.
- Bacteria: Bacillus thuringensis: This bacterium is highly effective against several insect-pests belonging to order Lepidoptera. They cause disease due to which insect turns black and die.
- Fungi: *Metarhizium anisopliae*, *Beauveria bassiana* and *Verticillium lecanii* are used against important pests like gram pod borer, tobacco caterpillar and sucking pests like thrips, aphids and mealy bugs. The fungi develop hyphae inside insect system resulting in the death of the insect due to mechanical congestion. The mode of action makes these organisms to perfectly suit to the needs of organic farming. In certain cases they produce toxins to kill the insect.
- Nematodes: *Heterorhabditi* sp and *Steimernema* sp: These nematodes harbor certain bacteria which act as toxins to insect systems. Protozoa
- *Variomorpha* sp: These were found effective against insect pests and can effectively incorporated as tools in organic farming.

## Cultural methods

Agronomic adjustments, necessary for higher yield, are at the same time are directed at prevention, mass multiplication and spread of pests by modifying the crop microclimate.

**Summer ploughing:** White-grub insect is serious in some areas only. The adult beetles emerge from the soil during June-July with the first showers of rain. Plough the field twice during May-June. It would help in exposing the beetles resting in the soil. Sow the crop early wherever possible.

**Sanitation:** It includes removal or destruction of breeding refuges and over wintering of pests. Seed material, farm yard manure etc carrying insect eggs or its stages of development should be carefully screened before their use. Destruction of alternate hosts minimizes pest population build up.

**Tillage and inter-cultivation**: Ploughing and inter-cultivation brings about unfavourable conditions for multiplication of pests as well as diseases and weeds. Quiescent stages (pupae) of harmful organisms will be exposed to dehydration or to predation by birds and other stages may be mechanically damaged or buried deep in the soil.

**Seed:** Always use clean seed and planting material.

**Cultivar selection**: Cultivars with high yield potential and quality without resistance to pests and diseases are the main causes of frequent epidemics and mass multiplication of pests and diseases. A large number of cultivars resistance/ tolerance to pest and diseases has been developed to suit different agro-ecosystems. Selection of such cultivars can bring down the losses considerably.

**Time of sowing:** As weather influences developmental rhythm of plansts as well as growth and survival of pests and diseases, serious setback occurs when the weather conditions are such as to bring about coincidence of susceptible growth stages with highest incidence of pests and diseases. Therefore, adjustment in sowing dates is often resorted to as an agronomic strategy to minimize the crop losses. Maize sown late suffers little borer damage, as by then the egg parasite *Trichogramma* is able to keep down the population of the pest. Rice may suffer less from borer attack if planted early (early June). Early maturing cotton cultivars have become popular in Punjab and Haryana as they escape pink bollworm. Avoid early sowing in June as early sown crop suffers heavily from Jassid in sesame.

**Alternate host:** Baru (*Sorghum halepense*) is the alternative host plant from which this mite spreads to the sugarcane crop. So destroy the weed, if growing near the sugarcane fields.

**Plant population:** Plant population per unit area influence crop microclimate. Dense plant canopy leads to high humidity build up congenial for pest and disease multiplication. Keeping the total plant population constant, inter and intra row plant population can be adjusted to minimize the humidity build up within the plant canopy.

**Manures and fertilizers**: Excessive nitrogen increases susceptibility of crop to sucking and leaf eating pests. Higher rates of nitrogen application than the recommended rate to hybrids without corresponding increase in phosphorus and potassium is the main factor for heavy pest and disease incidence. Balanced application of NPK helps the crop to tolerate pests and diseases considerably.

**Water management:** Irrigation can reduce the soil inhabiting pests by suffocation or exposing them to soil surface to be preyed upon by birds. Irrigating potato crop at tuber formation can minimize potato scab. Anthracnose of beans, early blight and charcoal rot of potato can be checked by furrow irrigation than sprinkler irrigation.

**Habitat diversification:** Many pests prefer feeding on a particular plant or others. This preference may be exploited to reduce the pest load on crop. Crop rotation, intercropping, traps cropping and strip cropping can bring down the pest load considerably.

**Planting methods**: Adopt suitable planting methods to reduce the insect-pest incidence. Ridge sowing reduces the incidence of pod borer in pigeon pea. (Agarwal *et al.,* 2003) and cut worm in sunflower (Bakhetia et al., 1995). The wider spacing (60 × 20 cm) showed less black aphid infestation in chickpea than closer spacing (30 × 10 cm).

**Trenching**: The insects like army worm, red hairy catterpiller, nymphs of grass hoppers can be checked by digging trenches around the field. The marching insects fall in the trench and these can be easily killed.

**Crop rotation:** The evaluations of suitable cropping systems to perform phyto-sanitary function in avoiding the carry over effects of crop specific pests and diseases to succeeding crops. Inclusion of green manure crop helps to encourage beneficial organisms. Crop rotations with non grass crops helps to reduce the borer population (Alvares, 1999). The weed population also reduced in green manured fields as the weeds are also incorpora ted into soil alongwith green manure crop.

**Habitat diversification**: Many pests prefer feeding on a particular plant or others. This preference may be exploited to reduce the pest load on crop. Crop rotation, intercropping, traps cropping and strip cropping can bring down the pest load considerably.

**Intercropping and mixed cropping**: Adopt intercropping and mixed cropping systems for organic management of pests. Intercropping of cowpea and chilli reduces pod borer population in soybean (Ramangouda, 2002). Incidence of *H.*

*armigera* in chickpea decreases when grown as intercrop with wheat, barley and mustard. The incidence of insect-pests is less on legumes intercropped with maize (Singh, 2001). Intercropping of pea with barley and linseed reduces the insect-pest incidence in peas. (Prasad *et al.,* 1987). Intercropping of coriander and garlic in sugarcane helps to manage sugarcane top borer (Verma et al., 1987). Intercropping of coriander and linseed in chickpea helps to manage chick pea pod borer (Kasyapa, 1998). Prefer mixed cropping of fodder maize with cowpea/ bajra/ sorghum (PAU 2020).

## Trap crops

Plant trap crops alongwith main crop to attract the pests and predators that feed on insect-pests. The planting of marigold around tomato crop helps in controlling *H. armigera* because *H. armigera* lays eggs on marigold which is xposed to *T. chilonis* (Kalyayan, 2007). The planting of napier bajra around maize fields helps to manage maize stem borer.

Bihar hairy caterpillar (*Bhaboo Kuta*) attacks cowpea from August to November. To keep cowpea free from attack of this pest, plant a row of til (Sesamum) around cowpea field at the time of its sowing. Females prefer to oviposit on til rows from where larvae in gregarious phase should be collected and destroyed mechanically.

**Lopping**: The lopping is recommended in tall basmati varieties to prevent lodging and it also decreases the stem borer incidence. The detopping of cotton seduces the *Helicoverpa* incidence.

**Water management:** Water management is best tool against some insects-pests. Irrigation application prevents termite damage to various crops. Drain out excess water to suppress brown plant hopper population in rice crop. Irrigating potato crop at tuber formation can minimize potato scab. Anthracnose of beans, early blight and charcoal rot of potato can be checked by furrow irrigation than sprinkler irrigation. Do not delay or withhold irrigation during April-June to prevent spider mite in brinjal.

**Manures and fertilizers**: Excessive nitrogen increases susceptibility of crop to sucking and leaf eating pests. Higher rates of nitrogen application than the recommended rate to hybrids without corresponding increase in phosphorus and potassium is the main factor for heavy pest and disease incidence. Balanced application of NPK helps the crop to tolerate pests and diseases considerably.

## Bird perches

- Bird perches are resting places for predatory birds to rest and to look for preys; such as insect pests of cotton, peanuts, and cowpeas. Predatory birds prefer to look for prey in field crops where they have places to rest.

- To make bird perches, use bamboo or wooden poles or tree branches. Erect either of these at regular intervals in the field.
- To have live bird perches within the field, plant *Setaria species* (foxtail cultivars). These plants are found to be attractive to predatory birds. The birds feed on their seeds. In cotton field, plant Setaria in every 9$^{th}$ or 10$^{th}$ row of cotton. Once the birds are on the field, they prey on cotton bollworms and other insects.

**Natural enemies**: Conserve and promote the natural enemies by providing artificial structures viz., nesting boxes for wasps and predatory birds; planting or retaining food or shelter plants on the boundary; placing bundles of rice residues for attracting spiders and other predators; establishing bird perches; using trap crops; Planting of tall sorghum varieties on borders and inside pigeon pea fields acts as bird perches which eats pod borer larvae (Saxsena *et al.,* 2010). A single wolf spider can eliminate atleast 20 brown plant hoppers per one day in rice crop (Alvares, 1999).

### Mechanical methods

**Hand picking:** Hand picking and destruction of eggs, gregariously feeding larvae and sluggish adults. Spreading blue or green cloth pieces in the evening hours in the crop field attracts spodoptera larvae, which can be collected and destroyed (Kalyayan, 2007). Egg masses and young larvae of tobacco caterpillar feeding gregariously should be collected along with leaves and destroyed by plucking the infested leaves or by pulling out the infested plants and burying them underground. The grown up caterpillars can be destroyed by crushing them under feet (PAU 2020-21). or by picking and putting them into kerosenized water (PAU 2021).

**Light traps**: Light traps can be used for trapping various insect-pests. When a thin gunny bag or polythene sheet smeared with greese or oil is placed near electric lamp. Use light-traps for the destruction of moths.

**Net**: Net can be used to protect nursery plants from insect vectors.

**Banding:** Stem banding with sticky or slippery material can be used to prevent insect-pests to reach the inflorescence of trees like mango mealy bugs.

**Stickers:** Use low cost yellow sticky traps @ 100 per ha during initial phase of cotton crop to check early infestation of whitefly.

**Rope**: Rice leaf folder can be managed by running a rope in rice fields.

### Behavioural methods

**Pheromones**: Pheromones are ectohormons secreted by an organism, which elicit behavioural responses from other members/sex of its own species. These are extremely selective, nontoxic, highly biodegradable and effective at low application rates. Synthetic sex pheromones are commercially available and are used for surveillance, monitoring and control of many Lepidopterous pests such as spotted bollworm, tobacco caterpillar, potato tuber moth, diamond back moth and leaf folder etc.

**Fairomones**: These are volatile compounds that evoke behavioural response adaptively favourable to the receiver. Fairomones are released either by the host plant or by the host insects. While former issued by the pest and natural enemies to locate their habitats, the later is used for prey finding and parasitisation/ preying. Fairomones from host plant can be effectively used to mass trap pest species as well as for monitoring. Use of fairomonal compounds to increase the efficiency of the predator *C. carnea* and the egg parasitoid *T chilonis* had been successfully demonstrated.

**Conservation of useful birds**: Predatory birds like owls, falcons, hawks, eagles, kites, etc. eat a large number of rats and mice. A single owl normally eats 4-5 rats a day. Insect eating birds like drongos, babblers, shrikes, lapwings, mynas, and many other small birds like sparrows and weaver birds feed a large number of insects to their young. A single pair of house sparrows feeds insects to their young about 250 times a day. Therefore, the useful birds should not be killed.

## Others

- Grow bajra, maize and jawar as barrier crops, being least preferred hosts for the mealybug.
- Do not grow bhindi, moong and arhar in the cotton crop and as border rows in order to reduce the incidence of American bollworm, spotted bollworms, jassid and whitefly.
- Bhindi, moong, dhaincha and castor are also the most preferred hosts of tobacco caterpillar, helping the pest to multiply and shift to cotton.
- The above pests on these crops grown in the vicinity of cotton fields, should be properly controlled in order to check their migration to the cotton crop.
- Do not throw the uprooted infested plants in cotton fields/water channels to check further spread of mealybug.
- Prevent movement of sticks from infested areas to new areas to avoid spread of mealybug.
- The trees/fruit plants near cotton fields harboring mealybug population should be sprayed with recommended bio-insecticides.
- Egg masses and young larvae of tobacco caterpillar feeding gregariously should be collected along with leaves and destroyed.

## Management of rat damage

A. **Mechanical Control:** i) During irrigation of vacant harvested fields rats coming out of flooded burrows should be killed with sticks. ii) Traps can be used to control rodents. Place 40 traps per ha at runways and activity sites of rodents. Kill the trapped rats by drowning in water and the interval between two trapping at the same location should not be less than 30 days.

B. **Environmental Control**: Weeds, grasses and bushes should be removed as these provide shelter and food to rodents. Highly infested bunds, water channels and field pavements should be periodically rebuilt to destroy permanent rat burrows. Waste lands along roads, canals, railway lines, other uncultivated areas and forest strip serve as hiding places for rodents. So, to protect the adjoining crops, rat control operations must be carried out in these areas also.

C. Biological Control Owls, kites, eagles, falcons, cats, mongoose, jackals, snakes and lizards are the natural predators for rats and mice. These should be protected.

D. **Acacia twigs:** Insert a thorny twig of *acacia* (*kikkar*) in each rat hole.

E. **Bird perches:** Erect 10-12 bird perches $ha^{-1}$ to attract owls at night and these owls feed on rats.

F. **Flour and cement balls:** Maize flour balls mixed with cement powder and sugar can be used for rat. Rats eat these balls and cement becomes hard in the stomach and cannot be excreted leading to their death. It also irritates them and they bite other rats in the hole and force them to run away.

## Harmful birds

Several fruits are damaged by birds at the bud stage and ripening stage. Parakeet is the major bird pest causing serious damage to guava, peach, pear, almond, grapes, mango and ber. House crows damage peach, plum and grapes. The major damage to grape is caused by mynas, especially the bank mynas.

## Management of bird damage

### A. Mechanical Control

1. Make false gun-shots at different intervals to scare the birds.
2. Frequent beating of drums and use of Gopia at different points in the orchard is very effective against the birds.
3. Covering the vines of grapes and isolated fruit trees with nylon nets prevents the bird damage.

4. Fixing of scare crows i.e. a discarded earthen pot painted to stimulate human like head supported with wooden sticks and clothed in human dress to give a human like appearance is one of the most effective traditional techniques to keep the birds away. Position, direction and the dress of the scare crow should be changed atleast at 10 days interval. The height of the scare crow should be 1 metre above the plant height.
5. Use automatic bird scarer by shifting their position periodically and supplementing their noise with actual gunfire's. The other simplest method is the use of rope-crackers. It involves tying of sets of small fire crackers at a distance of 6-8 inches apart on a rope and igniting it from the lower end. The explosions caused by fire crackers on catching fire at different intervals scare the birds feeding on fruits. Fix the rope-crackers in the centre of the orchard.

### B. Cultural practices

As far as possible sowing of maize and sunflower crops should be avoided in and around the orchards.

### C. Alarming calls playing of cassettes

(Available at Communication Centre, PAU) of distress or flock calls of parakeets and crows respectively in a tape-recorder at peak volume for 1/2 hr. twice each in the morning between 7.00 to 9.00 a.m. and in the evening at 5.00 to 7.00 p.m. respectively, with a pause of 1 hour, scare the birds or halt their activities in orchards. Use of distress or flock calls remain effective for 15-20 days. Better results can be obtained by using this technique in sequence or in combination with other methods as an integrated pest management. For covering larger area use of amplifier or additional speakers (as per requirements) can be done.

**Eco-friendly Management of Termites Through Earthen pot Based Traps in orchards**: Termites are the important soil borne insect-pests of pear, ber, peach, grape and amla under Punjab in nursery and grown up trees in the orchards. They feed on cellulose, particularly the wood. They are social polyphagous and polymorphic insect, living in large colonies comprising of queen, king, soldiers and workers nests called termitaria. These nests are constructed both above and below the ground. The workers class causes all the damage by feeding on roots and above ground parts of the trees. Peak activity period of termities in Punjab is April to June and September to October.

Eco-friendly management of termites in pear, ber, peach, grape and amla can be carried out by burying maize cobs without grains (*Gul*) filled 24-holed earthen pots of diameter 13 inch, with lid @ 35 per hectare in termite infested orchards during first week of April and again during first week of September. These earthen pots are buried in the soil keeping their neck outside the soil surface. The pots should be removed from soil after 20 days of installation and collected termites should be destroyed by dipping in water containing a few drops of diesel oil.

## Integrated pest management

Integrated pest management (IPM), which by definition is a pest management system that, in the context of associated environment and population dynamics, utilizes all the appropriate techniques to minimize the pest population at levels below those causing economic injury. Though several parasitoids, predators and pathogens of pests, antagonistic microorganisms were known to be effective for several decades, they were not commercially exploited because of quick knock down effect and easy availability of chemical pesticides instead of **biopesticides and IPM.** Steadily, there has been growing appreciation about the role of cultural and biological methods in pest control. Cultural and biological methods are the two major components in integrated plant protection.

## Trap cropping

Trap cropping is the planting of a trap crop to protect the main cash crop from a certain pest or several pests. The trap crop can be from the same or different family group, than that of the main crop, as long as it is more attractive to the pest. There are two types of planting the trap crops; perimeter trap cropping and row intercropping. Perimeter trap cropping (border trap cropping) is the planting of trap crop completely surrounding the main cash crop. It prevents a pest attack that comes from all sides of the field. It works best on pests that are found near the borderline of the farm. Row intercropping is the planting of the trap crop in alternating rows within the main crop.

### Advantages of trap cropping

1. Lessens the use of pesticide
2. Lowers the pesticide cost
3. Preserves the indigenous natural enemies
4. Improves the crop's quality
5. Helps conserve the soil and the environment

### Tips for successful trap cropping

- Make a farm plan. This will guide you on where the trap crops are to be sown or planted.

- Learn to know and identify the pests.
- Select a trap crop that is more attractive to the pest than the main crop. Ask for assistance from your local agriculturist.
- Monitor your plants regularly.
- Immediately control the pests that are found in the trap crop. Prune or remove the trap crops once the pest population is high, otherwise they will serve as the breeding ground and the pests will attack the rest of your farm.
- Be ready to sacrifice your trap crop as an early crop and destroy them once pest infestation is high.
- Always keep farm records.

  The use of living organisms to suppress the population of a specific pest organism, making it less abundant or less damaging than it would otherwise be.

## Natural enemies

- Natural enemies are living organisms that kill pests
- Decrease pest reproductive potential
- Compete with pest organisms

## Other liquid formulations

**Cow urine**: Foliar spray of cow urine diluted with water in 1:20 ratio is effective against pathogen and insect management besides the cow urine also acts as a growth promoter.

**Fermented curd water**: Fermented curd water (butter milk) is effective against management ofwhite fly, aphids, jassids and yellow rust of wheat.

**Dashpani extract**: Crush neem leaves(5 kg), Virtex regundo leaves (2 kg), Aristolochia leaves (2 kg), papaya leaves (2 kg), Tinospora cordifolia leaves (2 kg), Annona squamosa custard apple leaves (2 kg), Pongamia pinnata karanj leaves (2 kg), Ricinus communis Castor leaves (2 kg), Nerium indicum (2 kg), Calotropis procera leaves (2 kg), green chilli paste 2 kg , garlic paste 250 gm, cow dung 3 kg and cow urine 5 litres in 200 litres water ferment for 30 days, shake regularly three times a day. Extract after crushing and filter it. The prepared extract can be stored for six months and the above mentioned extract is sufficient for one acre.

**Neem-cow urine extract**: Crush five kg neem leaves in water, add 5 litres cow urine and 2 kg cow dung, ferment for 24 hours with intermittent stirring, filter, squeeze and dilute the extract to 100 litre. Use the extract for foliar spray against sucking pests and mealy bug. The extract is sufficient for management of sucking pests and mealy bugs in one acre.

**Neem-Akk extract:** Collect 200 litre of cow urine over a period of time and immerse 5 kg neem leaves and akk (Calotropis) leaves. Keep the mixture for 10 days and stir the mixture in morning and evening. Filter the extract and and spray on a prophylactic basis on the crops in 1:14 ratio.

**Herbal-5**: Take five pots with 10 litre cow urine in each pots Add 2 kg leaves of Calotropis (Akk), neem, bhang (Canabis), Dhathura and ashwagandha in each pot so that each pot receives one type of leaves. Keep these pots air tight under soil or wheat bhusa for 25 days. Mix all the ingredients of the pots and add 1 kg garlic and 1 kg chillies and boil it for half an hour. Use iron or copper utensils. Put 200 g copper wire if used utensil is made up of iron and 200 g iron nails if utrnsil is of copper. Filter the solution with cloth and use as spray @ 15 ml/litre of water.

**Asafoetida (hing)**: Mix 4 litre cow urine with 10 g Asafoetida (hing) in 10 litre of water. Its spray is effective against sucking pests.

**Mixed leaves extract**: Crush 3 kg neem leaves in 10 litres cow urine. Crush 2 kg custard apple leaves, 2 kg papaya leaves, 2 kg pomegranate leaves, 2 kg guava leaves in water. Mix the two and boil the same for 5 times at same interval till the extract solution becomes half. Keep for 24 hours, then filter, squeeze the extract. The extract can be stored for six months. Dilute 2.0 to 2.5 litres of extract with 100 litres water and it can be used against sucking pests, pod/fruit borer in one acre.

**Chilli-garlic extract**: Crush 1 kg Ipomea (besharam) leaves, 500 g hot chilli, 500 g garlic and 5 kg neem leaves in 10 litres cow urine. Boil the suspension 5 times till it is reduced to half. Then filter, squeeze the extract and store in glass or plastic container. 2-3 litres extract diluted to 100 litres is used for one acre. This is useful against leaf roller, stem/fruit/pod borers.

**Ginger-chilli-garlic extract:** Crush 10 kg garlic, 5 kg ginger and 5 kg chilli in 70 litres of water. Filter it and use 60 litre of extract for one hectare against leaf folder.

**Fermented whey**: Ferment whey in copper container for five days and use as foliar spray to control insect-pests.

**Ash**: Ash dust acts as repellant against red pumpkin beetle.

**Pest repellant:** Take 200 gm Artemisia/Stinging needle/Lavender leaves, 100 gm Hot chillies, 100 gm garlic and Egg yolk - 1 (acts as an emulsifier) and 100 ml cooking oil. Mix the materials in a container or put them directly in a mixer grinder. These materials are grinded until the contents are turned into a fine paste, then add 5 litres of water to the thick paste and mix it thoroughly. This stock solution can be diluted 10 times with water, stir and spray. The pest repellant spray is effective on vegetables and fruit trees

**Marigold extract**: Marigold (*Tagetes* sp) is an excellent pest repellent for most of the pests. When these plants are planted along with the main crop especially in vegetables help in managing soil borne insects, pathogens and nematodes due to the allelopathic effect.

Grind 5 kg of marigold plants along with the flowers into fine paste. Add 100 litres of water and allow it to ferment for a week. Add 100 grams of soap nut powder on the day of spraying which acts as emulsifier. Stir the solution for 15 minutes and spray during evening hours.

**Fenugreek formulation**: Mix 1.0 kg of fenugreek (*Trigonella foenum-graecum*) flour powder with 2.0 litre water and keep the mixture for one day. After one day, dilute the mixture to 40 litre and spray on one hectare area against fruit borer and tobacoo catterpiller.

**Broad spectrum formulation 1**: Mix 3 kg crushed neem leaves and one kg neem seed kernel powder with 10 litre cow urine In copper container. Seal the container and allow it to ferment for ten days. After ten days, boil the suspension, till its volume is reduced to half. Grind 500 g green chillies in one litre water, keep it over night. Crush 250 g garlic in another container and keep overnight. Boil chilli and garlic extract mixture next day and filter it. This is a broad spectrum bio insecticide and can be used on all crops against variety of insects by using 250 ml concentrate in 15 litres of water.

**Broad spectrum formulation 2:** Suspend 5 kg neem sed kernel powder, one kg *karanj* seed powder, 5 kg chopped leaves of *Ipomea* spp. (besharam) and 5 kg chopped neem leaves in 20 litre drum. Add 10 litres cow urine and fill the drum with water to make 150 litre. Seal the container and ferment it for 8-10 days. After that, mix the contents and distil these in a distiller. The distillate will act as a good pesticide and plant growth promoter

**Neem oil spray:** Neem oil acts as a good insect repellent and anti-feedant in variety of crops. Mix 3 litres of neem oil in 100 litres of water, stir well and add 100 gm of soap (acts as emulsifier) into solution. Spray the solution in evening hours.

**Homemade neem extract spray:** To control Pod sucking bug in moongbean, spray the crop at seed filling stage of the crop upon pest appearance using 3750 ml per hectare homemade neem extract using 250-325 litres of water per acre with manually operated knapsack sprayer.

**Preparation of homemade neem extract:** To prepare neem extract, boil 5.0 kg mixture of neem leaves and fruits in 10 litres of water for 30 minutes. Then, filter this material through muslin cloth and use the filterate for spraying at the recommended dose.

**Botanical insecticides:** Plants, during their long evolution, have synthesized a diverse array of chemicals to prevent the colonization by insects and other herbivores. They produce terpenoids, alkaloids, flavonoids, phenolic compounds, unusual amino acids etc. These chemicals exhibit repellent, feeding and oviposition deterrent and insecticidal properties besides interfering with their normal growth, development and reproduction. Several plant based insecticides such as nicotinoids, natural pyrethrins, rotenoids, neem products, etc have been used in the past for suppression of pest species. Among them, neem based products are ideally suited for practical utilization because of their greater effectiveness against diverse group of insect pests under field conditions, ready availability and simple methods for preparation of insecticidal products.

Neem has been evaluated against more than 500 species of insects at global level and most of these insects have been found to be susceptible. Neem seed extracts influence insect biology and behavioural ecology by interfering with feeding, development, mating, egg production and egg hatchability. Neem products give good control of caterpillars such as diamondback moth on Cole crops and gram pod borer on red gram, cotton, tomato etc.; grass hoppers, leaf hoppers, plant hoppers, leaf miners, whiteflies, scales and mealy bugs. Neem kernel powder at 1 in 2 per cent gives good control of pulse beetle and other storage pests. Inspite of its usefulness, the large-scale practical utilization of neem in pest management still faces several hurdles. There is an urgent need to characterize best ecotypes of neem tree for different environmental conditions. Intensive breeding and selection work need to be undertaken for economic production of high quality raw materials for commercial production of neem based insecticidal products. The development of appropriate techniques for harvesting, depulping, drying and storage of neem seeds is a pre-requisite for large-scale production of neem-based pesticides (Sannaveerappanavar, 2003).

The commercial products available in the market or crude extracts prepared locally vary widely in quality and quantity of different active components. To overcome these difficulties, guidelines approved by the Central Insecticides Board in 1991 stipulate that kernel based formulations and oil based formulations should contain at least 1500 and 300 ppm of azadirachtin, respectively. In recent years, neem products with and 50,000 ppm of azadirachtin have been introduced. Field and laboratory studies done at the University of Agricultural Sciences, Bangalore, have shown that the aqueous neem seed kernel extract (NSKE at 4-5%) is more effective and economical to use compared to commercial neem based products. Similarly, neem oil emulsions (2-4%) prepared by using soap can be effectively used against scales and mealy bugs on fruit trees such as guava, custard apple and sapota.

Other locally available plant products such as seed of pongamia and mohua have been tested for their efficacy against scales and mealy bugs. Microbial Insecticides:

**Microbial insecticides**: The insecticidal substances obtained from the microorganisms are microbial insecticides. The insecticidal crystal proteins (endotoxins) produced by the bacterium, *Bacillus thuringiensis* kurstaki are effective against lepidopteran pest species. These toxins are very specific in their action, easily bio-degradable and being stomach poisons, safer to non-target organisms. Presently many commercial products are available in the market for use against caterpillar pests. However, some pests like diamondback moth on Cole crops have developed resistance to certain Bt. Endotoxins. Other microbial products that have been recently introduced for pest management are the insecticidal toxins produced by soil derived from Streptomyces avermitilis possess potent anthelminthic, acaricidal and insecticidal properties at very low does. Like wise, spinosyns (Spinosad), toxins obtained from another Actinomycetes organism, Saccharopolyspora spinosa very good insecticidal property against lepidopteran pests like diamondback moth, gram pod borer; etc.

**Biological agents:** The natural enemies of insects viz. entomophagous parasitoids, predators and the insect Pathogens play an important role in keeping the pest populations at low levels, However, the use of broad-spectrum insecticides in agro-ecosystems has adversely affected these beneficial insects, often leading to out break of certain minor pest species. In organic farming, it is possible to encourage the activity of insect predator and parasitoids and if required, their population can be augmented through inundative releases. Some of the commonly used bio-control agents are: *Trichogramma* spp. Against cotton boll worms, tomato fruit borer. Sugarcane borers; *Goniozus nephantidis* and *Bracon brevicornis* against coconut black headed caterpillar, predatory coccinellid beetles against scales and mealy bugs on horticultural crops; the green lace wing, *Chrysoperla* sp. for the control of aphids and bollworms on cotton etc. Among the insect pathogens, certain viruses and fungi have been utilized in suppression of crop pests. Helicoverpa NPV, Spodoptera NPV are available to the farmers. Likewise, the fungal pathogens, *Metarrhizium* sp. and *Beaveria* sp. are commercially available to the farmers. Recently, entomopathogenic nematodes, *Steinernema* sp. and *Heterorhabditis* sp. have been made available to the farmers for use in pest management.

# 7

# Organic Diseases Management

A bio-pesticide is defined as a pesticide derived from natural materials. The biological control involves the use of one living organism to control another. This management technology has attained much attention in the recent times. We can also say that the use of microbial antagonists (including bacteria or fungus) to suppress plant disease pathogens.

The biological control was first coined by Harry Smith in relation to biological control of insects means suppression of insect population by native or introduced enemies. In generic terms, the population leveling process in which the population of one species lowers the number of another (Table 1) as compared to use of chemical fungicides in chemical methods.

**Table 1.** Difference between chemical pesticides and biological control agents:

| Chemical pesticides | Biological control agents |
|---|---|
| Implicated in ecological, environment and human health problems | Non-toxic to human and animals, not a water contaminant concern |
| Require yearly treatment | Once colonized may last for years |
| Broad spectrum | Host specific |
| Toxic to both beneficial and pathogenic species | Only effect one or few species |

## Requirements of successful bio-control

1. Highly effective bio-control strain must be obtained.
   a) Be able to compete and persist
   b) Be able to colonize and proliferate
   c) Be non-pathogenic to host plant and environment.
2. Inexpensive production and formulation of agent must be developed.
   a) Production must result in biomass with excellent shelf life.
   b) To be successful as agricultural agent must be
      i) Inexpensive
      ii) Able to produce in large quantities
      iii) Maintain viability

3. Delivery and application must permit full expression of the agent
    a) Must ensure agents will grow and achieve their purpose.

    The bio-pesticides are divided into three categories: biochemical pesticides, microbial pesticides, and plants containing added genetic material.

- Biochemical pesticides contain naturally occurring substances that control pests. Substances that control diseases in this category include potassium bicarbonate, hydrogen dioxide, phosphorous acids, plant extracts, and botanical oils.
- Microbial bio-pesticides contain microorganisms as the main active ingredient that function as biological control agents, affecting the pathogen directly or indirectly through the compounds they produce or by stimulating specific plant responses.
- Plant-incorporated protectants (PIPs) are the least common type of bio-pesticide. These are pesticidal substances produced by plants that contain genetic material added to the plant often through genetic engineering. Biopesticides also exist for the management of weeds, insects, & nematodes.

## Mechanism of biological control of plant pathogens

**Antibiosis**: Inhibition of one organism by another as a result of diffusion of a metabolic products such as antibiotic lytic agents, enzymes, volatile compound s and other toxic substances. Antibiotic production common in soil bacteris and fungi. Example: zwittermicin antibiotic production by Bacillus cereus against Phytophthora root rot in alfalfa, Marigold (Tagetes species) rootsreleases terthienyls, chemicals that are toxic to several species of nematodes and fungi.

**Nutrient competition**: Competition between micro-organisms for carbon, nitrogen, oxygen, iron and other nutrients and the organisms limit the growth of others.

**Destructive mycoparasitism**: The parasitism of one fungus by another by direct contact, cell wall degrading enzymes, producing antibiotics. Example: *Trichoderma harzianum* used as seed treatmen against pathogenic fungus.

## Mode of action

1. **Direct mechanism:** Direct lysis or killing of pathogen by biocontrol agents.
    a) **Antibiosis**: Antibiotic mediate suppression in which antibiotics are microbial toxins that can kill or poison other micro-organisms at low concentration.

i) **Volatile antibiotics:** Hydrogen cyanide, Aldehydes, Alcohols Ketones, Sulfides.

ii) **Non- Volatile antibiotics:** Polyketides (diacetylphloroglucinol DAPG and mupirocin), Heterocyclic nitrogenous compounds.

b) **Paracitism:** Direct utilization of pathogens as source nutrients.

i) **Mycoparaisitism:** Refers to association in which a parasitic fugus (hyperparasite) live as a parasite to another fubgus (hypoparasite). Also known as Hyperparasitism**,** when hyperparasites (biocontrol fungi) utilize hypoparasites (pathogenic fungi) as source of nutrients. Hyperparasites produce parasitizing hypae to acquire host nutrients. May also require cell wall degrading enzymes. e.g., *Pythium oligandrum* Vs *Fusarium culmorum*

2. **Indirect mechanism:** Exclusion of plant pathogen as a result of the presence, activity or products of biocontrol agent

i) **Competition:** Exclusion of pathogens by biological control agents via competition for space or nutrients.. Production of substances (such as siderophore) for nutrient (such as iron) acquisition which deprive pathogens of nutrients.

ii) **Induced systemic resistance**: Also known as systemic acquired resistance (ISR), resistance in plants to varieties of pathogens induced by the presence pr products of biocontrol agent.

**Induced systemic resistance may be:**

- Necrogenic pathogen (applied on leaf)
- Non-pathogenic bacteria(PGPR) (applied on root or seed)
- Metabolites of pathogenic or saprophytic bacteria.
- Microbial metabolites able to induce ISR: Polyacrylic acid, Ethylene, Salicylic acid, Acetyl salicytic acid, Amino acid derivatives, Harpin (*Erwin amylovora*).
- Stress can induce defense mechanism against pathogens.

**ISR defense responses**

- Physical thickening of cell wall by
- Lignification,
- Deposition of callose,
- Accumulation of antimicrobial low molecular weight substances (e.g., phytoalexins).

- Synthesis of various proteins (e.g., chitinases, glucanases, peroxidases and other pathogenesis related (PR) proteins.

## 1. Method of application of bio-pesticides

a) **Seed Treatment**: Prepare 5% jaggery solution by boiling 500 gm of jaggery in 10 litres of water for 15-20 minutes. Sufficient solution has to be prepared depending upon the quantity of seeds to be treated. The solution has to be cooled. Add bio-pesticide into solution and thoroughly mix the contents of solution. The general recommendation is 10 gm bio-pesticide per kg of seed. Put the seeds on a polythene sheet and the bio-pesticide solution is poured over the seeds and seeds are mixed thoroughly with solution. Dry the seeds in shade and sown immediately.

b) **Nursery bed**: One kilogram of bio-pesticide is mixed with 100 kilograms of sand. The soil mixture can be used as a nursery soil or the same can be mixed with nursery bed of 10 square metres.

c) **Soil drenching**: Prepare a solution by adding 10 gram of bio-pesticide in a litre of water and stir the solution well. Using a rose can to drench the soil with the solution.

d) **Seedling dip**: Prepare a solution by adding 10 grams of bio-pesticide in a litre of water. Dip the seedlings in this solution for half an hour.

e) **Soil application**: A mixture is prepared by adding 5 kilograms of bio-pesticide with 100 kilograms of well rotten FYM. Broadcast the mixture on soil before ploughing. Apply light irrigation, if the mixture is applied in standing crop.

f) **Foliar spray** : Mix one kg bio-pesticide in 150 litres of water and spray directly on the crop. The solution is sufficient for one acre.

## Requirements of successful biocontrol agents

1. Able to compete and persist
2. Able to colonize and peroliferate
3. Must be non-pathogenic to host and environments.
4. Must have excellent shelf life.
5. Must be inexpensive.
6. Must be able to produce in large quantities.
7. Able to maintain viability.
8. Delivery and application methods must support product establishment.

## Importance of biological control

1. Alternative method of weed control.
2. Can be used where other method is not applicable.
3. Biocontrol agents are nontoxic to man and the environment
4. Act on selective target organism.
5. Biocontrol agents are self sustaining- easy adaption.
6. Diversify mode of actions.
7. Reduced possibility of including resistance in pathogens.
8. Cost effective
9. Long term effects.

## Limitations of biological control

1. Expensive/difficult to develop
2. Handling require training
3. High selectivity/host specificity
4. Variability in effectiveness
5. More susceptible to environmental conditions.
6. Short shelf life of formulations.
7. Slow action.
8. Storage problem.

## Commercial biological control

**Kodik-**a successful biological control product for suppression of soil borne plant pathogens of cotton. *Bacillus subtilis strain GB03 is a*ctive microbe and its mode of action is antibiosis.

**Mycostop:** Biofungicide for vegetables and ornamental crops. Streptomyces griseoviridis strain K61 and its mode of action is competition(Hyperparasitism, Antibiosis).

## Plant pathogen control by Trichoderma

Trichoderma species are available in all agricultural soils.

Antifungal properties (Mycoparasitism and Nutrient competition) are known since 1930s.

Agriculturally used as bio-control agent and as a plant growth promoter.

## Genetic modification

The wild strains-Heterokaryotic contain nuclei of dissimilar genotypes within a single organism.

Biocontrol strains-Homokaryotic contains nuclei which are similar or identical. Allows genetic distinction and non-variability which is an important for quality control.

Most strains have innate resistance to some agricultural chemicals.

Strains available for commercial use are selected or modified for resistance to specific chemicals.

## How it is applied

- Favoured by presence of high levels of plant roots.
- Some are highly rhizosphere competent
  - ii) Capable of colonizing the expanding root surface
  - iii) Can be used as soil or seed treatment
- Action against pathogenic fungi
  1. Attachment to host hyphae by coiling
     - a) Lectin-carbohydrate interaction
  2. Penetrate the host cell walls by secreting lytic enzymes
     - a) Chitinases
     - b) Proteases
     - c) Glucanases
- Some strains colonizes the root with mycoparasitic properties
  - a) Penetrate the root tissue
  - b) Induce metabolic changes which induce resistance
  - c) Accumulation of antimicrobial compounds.
- Commercial availability
  - a) Seed coating, seed pieces, transplant starter
  - b) Protects roots from diseases caused by Pythium, Rhizoctonia and Fusarium
  - c) Interacts with the Rhizosphere, near the root haits and increase the available from nutrients needed by plants.
- Transgenes

a) Biocontrol molecules contain large number of genes which allow biocontrol to occur

b) Cloned several genes from Trichoderma as transgenes

   i) Produce crops which are resistant to plant diseases.

## Integrated plant protection

Plant protection (against pests, diseases and weeds) determines the effectiveness of other inputs in crop production. Exclusive reliance on pesticides, fungicides and herbicides resulted in pesticide and herbicide resistance, pest resurgence, residues and environmental pollution. This led to the development of integrated plant protection strategies, which are components of sustainable agriculture with a sound ecological foundation. Integrated plant protection should be understood as an ideal combination of agronomical, biological, chemical, physical and other methods of plant protection against entire complex of pests, diseases and weeds in a specified farming ecosystem, with the object of bringing down their infestation to economically insignificant levels with minimum interference on the activity of natural beneficial organisms. The essence of integrated plant protection concept lies in the harmonious integration of compatible multiple methods use singly or in combination against insect pests, pathogens and weeds.

## Integrated disease management

For mitigating the loses due to diseases, several methods such as fungicides, organ mercurial, chemotherapy, thermotherapy, cultural methods and host resistance are employed. However, no single method is effective in controlling a disease. Therefore, integrated disease management (IDM) became imperative for effective disease control. Integrated disease management in organic farming combines the use of various measures. The usefulness of certain measures depends on the specific crop-pathogen combination. In many crops, preventative measures can control diseases without the need of plant protection products. However, for certain disease problems, preventative measures are not sufficient. For example, organic apple production strongly depends on the multiple uses of plant protection products.

All the cultural methods discussed under IPM hold well for IDM also. Broad based tentative IDM components are being adopted for disease control. However, all these components are not feasible for any specific ecosystem or any specific disease. For many other diseases the role of host resistance, cultural methods and chemical methods are integrated. Solar heat therapy (drying the seed in hot sun after harvest and again before sowing) is a common practice in our agriculture. Among mechanical methods for prevention and against spread of diseases uproot and burn is the age old and the best method so far. It is better to prevent and control vectors against spread of diseases. Disease affected plants are to be uprooted and burnt and alternate and collateral host-crops, grasses, stubbles etc. destroyed. Disease can affect any part of a plant. Disease may be fungal, bacterial and viral. Viral diseases are more serious than fungal and bacterial.

Disease management in organic cropping systems combines various components which can be divided into strategic preventative measures, tactical preventative measures and control measures. For each crop-pathogen relationship and cropping system such components will contribute to different extent to disease management (Termorshuizen, 2002). The development of integrated disease management systems depends on thorough knowledge of the cropping systems as well as of the pathogen and can only be achieved by interdisciplinary research.

### Pathogen characteristics and disease management

Host-specificity and mobility are the two main characteristics of pathogens determining the choice of disease management measures (Wijnands et al., 2000). Strictly host-specific pathogens which are not mobile can be controlled by using cropping systems with low frequencies of the susceptible crop. Examples are cyst nematodes of potato or sugar beet. Pathogens which are not host-specific and not mobile can be controlled by the choice and sequence of crops grown in the rotation supported by preventative measures increasing soil suppressiveness and plant health. Examples are the soil borne pathogens *Sclerotinia sclerotiorum* and *Rhizoctonia solani*. Host-specific pathogens with high mobility such as *Phytophthora infestans* in potato cannot be controlled by crop rotation.

Preventative measures are sanitation in a cropping area and the choice of crop structure and planting date in combination with resistant varieties. In many situations also control measures such as applications of plant protection products may be needed to achieve sufficient yield. Also pathogens which are not host-specific but highly mobile cannot be controlled by crop rotation. Disease prevention depends on strengthening the crop, escaping the disease by choosing proper seeding dates and creating an open crop structure. Disease control by using crop protection products may be needed in many cases. Example for a mobile pathogen with a broad host range is *Botrytis cinerea* causing grey mould in various crops such as beans, peas, strawberries, grapes and many other crops. How differently various measures contribute to disease management in different crop-pathogen relationships will be illustrated by the comparison of two systems. In wheat, various *Fusarium* spp. can cause *Fusarium* Head Blight (FHB) leading to a decrease of yield and, more important, the production of mycotoxins in the grain. *Fusarium* sp have a broad host range and also can survive saprophytically.

Mobility of spores of most *Fusarium* sp is low. In apple, *Venturia inaequalis* can cause apple scab on leaves and fruit resulting in reduced yields and quality of fruit. The pathogen is strictly host-specific and can survive only on apple tissues. The mobility of spores is low.

## Preventative measures

Many measures for preventative disease control have a long-term strategic character. Various aspects of the farm management and the long-term cropping system as well as of the location including the farm neighbourhood have impact on diseases of crops and thus should generally be considered in disease management.

**Soil conditions:** Optimize soil conditions through agronomic management.

**Soil structure, soil suppressiveness, biological soil disinfection, and catch crops:** These measures are important for managing soil borne diseases but will have no direct effect on the above-ground development of *Fusarium* spp. and *V. inaequalis*.

**Avoidance of pathogen sources in neighbourhood of field and crop rotation in neighbouring field:** Since most *Fusarium* spores travel only a few centimetres, sources in the crop neighbourhood will not cause epidemics. Ascospores of *V. inaequalis* produced in neighbouring orchards may reach the crop. Abandoned orchards and orchards with high apple scab pressure should not be found in the neighbourhood of an apple orchard.

**Disposal of crop residues:** Ensure proper disposal and decomposition of crop stubbles and off-farm wastes.

**Dispose of diseased plants:** In sugarcane, remove the smut whips gently (without shaking) after putting them inside a closely woven drill bag. Then uproot the entire diseased clumps and bury them deep. Immerse the bag used for collecting the smut whips in boiling water for 5 minutes after every roguing of the crop. Do not ratoon the smut infected crop.

Phyllody in sesame is caused by a Mycoplasma like organism (MLO). The flowers are modified into leaf like structures and do not bear pods. It is transmitted by Jassid (*Orosius albicinctus*). Therefore, control of jassid is required. Also rogue out the diseased plants to prevent further spread.

**Removal of crop residues from field**: *Fusarium* sp threatening wheat crops are surviving primarily in stubble of cereals including maize. Removing this potential inoculum sources is not feasible, although physical removal especially of maize stubble may have a significant impact on disease development. In apple, removal of fallen leaves as the principle inoculum source of apple scab in spring is an interesting option. Removal of leaves by using specially designed vacuum cleaners has been demonstrated. However, mechanization is difficult, cost intensive and application depends much on orchard circumstances.

**Biological crop residue treatments**: Microbial decomposition of crop residues is a natural process which can be supported by adding stimulating nutrients or selected micro-organisms. Also earthworms can be protected and stimulated to consume plant residues on the soil surface. In arable crops, stimulation of resident microbial populations on residues may be achieved by creating a suitable microclimate, e.g. by using mulches.

**Pull and Burn:** Adopt 'Pull and Burn' method to manage diseases and their spread.

**Biodiversity:** Maintain biodiversity at the farm with wild plants and trees to ensure natural checks and balances in disease management. The biodiversity in organic ecosystems helps in checking plant diseases spread.

Main inoculum source of FHB are crop residues of preceding diseased crops. The best documented example is the high risk of FHB when wheat is grown after maize. Maize stubble are often colonized by the same *Fusarium* sp affecting wheat and such *Fusarium* sp can survive and multiply on maize stubble for several years. Avoiding growth of maize in rotation with wheat will substantially reduce risks of FHB epidemics. Rotation schemes with cereals grown after cereals should generally be avoided. In the perennial apple production crop rotation is no issue.

Avoid growing of brinjal, after the susceptible crops like pea cauliflower, carrot and binjal, instead tomato and chilli can be grown. This will help in reducing the disease and spread of inoculum of White rot (*Sclerotinia sclerotiorum*)in the field and avoid growing of main season brinjal crop (July-February) in the White rot disease prone fields.

**Alternate hosts:** Destroy alternate hosts. Weeds like Gutpatna, Bathu, Jangli Palak and wild castor are the alternate host plants on which the Bihar hairy caterpillar (*Bhabu Kuta*) feeds in gregarious phase and later on migrates to berseem. Destroy these weeds growing adjoining to berseem fields before its sowing. The attacks of Gram caterpillar (*Helicoverpa armigera*) is serious on berseem seed crop. Avoid raising of berseem seed crop adjoining to tomato, gram, late sown wheat, sathi moong, sathe mash and sunflower on which the pest multiplies and later shifts to berseem. Control the weeds, particularly itsit/chapatti as it acts as an alternate host for the tobacco caterpillar in summer moong/mash. Destroy the collateral host Takri grass (Digitaria sanguinalis) from the maize field to manage Brown stripe downy mildew (*Sclerophthora rayssiae* var zeae).

**Eradicate weeds**: Eradicate weeds like kanghi buti, peeli buti, puth kanda, congress grass, itsit growing on field bunds, waste lands, road side and irrigation channels/canals to avoid further spread of whitefly, mealybug, tobacco caterpillar and spotted bollworm to cotton fields. Regular surveillance on alternate host crops like brinjal, cucurbits (cucumber, long melon, chappan kadu), tomato, chilli, okra from February onwards and on cotton, moong from April onwards should be carried out for timely management of whitefly on these crops.

Leaf curl disease is caused by whitefly transmitted virus. The disease can be reduced by avoid growing American cotton in and around citrus orchards and adjoining bhindi crop. Follow clean cultivation and destroy Kanghi buti (*Abutilon sp.*) and Peeli buti (*Sida sp.*) which act as collateral hosts.

**Root knot**: This disease occurs in patches and is caused by the nematode *Meloidogyne arenaria* and *M. javanica*. The affected plants show poor growth with chlorotic leaves. The root knots become stubby. The plants affected early in the season become stunted and bushy. Disease can be checked by exposing the soil to the sun during May and June to reduce nematode population. Practise green-manuring wherever possible or add organic manures to the soil.

## Cultural measures

Tactical preventative measures deal with the planning and realization of a certain crop. Typical measures are the choice of variety, seed quality, seeding time and crop structure.

**Soil solarization:** Soil solarization by covering the moist soil with polythene sheet during hot summer raises the temperature of soil by about 13 °C higher than the outside reaching 50-55 °C at 8 cm soil depth and thus kills fungi, bacteria and nematodes (Chand, 2007).

**Hot water treatment:** Hot water treatment in sugarcane reduces incidence of red rot.

**Water soaking:** To prevent Loose Smut in wheat, soak the wheat seed in ordinary water from 8 a.m. to 12 noon on any calm and sunny day during May/June. After 4 hours soaking, spread out the moist seed in the sun in a thin layer on cemented floor (pucca), on tarpaulin or sheets of cloth. Dry the seed completely and store in a dry place till sowing. A coryniform bacterium (*Clavibacter xyli*) has been found to be associated with the Ratoon Stunting disease. The moist hot air treatment of seed canes at 54 °C for 4 hours is effective in destroying the causal organism. In sugarcane, leaf scald disease is caused by the bacterium *Xanthomonas albilineans* whereas grassy shoot disease is caused by mycoplasma like bodies. The moist hot air treatment of the seed-canes at 54°C for 4 hours inactivates these causal organisms. (PAU. 2020-21).

For prevention of ergot, immerse seed in 10 per cent salt solution and remove the sclerotia and smut-balls by skimming. Then wash the seed in ordinary water and dry it thoroughly (PAU 2021).

Bacterial stalk rot (*Dickeya zeae*): Water soaking and rotting of basal stem especially the leaf sheaths followed by rapid rotting of basal internodes. The rind loses its natural green colour and gives appearance as if boiled in water. The rotten stalks emit a characteristic fermenting odour and may break over from the second or third basal internode. The infected plants wilt. Destroy the diseased plant debris. Excessive rains and poor drainage favours the disease, so keep the fields well drained. Timely watering particularly during flowering and fruiting stages to cotton helps to mitigate the intensity of against tirak disease of groundnut. Prevent drying of soil by regular irrigation from tuberization to maturity to manage Common scab (*Streptomyces scabies*) in potato.

**Green manuring:** Practice green manuring with sunnhemp/sesbania/coeppeas as it reduces incidence of diseases.

**Inter and mixed cropping:** Intercropping and mixed cropping like Barley/ wheat+chickpea mixture reduces the spread of rusts. Wheat/Barley+chickpea intercropping systems have significant control of *Ascochyta* blight.

**Crop rotation:** Tailor suitable crop rotations which are helpful in checking the incidence of diseases. Crop rotation in groundnut manages soil born diseases. Rotate gram with non-susceptible crops such as wheat and barley to prevent foot rot and stem rot (PAU 2020-21).

**Selection of crops and varieties:** Select locally adapted and resistant crops and recommended crop varieties.

**Clean seed:** Use clean disease free seed and planting material.

**Tillage:** Adopt suitable cultivation methods like summer ploughing, deep ploughing etc. Deeper sowing in chickpea reduces the incidence of *Ascochyta* blight.

**Planting method:** Ridge sowing helps to reduce diseases in many crops. Wider spacing in groundnut and deep sowing in chickpea are effective against diseases.

**Tillage:** Primary inoculums of FHB are crop residues left on the soil after tillage. Using reduced tillage systems will increase FHB risks since much more residues will be present on the soil surface. In apple orchards, tillage is not an option. The summer ploughing helps to decrease the incidence of insect-pests and weeds.

**Crop structure**: Crop structure affects microclimatic conditions within the canopy and determines the distance pathogen spores have to spread to reach susceptible host tissue. A dense wheat crop will favour pathogen sporulation on the soil, but may block spore flights of *Fusarium* sp. Depending mainly on splash dispersal during rainfalls, vertical leaf positions may also block spore flights. Ears on taller plants may have a better change to escape infections which may have a moderate effect on FHB. The canopy structure of apple trees is managed to obtain sufficient yield and possibilities to create more open canopies are limited. Spores of *V. inaequalis* are very much adapted to infect trees and spread from orchards floors and within canopies. Possibilities to manage the apple scab by crop structure are low.

**Sowing time:** Adjust sowing or planting times and crop geometries so as to create an unfavourable environment for the potential plant pathogens. Late sown chickpea (end November to December) reduces root rot, wilt and blight diseases. Early planting of mustard helps in managing the white rust. The incidence of yellow vein mosaic in okra is less if sown during February-March (Chand, 2007). Sow chick pea crop during October 25 to November 10 to avoid stem rot and foot rot (PAU 2020-21). Do early sowing to reduce damage by early shoot borer in sugarcane.

**Hoeing:** For management of Gummosis/Foot Rot (*Phytophthora parasitica*) in Citrus, avoid deep hoeing and avoid injuries to the trunk bark and roots.

**Nutrition:** Supply balanced nutrition through crop rotations, organic manures and composts to produce hardy crop plants, tolerant to diseases.

**Mustard cake:** For the management of root knot nematode in rice, apply mustard cake @ 40 g per square metre (1.0 kg per marla)10 days before sowing of nursery with last preparatory tillage operation after rauni. Also, practice puddling for nursery sowing.

**Caution:** Exercise waiting period of 10 days between application of Mustard cake and sowing of nursery.

**Organic manure:** In disease prone areas use of FYM decreases the disease incidence of Stalk rot of cauliflower seed crop (*Sclerotinia sclerotiorum*) and the number of sclerotia of the fungus. In scab infested fields, do green manuring with 7-8 week old sunhemp or summer moong before rice or do green manuring before sowing of potatoes.

**Irrigation:** For management of Gummosis/Foot Rot (*Phytophthora parasitica*) in Citrus, avoid flood irrigation. Do not allow irrigation water to stay in direct contact with the tree trunk. Avoid deep hoeing. Avoid injuries to the trunk bark and roots.

Irrigation water management helps controlling some diseases in crops like rice. Drip irrigation in sugarcane reduces the red rot in sugarcane. To manage Sclerotinia Stem rot disease in rape seed and mustard, avoid irrigation to the rapeseed-mustard crop during 25$^{th}$ December to 15$^{th}$ January (PAU 2020-21).

The Stem rot (*Sclerotinia sclerotiorum*) disease symptoms are visible in berseem during January and February when temperature is low. The fungus which is present as sclerotia in the soil attacks the basal portion of the berseem stem and causes it to rot. It produces white cottony mycellium which begins to grow on the dead organic matter on the surface of the soil. The field affected by the pathogen should be heavily flooded during summer months so that hard, black sclerotia, present in the soil may be destroyed. Cut the crop and expose the soil to the sun. Do not pond water in the field continuously to prevent BLB in rice.

**Time of harvesting:** Terminate the crop as early as economically feasible. For this purpose give last irrigation to cotton by end of September. It would reduce bollworms damage and their carryover. Destroy all trash collected during the ginning process. Remove all seed from the ginneries by the end of March.

## Others

- Do not grow rice nursery and crop under shade and near wheat straw stack (Kup) to prevent rice crop against Bacterial blight.
- Do not apply excessive dose of nitrogen. Nitrogen should not be applied beyond six weeks after transplanting (except when LCC is used) to manage rice crop against Bacterial blight.
- Avoid excessive use of nitrogeneous fertilizers and field should be free from weeds and debris.
- After the last picking, allow sheep, goats and other farm animals into cotton fields to feed or plant debris and un-opened bolls.
- Do not stalk the cotton sticks under shade or in the field. Beet the sticks on ground to dislodge the pink bollworm larvae surviving the unopened bolls. Stalk the cotton stick vertically.
- Sterility mosaic is a viral disease transmitted by an eriophyid mite (*Aceria cajani*) ao do not allow any arhar plant standing around sugarcane, cotton and other fields during winter season in order to check the primary source of infection.
- Solarize water saturated nursery bed using transparent polythene sheet (50 micron) for 40 days in the months of May-June for the control of root knot nematode in nursary beds.
- Incorporate 40 days old Toria and Taramira crops into tomato nursery beds 10 days before sowing and turn the soil 3-4 times before sowing of tomato.
- Grow garlic in root knot nematode infested fields in rotation with other vegetable crops.

- Green manure with sunhemp (50 day old crop) or marigold (60 days old crop) in root knot nematode infested fields.
- Do not green manure with Dhaincha in root knot nematode infested fields.

**Removal of weeds**: Keep the field free from Rewari weed which serves as a source of inoculums of Rust (*Uromyces viciae fabae*) in peas.

**Alternate crops:** Avoid growing of pea after the susceptible crops like brinjal, cauliflower, carrot and pea, instead tomato and chilli can be grown. This will help in reducing the White rot (*Sclerotinia sclerotiorum*) disease and spread of White rot inoculum in the field.

## Microbial biofungicides

The biological methods in disease management involve the use of different fungi against the pathogenic organisms (Table 2). The commonly available biofungicides are:

***Trichoderma viride:*** *Trichoderma viride* is used to control root and stem rots, wilts, blights and other fungal diseases. It is effective against plant pathogens. Its metabolites also stimulate seed germination, plant growth and early flowering.

***Pseudomonas fluorescens*:** It is effective against root and stem rots, damping off, blights, downy mildews and powdery mildews and other fungal diseases. It grows optimally at 25-30°C. It enters the plant vascular system and reaches the various parts of the plant system and acts as systemic control agent against various fungal and bacterial pathogens. It also acts as a plant growth promoter and stimulates seed germination.

## Inorganic substances

Sulphur and copper can be used for disease management on organic farms in a restricted manner i.e. with prior approval of the certifying agency. The maximum use of copper for disease management is limited to 8 kg $ha^{-1}$ $year^{-1}$. Bordeaux mixture can also be used in a restricted manner.

**Preparation of Bordeaux mixture:** Dissolve 1 kg copper sulphate in 10 litre of water in a copper or plastic vessel. In another vessel, slake 1 kg fresh quick lime by adding small quantity of, preferably warm, water. When slaking is over, add 5 litre of more water to it and stir well to get a uniform suspension of lime. Transfer the lime suspension, through a sieve, to a vessel containing 85 litre of water and stir well. Then add already prepared 10 litre of the copper sulphate solution to the lime solution with constant stirring. The Bordeaux mixture with proper composition will turn red litmus to blue and turmeric powder to orange red.

**Table 2.** Disease management of crops with bio-pesticides

| Crop | Disease | Bio-pesticide/bio-product |
|---|---|---|
| Cotton | Wilt and leaf spots | *Trichoderma* and *Pseudomonas* |
| Sugarcane, Basmati rice, Maize | Borer | *Trichogramma* |
| Potato | Early and Late blight | *Trichoderma, Pseudomonas, Mentha oil* |
| Basmati rice | Foot root | *Trichpderma* |

**Physical treatments:** *Fusarium* spp. on seeds can be controlled by warm water treatments. However, the effect on FHB will be limited.

**Natural compounds and biocontrol agents as plant protection products**: The control of FHB does not depend on plant protection products since preventative measures such as rotation, and tillage can be used. In apple, preventative measures such as removal of fallen leaves can delay the outbreak of epidemics. However, epidemics need to be controlled by multiple applications of plant protection products such as copper. Environmentally friendly new products are strongly needed.

Biological based technologies are most effective when integrated with physical and chemical approaches in the process towards sustainable ecological based plant diseases management. The important measures are:

- Use of antagonists is equivalent to natural enemies used in control of insects.
- Seeds treated with antagonists can be used as feed and food.
- Use of non-pathogenic materials like *F. oxysporum* (Fusariclean) and *F. fluorescens* (biocoat) for fusarium wilt in vegetables and flower crops.
- 2,4-diacetyl fluoroglucinol for all diseases of wheat.
- Biosave 10 and biosave 11 (based on strains of *Pseudomonas syringal* against storage rot of vegetables and fruits.
- Use of root-knot nematode.
- Use of molecular tools to assure pathogen free planting materials.

**Indigenous formulations**: These formulations are prepared from the locally available materials. These are generally based on the indigenous traditional knowledge and are used by the organic growers. They need to be tested for their efficacy in disease management.

**Beejamrit**: Take 5 kg fresh cow dung in a cloth bag and suspend it in a container filled with water to extract the soluble ingredients of cow dung. Suspend 50 g lime in 1 litre water, separately. After 12-16 hours squeeze the bag to collect extract and add 5 litre cow urine, 50 g virgin forest soil, lime water and 20 litre water. Incubate the mixture for 8 - 12 hours and then filter the contents. The filtrate is used for seed treatment (Yadav, 2010).

**Seed treated with hing (Asafoetida)** : Gram seed treated with hing (Asafoetida) @ 10 gm/10 Kg seed control pod borer & wilt.

**Seed treatment with butter milk:** Soaking of seed in buttermilk, (@ 5 litre/40 kg seed) control wilt disease in Pigeonpea and gram.

**Garlic extract**: Garlic extract (10%) checks growth of R. bataticola and Fusarium oxysporum of chickpea (Gaur and Sharma, 2010).

**Cow milk**: Cow's milk was reported to acts as an excellent sticker and spreader due to presence of casein protein has excellent spreader and sticker property. It can be used @ 10% aqueous suspension for effectively controlling powdery mildew. Milk sprays induced systematically acquired resistance in chilli against leaf curl, a viral disease (Kumar *et al.,* 2002).

**Cow dung extract**: Paddy blast can be controlled by spraying extract of 3 kg cow dung in 10 litre of water (Alvares, 1999).

**Bael leaf extract**: Boil 100 g of bael leaves in 2 litre water and dilute it with 100 of water and few litre of cow urine. It is used to control diseases of rice (Alvares, 1999). **Cow urine**: Cow urine can be used to control sclerotinia rot caused by Sclerotinia sclerotiorum of cucumber (Basak et al, 2002).

**Cow dung ash**: It is dusted on plants to protect them from certain fungal diseases. The possible effect of ash on plant growth or disease management may be due to composition of cow dung ash as it contains K, Ca, Mg , Fe, Mn, Zn and Cu.

**Cow urine+buttermilk**: Cow urine+buttermilk (1 litre of buttermilk cow+ 1 litre of cow urine +8 litre of water) is used for control of rusts.

**Papaya leaves**: Prepare solution from 2 kg fresh leaves of papaya in 3-4 liters of water and keep in overnight. After filtration, this is diluted with 50-60 litres of water and 250 ml soap solution added to it, is effective to control brown spot disease of rice.

# 8

# Organic Production Requirements

The use of synthetic pesticides, weedicides and agro-chemicals led to contamination of products and the quality of the produce is under question. In order to check the pesticide residue laboratories were set up, to test the pesticide contamination in food and drink, but it did not prevent terrestrial and aquatic ecosystem on land and water. Thus, the clean and uncontaminated food may be obtained by growing than in places, which is not contaminated and not applied with toxic chemicals. The standards are set of requirements which makes the food products to be grown under specified conditions, using only permissible inputs, following organic principles during growing, harvesting, processing, packing and transportation and the same came to be known as 'Organic Standards'. Organic standards typically contain lists of materials that are permitted as farm and processing inputs, such as fertilizers, pesticides and food additives. All other materials should be considered as prohibited unless the relevant certification programme approves their use. Organic standards generally emphasis the use of good management practices to minimise the need for inputs wherever possible. Organic standards address such broader aspects as biodiversity, native vegetation retention, waterway management, animal husbandry, ethics and waste management.

**Organic production standards:** In order to assure the consumer that a product is produced organically, a kind of quality control is needed. The organic quality is based on standards, inspection, certification and accreditation. All organic food is produced and handled according to strict rules called 'organic standards'. These standards cover all aspects of food production from animal welfare and wildlife conservation, to not allowing artificial food additives. All organic forms are visited at least once a year by a certifying inspector to check that standards are met. Organic standards do not define a quality status which can be measured in the final product (E.g. quality of pesticide residues, heavy metals, etc.). They define the way of production (E.g. that no chemical pesticides and fertilizers are used). There are organic standards at the national as well as international level.

## IFOAM (International Federation of Organic Agriculture Movement) Basic Standards

- IFOAM started during 1970, which now has more than 750 member organisations in over 100 countries.
- Took the lead in setting standards for organic agriculture.
- Several countries and organisations have their own standards but largely follow IFOAM standards.
- Meanwhile, the FAO and WHO jointly brought out guidelines during 1999, known as Codex Alimentarius Commission' guidelines for production, processing, labelling and marketing of organically produced food (ACIGL-32-1999).
- India – NPOP (National programme for organic production) during 2000.
- IFOAM basic standards are most important organic standards which also describes the principle of organic farming.
- These are the mother or organic standards.
- They are standards for standard setting on the national of international level.
- They are not for certification.
- It provides framework for certification bodies and standard setting organisations to develop their own certification standards.
- They are regularly reviewed and updated by the IFOAM members from all over the world.

## Aims and Activities of IFOAM

- Information exchange about all facts or organic agriculture.
- Promotion of the worldwide development of organic agriculture.
- Exchange of knowledge and experience among members and the organic movement as a whole.
- Representing the organic movement in international institutions.
- Continuously revising the IFOAM basic standards and the norms for accreditation.
- Developing a harmonised international organic guarantee system from the basic standards to the IFOAM accreditation programme.

**Standards :** Standards are rules of production for organic agriculture. They determine the production process within the ecological and social environment through which the product emerges. There are standards at various levels.

**International Standards:** International standards are those standards for organic agriculture approved by international bodies and recognised by legal authorities. E.g. Codex Alimentarius Commission guidelines.

**IFOAM Basic Standards:** They were first published in 1980. Since then they have been subjected to biennial review and publication. These basic standards define organic products grown, produced and handled. They reflect the current state of organic production and processing methods.

**Codex Alimentarius:** The codex guidelines for organically produced food will be regularly reviewed at least every four years based on given codex procedure.

**Regional /Supranational Standards:** Different regions in the world are involving regional or supranational standards for organic agriculture. E.g. European Union Council's regulations.

**European Union Council's Regulations:** The European Union regulation on organic production lays down minimum rules governing the production, processing and import of organic products, including inspection procedures, labelling and marketing for the whole of Europe. Each European country is responsible for enforcement and for its own monitoring and inspection system. Applications, supervision and sanctions are dealt with at regional levels.

National Standards: National standards are basic organic agriculture standards prepared by respective countries on the basis of which detailed standards are prepared by certification agencies and statutory boards for the development of crops.

### Some of the national standards are

1. USDA Standards
2. Canadian Organic Standards
3. Australian Organic Standards
4. Organic standards

### Codex Alimentarious Guidelines

In 1991, the Codex Alimentarius Commission, a joint FAO / WHO food standards programme began developing guidelines for the production, processing, labelling and marketing of organically produced foods. The guidelines include general sections describing the organic production concept and the scope of the text, description and definitions, labelling and claims (including, products in transition/ conversion), rules of production and preparation including criteria for the substances allowed in organic production; inspection and certification systems; and import control. Guidelines for the production, processing, labelling and marketing of organically-produced foods adopted during 1999. During 2001, it included sections ensuring livestock and livestock products and bee keeping and bee products. In earlier guidelines (IFOAM), only organic farming and processing were included, but in codex labelling and marketing are included codex guidelines are widely adopted throughout the world.

### Codex guidelines for organic production system

1. Enhance the biological diversity within the whole system.
2. Increase the soil microbial activity. E.g. flora, fauna, etc.
3. Maintain long-term soil fertility.
4. Recycle wastes of plant and animal origin in order to return nutrients to the land. Thus, minimising the use of non-renewable resources.
5. Rely on renewable resources in locally organised agricultural ecosystems.
6. Promote the healthy use of soil, water and air as well as minimise all forms of pollution that may result from agricultural practices.
7. Handle agricultural products with emphasis on careful processing methods in order to maintain the organic integrity and vital qualities of the products at all stages.
8. Become established on any existing farm through a period of conversion the appropriate length of which is determined by site specific factors, such as history of land, type of crops and livestocks to be produced.

### European Union Regulations

The European Union was one of the first to set up a policy on organic farming by adopting EU Council regulation. With this regulation, the council created a community frame work defining in detail the requirements for agricultural products or food stuffs bearing a reference to organic production methodology. The regulation is set up primarily as a labelling regulation, meant to regulate the internal market for organic products but it also describes the organic production standards and inspection and supervision requirements. It deals with all agricultural products and aspects of primary food production and food processing.

### National Standards for Organic Production (NSOP)

In India, standards for organic agriculture were announced in May, 2001 and the National Programme on Organic Production (NPOP) is administered by Agricultural and Processed Food Products Export Development Authority

(APEDA) using the IFOAM basic standards under the Ministry of Commerce. It includes definite principles, basic standards of production, documentation, inspection and certification guidelines. The Government has set the frame conditions in which the organic sector of a country operator which include content and legal status of organic standards, the regulations concerning the use of organic claims and labels, the legislation on consumer protection and the accreditation system. As per the national accreditation policy, all certifying agencies operating in India are to obtain accreditation from any one of the accrediting agency appointed by the Government of India (E.g. Spice Board, Coffee Board, Tea Board, APEDA, etc.).

The important organic production requirements as per national standards for organic production developed by APEDA are reproduced below:

## Crop production and animal husbandry in general

### Conversion requirements

Organic agriculture means a process of developing a viable and sustainable agro-ecosystem. The time between the start of organic management and certification of crops and/or animal husbandry is known as the conversion period. The whole farm, including livestock, should be converted according to the standards over a period of three years.

### Conversion period in organic farming

- It is the time between the start of organic management and certification of crops.
- Conversion period varies from 1 to 3 years depending on the current usage of chemical fertilizers, pesticides and false usage of lands.
- It is determined by certification agencies, while deciding the period, ecological regions are considered.

For a sustainable agro-ecosystem to function optimally, diversity in crop production and animal husbandry must be arranged in such a way that there is interplay of all the elements of the farming management. Conversion may be accomplished over a period of time. A farm may be converted step by step.

### Maintenance of organic management

- The standards requirements should be met during the conversion period.
- If the whole farm is not converted, the certification programme should ensure that the organic and conventional parts of the farm are separate and inspectable.

- Converted land and animals should not get switched back and forth between organic and conventional management.

## Landscape

- Organic farming should contribute beneficially to the ecosystem.
- Areas which should be managed properly and linked to facilitate biodiversity:
- Extensive grassland such as moorlands, reed land or dry land
- In general all areas which are not under rotation and are not heavily manured
- Extensive pastures, meadows, extensive grassland, extensive orchards, hedges, hedgerows, groups of trees and/or bushes and forest lines
- Ecologically rich fallow land or arable land
- Ecologically diversified (extensive) field margins
- Waterways, pools, springs, ditches, wetlands and swamps and other water rich areas which are not used for intensive agriculture or aqua production
- Areas with ruderal flora.
- The certification programme shall set standards for a minimum percentage of the farm area to facilitate biodiversity and nature conservation.
- The certification programme shall develop landscape and biodiversity standards.

## Crop production

## Choice of crops and varieties

All seeds and plant material should be certified organic. When organic seed and plant materials are available, they shall be used. The certification programme shall set time limits for the requirement of certified organic seed and other plant material. When certified organic seed and plant materials are not available to the farmer then chemically untreated conventional materials shall be used. The use of genetically engineered seeds, pollen, transgene plants or plant material is not allowed under organic farming.

## Duration of conversion period

The establishment of an organic production system and building of soil fertility requires an interim period, the conversion period. The conversion period may not always be of sufficient duration to improve soil fertility and re- establish the balance of the ecosystem but it is the period in which all the actions required to reach these goals are started.

**The duration of the conversion period must be adapted to**

- The past use of the land
- The ecological situation

Plant products produced can be certified organic when the national standards requirements have been met during a conversion period of at least two years before sowing or in the case of perennial crops other than grassland, at least three years (thirty-six months) before the first harvest of products. The accredited inspection and certification agency may decide in certain cases (such as idle use for two years or more) to extend or reduce the conversion period in the light of previous status of the land but the period must equal or exceed twelve months.

The conversion period can be extended by the certification programme depending on, e.g., past use of the land and environmental conditions.

The certification programme may allow plant products grown by the farmer to be sold as "produce of organic agriculture in process of conversion" or a similar description during the conversion period of the farm.

For the calculation of inputs for feeding, the feed produced on the farm unit during the first year of organic management, may be classified as organic. This refers only to feed for animals which are themselves being reared within the farm unit and such feed may not be sold or otherwise marketed as organic. Feed produced on the farms in accordance with the national standards is to be preferred over conventionally grown / brought-in feeds and the inputs managed at own farm will help to enhance net profit by reducing input cost.

**Diversity in crop production**

The basis for crop production under organic farming shall take into consideration of soil structure and fertility and surrounding ecosystem and to provide a diversity of species in order to minimize nutrient losses.

**Diversity in crop production can be achieved by a combination of**

- A versatile crop rotation with inclusion of legumes
- An appropriate coverage of the soil during the year of production which diverse plant species

Where appropriate, the organic produce farm shall require sufficient diversity to save the organic crop from pressure of insects, weeds, diseases and other pests, while maintaining or increasing soil organic matter, microbial activity and general soil health. For non perennial crops, this is normally, but not exclusively, achieved by means of crop rotation particularly leguminous crops.

### Fertilisation policy in organic farming

1. Biodegradable materials are encouraged as sufficient quantities of biodegradable material of microbial, plant or animal origin available at organic farm should be returned to the soil to increase or at least maintain its soil health and the biological activity within it. They set limitations for use of biodegradable material. Excess use of it also pollute environment. Poultry manure, if it is produced outside the farm, is avoided.
2. Fertilization management should minimize nutrient losses. Accumulation of heavy metals and other pollutants shall be prevented.
3. Carbon based materials should form the basis for nutrition. Non synthetic mineral fertilizers and brought in bio-fertilizers shall be regarded as supplementary and not a replacement for nutrient recycling.
4. Adequate pH levels should be maintained in the soil.
5. The certification programme shall set limitations to the total amount of biodegradable material of microbial, plant or animal origin brought onto the farm unit, taking into account local conditions and the specific nature of the crops.
6. The certification programme shall set standards which prevent animal runs from becoming over-manured where there is a risk of pollution.
7. Brought-in material (including potting compost) shall be in accordance with standards.
8. Manures containing human excreta should not be directly used on crops which are used for direct consumption.
9. Mineral fertilizers shall be applied in their natural composition and shall not be rendered more soluble by chemical treatment. Chilean nitrate and all synthetic nitrogenous fertilizers, including urea, are prohibited.
10. Some of the mineral fertilizers have restricted use and can be used as supplementary to manures. E.g. Lime, gypsum, rock phosphate, KNO3 are permitted.

- **Permitted**: FYM, urine, crop residues, mulches, cover crops, poultry manure, biofertilizers, BD preparation, vermicompost, botanical extracts, etc.
- **Restricted:** Use of Blood meal, bone meal, compost, city waste, FYM from other farm.
- **Restricted minerals**: NaCl, KSO4, gypsum, MgSO4, lime, rock phosphate.

## Pest, disease and weed management including growth regulators

Organic farming systems should be carried out in a way which ensures that losses from pests, diseases and weeds are minimized. Emphasis is placed on the use of a balanced fertilizing programme, use of crops and varieties well- adapted to the environment, fertile soils of high biological activity, adapted suitable cropping systems, companion planting, green manures, etc. Growth and development should take place in a natural manner.

- Noxious animal, insects, pests, disease, weed, etc., are to be controlled by Cultivation methods (suitable crop /varieties, time of planting and other cultural management) Physical method (using light, heat, sound, etc., or manual or mechanical methods) Biological method (predators, parasites, etc.) In case of critical agricultural chemicals noted in the standards are to be used.
- The natural enemies of pests and diseases should be protected and encouraged through proper habitat management of hedges, nesting sites etc. Pest management should be regulated by understanding and disrupting the ecological needs of the pests. An ecological equilibrium should be created to bring about a balance in the pest predator cycle.
- Products used for pest, disease and weed management, prepared at the farm from local plants, animals and micro-organisms, are allowed. If the ecosystem or the quality of organic products is likely to be jeopardised, the procedure to evaluate additional inputs to organic agriculture and other relevant criteria shall be used to judge if the product is acceptable. Branded products must always be evaluated. Ionising radiation should not be executed for the disease and pest control, preservation of foods, removal of pathogens and sanitation. Products which can be used under the supervision are biopesticides (Bt, trichoderma, NPV, pseudomonas, etc.), Bordeaux mixture, sulphur, soft soap and most plant based products (neem, rotenone, pyrethrum, etc.). Maximum use of copper under NPDP @ 8 kg/ha/year.
- Thermic weed control and physical methods for pest, disease and weed management are permitted.
- Thermic sterilization of soils to combat pests and diseases is restricted to circumstances where a proper rotation or renewal of soil cannot take place. Permission may be given by the certification programme only on a case by case basis.
- All equipments from conventional farming systems shall be properly cleaned and free from residues before being used on organically managed areas.
- The use of synthetic herbicides, fungicides, insecticides and other pesticides is prohibited.

- The use of synthetic growth regulators and synthetic dyes is prohibited.
- The use of genetically engineered organisms or products is prohibited.
- Accredited certification programmes shall ensure that measures are in place to prevent transmission of pests, parasites and infectious agents.

**Contamination control**

- Buffer zone should be maintained to minimize contamination from outside and from within the farm. Three tier buffer zone can be maintained by planting *Leucinia*, *Sesbania aculeate* and *Crotolaria juncea* in the Northern India or a row of *Cajanus cajan*can also be planted as buffer plantation to avoid the contamination of pesticides.
- For protected structure coverings, plastic mulches, fleeces, insect netting and silage rapping, only products based on polyethylene and polypropylene or other polycarbonates are allowed. These shall be removed from the soil after use and shall not be burnt on the farmland. The use of polychloride based products is prohibited.
- In case of risk of contamination, the certification programme should set limits for the maximum application levels of heavy metals and other pollutants.
- In case of any suspicion of contamination, the certification programme shall ensure analysis of the relevant products to determine the level of contamination from the possible sources of pollution (soil and water).

**Soil and water conservation**

- Soil and water resources should be handled in a sustainable manner.
- Relevant measures should be taken to prevent erosion, salination of soil, excessive and improper use of water and the pollution of ground and surface water.
- Clearing of land through the means of burning organic matter, e.g. slash-and burn, straw burning shall be restricted to the minimum.
- The clearing of primary forest is prohibited. Relevant measures shall be taken to prevent erosion.
- Excessive exploitation and depletion of water resources shall not be allowed.
- The certification programme shall require appropriate stocking rates which do not lead to land degradation and pollution of ground and surface water.
- Relevant measures shall be taken to prevent salination of soil and water.

## Collection of non cultivated material of plant origin and honey

- The act of collection should positively contribute towards the maintenance of natural areas.
- When harvesting or gathering the products, attention should be paid to maintenance and sustainability of the agro-ecosystem.
- Wild harvested products shall only be certified organic if derived from a stable and sustainable growing environment of organic area. Harvesting or gathering the product shall not exceed the sustainable yield of the ecosystem, or threaten the existence of plant or animal species.
- Products can only be certified organic if derived from a clearly defined collecting area, which is not exposed to prohibited substances/pesticides, and which is subject to inspection.
- The collection area shall be at an appropriate distance from conventional farming, pollution and contamination.
- The operator managing the harvesting or gathering of the products shall be clearly identified and be familiar with the collecting area in question.

## Animal husbandry

### Animal husbandry management

Management techniques in animal husbandry should be governed by the physiological and ethological needs of the farm animals in question. This includes:

- That animals should be allowed to conduct their basic behavioural needs.
- That all management techniques, including those where production levels and speed of growth should be concerned, for the good health and welfare of the animals.

For welfare reasons the herd or flock size should not adversely affect the behavioural patterns of the animal.

The certification programme shall ensure that the management of the animal environment takes into account the behavioural needs of the animals and provides for:

- Sufficient free movement
- Sufficient fresh air and natural daylight according to the needs of the animals
- Protection against excessive sunlight, temperatures, rain and wind according to the needs of the animals
- Enough lying and/or resting area according to the needs of the animal. For all animals requiring bedding, natural materials shall be provided.

- Ample access to fresh water and feed according to the needs of the animals
- Adequate facilities for expressing behaviour in accordance with the biological and ethological needs of the species.

No compounds used for construction materials or production equipment shall be used which might detrimentally affect human or animal health.

All animals shall have access to open air and/or grazing appropriate to the type of animal and season taking into account their age and condition, to be specified by the certification programme.

### The certification programme shall allow exceptions in cases where

- The specific farm or settlement structure prevents such access provided animal welfare can be guaranteed
- Areas where feeding of animals with carried fresh fodder is a more sustainable way to use land resources than grazing, provided animal welfare is not compromised.
- Restrictions shall always include a time limit which shall be set for each exception.
- Poultry and rabbits shall not be kept in cages.
- Landless animal husbandry systems shall not be allowed.
- When the natural day length is prolonged by artificial lighting, the certification programme shall prescribe maximum hours respective to species, geographical considerations and general health of animals.
- Herd animals shall not be kept individually.
- The certification programme may allow exceptions, e.g., male animals, smallholdings, sick animals and those about to give birth.

### Length of conversion period

- The establishment of organic animal husbandry requires an interim period, the conversion period.
- The whole farm, including livestock, should be converted according to the standards set out in this document. Conversion may be accomplished over a period of time.
- Replacement poultry should be brought onto the holding at the start of the production cycle.
- Animal products may be sold as "product of organic agriculture" only after the farm or relevant part of it has been under conversion for at least twelve

months and provided the organic animal production standards have been met for the appropriate time.

- The certification programme shall specify the length of time by which the animal production standards shall be met. With regard to dairy and egg production, this period shall not be less than 30 days.
- Animals present on the farm at the time of conversion may be sold for organic meat if the organic standards have been followed for 12 months.

## Brought-in animals

All organic animals should be born and raised in the organic holding.

Organic animal husbandry should not be dependent on conventional raising systems.

When trading or exchanging livestock, it should preferably take place between organic farms or as part of a long term cooperation between specific farms.

When organic livestock is not available, the certification programme shall allow brought-in conventional animals according to the following age limits:

- 2 day old chickens for meat production
- 18 week old hens for egg production
- 2 week old for any other poultry
- Piglets up to six weeks and after weaning
- Calves up to 4 weeks old which have received colostrum and are fed a diet consisting mainly of full milk. Certification programmes shall set time limits (not exceeding 5 years) for implementation of certified organic animals from conception for each type of animal.

Breeding stock may be brought in from conventional farms at an annual rate not exceeding 10% of the adult animals of the same species in the organic farm. For brought-in breeding stock the certification programme shall allow a higher yearly maximum than 10% in the following cases and with specific time limits.-

- Unforeseen severe natural or manmade events
- Considerable enlargement of the farm
- Establishment of a new type of animal production on the farm
- Small holdings

## Breeds and breeding

- Breeds should be chosen which are adapted to local conditions. Breeding goals should not be at variance with the animal's natural behaviour and should be directed towards good health.

- Breeding shall not include methods which make the farming system dependent on high technological and capital intensive methods. Reproduction techniques should be natural.
- The certification programme shall ensure that breeding systems are based on breeds that can both copulate and give birth naturally.
- Artificial insemination is allowed.
- Embryo transfer techniques are not allowed in organic agriculture.
- Hormonal heat treatment and induced birth are not allowed unless applied to individual animals for medical reasons and under veterinary advice.
- The use of genetically engineered species or breeds is not allowed.

### Mutilations

The animals' distinctive characteristics should be respected. Species shall be chosen which do not require mutilation.

Mutilations shall be allowed only in exceptional cases and shall be kept to the minimum. Mutilations are not allowed. The certification programme shall allow the following exceptions:

- Castrations
- Tail docking of lambs
- Dehorning
- Ringing
- Mulesing

Suffering shall be minimized and anaesthetics used where appropriate.

### Animal nutrition general principles

The livestock should be fed 100% organically grown feed of good quality. All feed shall come from the farm itself or be produced within the region. The diet shall be offered to the animals in a form allowing them to execute their natural feeding behaviour and digestive needs.

The diet should be balanced according to the nutritional needs of the animals. Products from the organic feed processing industry shall be used. Colouring agents shall not be used in organic livestock production.

The certification programme shall draw up standards for feed and feed ingredients.

The prevailing part (at least more than 50%) of the feed shall come from the farm unit itself or shall be produced in co-operation with other organic farms in the region.

The certification programme shall allow exceptions with regard to local conditions under a set of time limit for implementation.

For the calculation purpose only, feed produced on the farm unit during the first year of organic management, may be classed as organic. This refers only to feed for animals which are themselves being reared within the farm unit and such feed may not be sold or otherwise marketed as organic.

Where it proves impossible to obtain certain feeds from organic farming sources, the certification programme shall allow a percentage of feed consumed by farm animals to be sourced from conventional farm. The maximum percentages of such feeds are given in the following table and shall be calculated in terms of the average diet for each animal category. These maximum percentages shall be followed the whole year round:

- Ruminants (dry matter intake) 15% Non-ruminants (dry matter intake) 20%
- These percentages will be reduced within 5 years to Ruminants (dry matter) 10%
- Non-ruminants (dry matter) 15%
- The certification programme shall allow exceptions to these percentages, with specific time limits and conditions in the following cases:
- Unforeseen severe natural or man-made events
- Extreme climatic or weather conditions
- Areas where organic agriculture is in early stages of development

**The following products shall not be included nor added to the feed given to farm animals**

- Synthetic growth promoters or stimulants
- Synthetic appetizers
- Preservatives, except when used as a processing aid
- Artificial colouring agents
- Urea
- Farm animal by-products (e.g. abattoir waste) to ruminants
- Droppings, dung or other manure (all types of excreta) even if technologically processed
- Feed subjected to solvent (e.g. hexane), extraction (soya and rape seed meal) or the addition of other chemical agents
- Pure amino acids
- Genetically engineered organisms or products thereof

- This covers both organic and conventional feeding stuffs.
- Vitamins, trace elements and supplements shall be used from natural origin when available in appropriate quantity and quality.
- The certification programme shall define conditions for use of vitamins and minerals from synthesized or unnatural sources.

**All ruminants shall have daily access to roughage. The following fodder preservatives shall be used**

- Bacteria, fungi and enzymes
- By-products of food industry (e.g. molasses)
- Plant based products

Synthetic chemical fodder preservatives shall be allowed in special weather conditions. The certification programme shall specify conditions for use of substances from synthesized or unnatural sources e.g. acetic, formic and propionic acid, vitamins and minerals.

The certification programme shall set minimum weaning times taking into account the natural behaviour of the relevant animal species.

Young stock from mammals shall be raised using systems which rely on organic milk, preferably from their own species.

In emergencies the certification programme shall allow the use of milk from nonorganic farming systems or dairy based milk substitutes so long as they do not contain antibiotics or synthetic additives.

### Veterinary medicine

Management practices should be directed to the wellbeing of animals, achieving maximum resistance against disease and preventing infections.

Sick and injured animals shall be given prompt and adequate treatment.

Natural medicines and methods, including homeopathy, ayurvedic, unani medicine and acupuncture, shall be emphasized.

When illness does occur the aim should be to find the cause and prevent future outbreaks by changing management practices.

Where appropriate, the certification programme should set conditions based on the farm's veterinary records to minimize the use of medicines.

The certification programme should make a list of medicines specifying the withholding periods.

The well-being of the animals is the primary consideration in the choice of illness treatment. The use of conventional veterinary medicines is allowed when no other justifiable alternative is available.

Where conventional veterinary medicines are used, the withholding period shall be at least double the legal period. Use of the following substances is prohibited:

- Synthetic growth promoters
- Substances of synthetic origin for production, stimulation or suppression of natural growth
- Hormones for heat induction and heat synchronization unless used for an individual animal against reproductive disorders, justified by veterinary indications

Vaccinations shall be used only when diseases are known or expected to be a problem in the region of the farm and where these diseases cannot be controlled by other management techniques. The certification programme shall define conditions for such cases. Legally required vaccinations are allowed. Genetically engineered vaccines are prohibited.

**Transport and slaughter**

Transport and slaughter should minimize stress to the animal. Transport distance and frequency should be minimized.

The transport medium should be appropriate for each animal. Animals should be inspected regularly during transport.

Animals should be watered and fed during transport depending on weather conditions and duration of the transport. Stress to the animal shall be minimized, especially taking into consideration:

- Contact (by eye, ear or smell) of each animal with dead animals or animals in the killing process
- Existing group ties
- Resting time to release stress

Each animal shall be stunned before being bled to death. The equipment used for stunning should be in good working order. Exceptions can be made according to cultural practice. Where animals are bled without prior stunning this should take place in a calm environment.

Throughout the different steps of the process there shall be a person responsible for the well-being of the animal. Handling during transport and slaughter shall be calm and gentle. The use of electric sticks and such instruments are prohibited.

The certification programme shall set slaughter and transportation standards that will take into consideration:

- Stress caused to the animal and person in charge
- Fitness of the animal
- Loading and unloading
- Mixing different groups of animals or animals of different sex
- Quality and suitability of mode of transport and handling equipment
- Temperatures and relative humidity
- Hunger and thirst
- Specific needs. of each animal
- No chemical synthesized tranquillizers or stimulants shall be given prior to or during transport. Each animal or group of animals shall be identifiable during all steps.
- Where the transport is by axle, the journey time to the slaughterhouse shall not exceed eight hours.
- Certification programmes may grant exceptions on a case to case basis.

**Bee keeping**

The collection area should be organic and/or wild and should be as varied as possible to fulfill the nutritional needs of the colony and contribute to good health.

The feed supplied should be fully organic.

Bee keeping is considered to be part of animal husbandry. The general principles therefore also apply to bee keeping.

The feeding of colonies shall be seen as an exception to overcome temporary feed shortages due to climatic conditions.

The foundation comb should be made from organic wax.

When bees are grown in wild areas, consideration should be taken of the indigenous insect population. Hives shall be situated in organically managed fields and/or wild natural areas.

Hives shall not be placed close to fields or other areas where chemical pesticides and herbicides are used.

Exceptions can be made by certification bodies on a case to case basis.

Feeding shall only take place after the last harvest before the season when no foraging feed is available.

Each bee hive shall primarily consist of natural materials. Use of construction materials with potentially toxic effects is prohibited.

Persistent materials may not be used in beehives where there is a possibility of permeation of the honey and where residues may be distributed in the area through dead bees.

- Wing clipping is not allowed
- Veterinary medicine shall not be used in bee keeping.

When working with the bees (e.g. at harvest) no repellent consisting of prohibited substances shall be used. For pest and disease control and for hive disinfection the following products shall be allowed:

- Caustic soda
- Lactic, oxalic, acetic acid
- Formic acid
- Sulphur
- Etheric oils
- Bacillus thuringiensis

## Plant protection

- Natural methods are to be adopted. Preventive, cultural thermophysical, etc., promote natural predators and bioagents.
- Prohibited: Synthetic products, GM products.Restricted: Most of the plant products. Even neem oil, biopesticides (NPV, fugae, bacterial, etc.).
- Not allowed: Alcoholic product based plant pesticides, soft soap based pesticides even sterilised insects.
- Allowed: Most of homeopathic, herbal and BD preparation, pheromone traps and mechanical traps-allowed / permitted.
- Until now, the Indian standards are only compulsory for products to be exported. It is planned to apply the same standards also for the domestic market.

## NSOP guidelines

## Formation of organic farmers group

- Farmers with similar farming and production should be brought together preferably in the same village in contiguous areas.

## Plant and planting material

- All seed and planting material essentially used from the same farm or other organic farming farms which are adopted to local soil and climatic conditions.

## Maintenance of Organic Management

- The standards requirements should be met during the conversion period.
- If the whole farm is not converted, the certification programme should ensure that the organic and conventional parts of the farm are separate and inspectable.
- Converted land and animals should not get switched back and forth between organic and conventional management.

## Soil Healths and Water Quality

- Enhancement of biodiversity.
- No burning of vegetation.
- Recycling, generation and addition of OM and nutrients.
- Measures to be taken to avoid land degradation and pollution of water resources and should not exploit it.
- Manures to avoid salanisation.
- Rain water recycling where possible.
- No clearing of forest.
- Measures to prevent soil and water erosion.

## Biodiversity

- Minimum crop rotation requirements / variety.
- Should manage weeds, insects, diseases and other pests while maintaining soil fertility, OM and microbial activity.
- Standards for orchards / plantation floor cover and/or diversity or refuse plantings in the orchard should be set.

## Soil productivity

- Compost prepared from agricultural products produced in the field. Native soil MO are to be utilised.
- Legumes, GM or deep rooting plants in rotation programme.
- Use mineral fertilizer (ground rock, lime, dolomite, etc.) microbial fertilizers, biodynamic preparations and botanical preparations.
- No synthetic / chemical fertilizers are permitted.
- Use of fertilizers and soil improvement materials noted only in the organic standards. Maximum amount of manures may not exceed 170 kg N/year/ha.

## Pest, disease and weed management

- Genetically engineered cultivars or plant materials are not permitted in organic production.
- The seed for raising a crop should either be organically produced or if organic seed is not available, conventional seed without any chemical treatment may be used.
- Whole farm including the livestock should be converted to organic in a step by step manner.
- If the whole farm is not converted, the certification programme shall ensure that the organic and conventional parts of the farm are separate and inspectable.
- Before products from a farm/project can be certified as organic, inspection shall be carried out during the conversion period.
- To ensure a clear separation between organic and conventional production, the certification programme (agency) shall inspect, where appropriate, the whole production system.
- Plant products produced can be certified organic when the national standards requirements have been met with during the conversion period of at least two years before sowing for annual crops or in the case of perennial crops other than grassland, at least three years before the first harvest of products.
- Biodegradable material of microbial plant or animal origin shall form the basis of the fertilization programme.
- Manures containing human excreta (faeces and urine) cannot be used on vegetation for human consumption.
- Mineral fertilizers shall only be used in a supplementary role to carbon based materials. Permission for use shall only be given when other fertility management practices have been optimized.
- Chilean nitrate and all synthetic nitrogenous fertilizers, including urea, are prohibited.
- Mineral fertilizers shall be applied in their natural composition and shall not be rendered more soluble by chemical treatment.
- Products used for pest, disease and weed management, prepared at the farm from local plants, animals and micro- organisms, are allowed.
- The use of synthetic herbicides, fungicides, insecticides and other pesticides is prohibited.

- In case of reasonable suspicion of contamination the certification programme shall make sure that an analysis of the relevant products and possible sources of pollution (soil and water) shall take place to determine the level of contamination.
- For protected structure coverings, plastic mulches, fleeces, insect netting

# 9

# National Program for Organic Production and Certification

**National Program for Organic Production (NPOP):** Organic products are grown under a system of agriculture without the use of chemical fertilizers and pesticides with an environmentally and socially responsible approach. This is a method of farming that works at grass root level preserving the reproductive & regenerative capacity of the soil, good plant nutrition, and sound soil management, produces nutritious food rich in vitality which has resistance to diseases.

India is bestowed with lot of potential to produce all varieties of organic products due to its various agro climatic regions. In several parts of the country, the inherited tradition of organic farming is an added advantage. This holds promise for the organic producers to tap the market which is growing steadily in the domestic and export market.

As per the available statistics, India's rank in terms of World's Organic Agricultural land was 9th and in terms of total number of producers was 1st as per 2018 data (Source: FIBL & IFOAM Year Book 2018).

The Government of India has implemented the National Programme for Organic Production (NPOP). The national programme involves the accreditation programme for Certification Bodies, standards for organic production, promotion of organic farming etc. The NPOP standards for production and accreditation system have been recognized by European Commission and Switzerland for unprocessed plant products as equivalent to their country standards. Similarly, USDA has recognized NPOP conformity assessment procedures of accreditation as equivalent to that of US. With these recognitions, Indian organic products duly certified by the accredited certification bodies of India are accepted by the importing countries.

### Area

As on 31st March 2018, total area under organic certification process (registered under National Programme for Organic Production) is 3.56 million Hectare (2017-18). This includes 1.78 million ha (50%) cultivable area and another 1.78 million

Hectare (50%) for wild harvest collection. Among all the states, Madhya Pradesh has covered largest area under organic certification followed by Rajasthan, Maharashtra and Uttar Pradesh. During 2016, Sikkim has achieved a remarkable distinction of converting its entire cultivable land (more than 76000 ha) under organic certification.

### Production

India produced around 1.70 million MT (2017-18) of certified organic products which includes all varieties of food products namely Oil Seeds, Sugar cane, Cereals & Millets, Cotton, Pulses, Medicinal Plants, Tea, Fruits, Spices, Dry Fruits, Vegetables, Coffee etc. The production is not limited to the edible sector but also produces organic cotton fiber, functional food products etc.Among different states Madhya Pradesh is the largest producer followed by Maharashtra, Karnataka, Uttar Pradesh and Rajasthan. In terms of commodities Oil seeds are the single largest category followed by Sugar crops, Cereals and Millets, Fiber crops, Pulses, Medicinal, Herbal & Aromatic plants and Spices and Condiments.

### Exports

The total volume of export during 2017-18 was 4.58 lakh MT. The organic food export realization was around INR 3453.48 crore (515.44 million USD). Organic products are exported to USA, European Union, Canada, Switzerland, Australia, Israel, South Korea, Vietnam, New Zealand, Japan etc.In terms of export value realization Oilseeds (47.6%) lead among the products followed by Cereals and millets (10.4%), Plantation crop products such as Tea and Coffee (8.96%), Dry fruits (8.88%), Spices and condiments (7.76%) and others.

The Government of India has implemented the National Programme for Organic Production (NPOP). The national programme involves the accreditation programme for Certification Bodies, standards for organic production, promotion of organic farming etc. The NPOP standards for production and accreditation system have been recognized by European Commission and Switzerland for unprocessed plant products as equivalent to their country standards. Similarly, USDA has recognized NPOP conformity assessment procedures of accreditation as equivalent to that of US. With these recognitions, Indian organic products duly certified by the accredited certification bodies of India are accepted by the importing countries.

- NPOP is implemented under FTDR Act under the Ministry of Commerce for exports.
- NPOP is implemented under AGMARK by Ministry of Agriculture for the domestic market.

## Major stakeholders

- Progressive farmers / NGOs
- Organic Enterprises – Exports and retail stores
- MoC – APEDA
- MoA – NCOF / / DMI /FSSAI
- States

India produced around 1.35 million MT (2015-16) of certified organic products which includes all varieties of food products namely Sugarcane, Oil Seeds, Cereals & Millets, Cotton, Pulses, Medicinal Plants, Tea, Fruits, Spices, Dry Fruits, Vegetables, Coffee etc. The production is not limited to the edible sector but also produces organic cotton fiber, functional food products etc. Among all the states, **Madhya Pradesh** has covered largest area under organic certification followed by Himachal Pradesh and Rajasthan.

## National Program on Organic Production (NPOP)

- Launched during 2001
- Internationally recognized
- Equivalence with EU and Switzerland
- USDA recognized conformity Assessment system
- Equivalence with Canada, Japan, South Korea and Taiwan - expected soon
- 28 Accredited certification bodies
    - 10 in public sector
    - 18 in Private sector
    - Offers certification in
    - Crop production
    - Wild harvest collection
    - Livestock
    - Apiculture
    - Aquaculture
    - Food Processing

## Operational Structure of NPOP

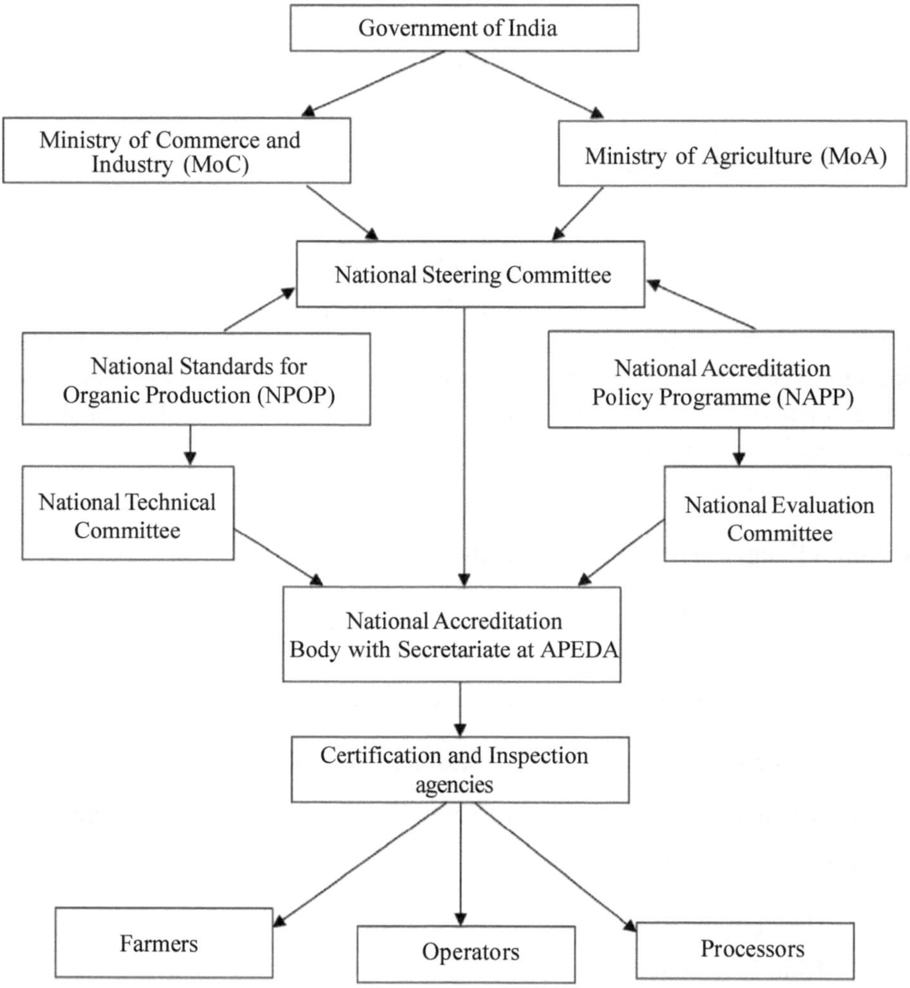


**Organic certification**: Certification means the agricultural farm and the farmer's methods inspected by an organic certifying agency to ensure that they comply with guidelines on organic farming. Each certifying agency has a code of standards, which is available to interested people. Certification is a procedure by which a third party gives written assurance that a product, process or services is in conformity with certain organic standards. Certification is the key to the National Organic Programme. The certification process focuses on the methods and production practices used in production.

Organic certification is a procedure to verify the compliance of organic standards. It is a certification process for organic food producers and other organic agricultural products. Any business directly involved in food production can be certified, including seed suppliers, farmers, food processors, retailers and restaurants. A lesser known counterpart is certification for organic textiles (or Organic clothing) that includes certification of textile products made from organically grown fibres. Requirements vary from country to country and generally involve a set of production standards for growing, storage, processing, packaging and shipping that include:

- Avoidance of synthetic chemical inputs (e.g. fertilizer, pesticides, antibiotics, food additives), irradiation, and the use of sewage sludge;
- Avoidance of genetically modified seed;
- Use of agriculture area which is free from prohibited chemical inputs for a number of years (often, three or more);
- For livestock, adhering to specific requirements for feed, housing, and breeding;
- Keeping detailed written production and sales records (audit trail);
- Maintaining strict physical separation of organic products from non-certified products;
- Undergoing periodic on-site inspections.
- In some countries, certification is overseen by the government and commercial use of the term organic is legally restricted. Certified organic producers are also subject to the same agricultural, food safety and other government regulations that apply to non-certified producers.

## Purpose of certification

Organic certification addresses a growing global demand for organic food. It is intended to assure quality of organic product and prevent fraud, and to promote commerce. The certification of organic farmed crops/products was not necessary in the early days of the organic movement, when small farmers would sell their produce at premium price directly at farmers' markets, as organics have grown in popularity and presently most of the human beings are health conscious and have capacity to pay premium price for organic products. Now-a-days, more and more consumers are purchasing organic food through traditional channels, such as supermarkets. As such, consumers must rely on third-party regulatory certification.

For organic producers, certification identifies suppliers of products approved for use in certified operations. For consumers, "certified organic" serves as a product assurance, similar to "low fat", "100% whole wheat", "sugar less", "cholesterol free", or "no artificial preservatives".

The main aim of certification is to regulate and facilitate the sale of organic products to consumers. Individual certification bodies have their own service marks, which can act as branding to consumers—a certifier may promote the high consumer recognition value of its logo as a marketing advantage to producers.

**Accreditation:** Accreditation means a procedure adopted by the National Accreditation Body for ascertaining the competence of a Certification Body to certify organic farms, products and processes as per the National Standards for Organic Products. Organic Accreditation authorizes private, foreign, or state organizations to certify farms or processing facilities.

In March 2000, the Ministry of Commerce launched NPOP (National Programme for Organic Production) design to establish National Standards for Organic Products, which could then be sold under the logo 'India Organic'. For proper implementation of NPOP, NAPP (National Accreditation Policy and Programme) has been formulated with Accreditation regulation announced in May 2001. These make it mandatory that all certification bodies whether internal or foreign operating in the country must be accredited by an Accreditation agency. An agricultural product can only be exported as organic product if a certification body duly accredited by APEDA certifies it.

## Third-party certification

- Organic certifying agent evaluates producers, processors and handlers to determine whether they conform to an established set of operating guidelines called organic standards. Those who conform are certified by the agent and allowed to use a logo, product statement or certificate to document their product as certified organic.
- In third party certification, the farm or the processing of the agriculture produce is certified in accordance with national or international organic standards by an accredited organic certification agency. To certify a farm, the farmer is typically required to engage in a number of new activities, in addition to normal farming operations:
- Study the organic standards, which cover in specific detail do's and don't's allowed for every aspect of farming, including storage, transport and sale.

**Compliance**-Farm facilities and production methods must comply with the standards, which may involve modifying facilities, sourcing and changing suppliers, etc.

**Documentation**-Extensive paperwork is required, detailing farm history and current set-up, and usually including results of soil and water tests.

**Planning**-a written annual production plan must be submitted, detailing everything from seed to sale: seed sources, field and crop locations, fertilization and pest control activities, harvest methods, storage locations, etc.

**Inspection**-annual on-farm inspections are required, with a physical tour, examination of records, and an oral interview.

**Fee**-an annual inspection/certification fee is to be paid by farmer to the certification agency for inspection and lab analysis. There are financial assistance programs for qualifying certified operations.

**Record-keeping**-written, day-to-day farming and marketing records, covering all activities, must be available for inspection at any time.

- In addition, short-notice or surprise inspections can be made, and specific tests (e.g. soil, water, plant tissue) may be requested.

For first-time farm certification, the soil must fulfill basic requirements of being free from use of prohibited inputs (synthetic chemicals, etc.) for a number of years. A conventional farm must adhere to organic standards for this period, often two to three years. This is known as being in transition. Transitional crops are not considered fully organic.

Certification for operations other than farms follows a similar process. The focus is on the quality of ingredients and other inputs, and processing and handling conditions. A transport company would be required to detail the use and maintenance of its vehicles, storage facilities, containers, and so forth. A restaurant would have its premises inspected and its suppliers verified as certified organic.

## How to get certification

- Application is made to the certification agency
- Screening of application by the certification agency
- Cost estimates for certification are sent for acceptance
- Acceptance of cost by the grower
- Signing of agreement
- Certification agency seeks detailed annual production plan detailing from seed to sale, crop rotations, pest management programme, storage locations and supply copy of the organic standards i.e. do's and don'ts of organic farming to the grower.
- Grower pays the fees
- On farm scheduled inspections- physical tour of farm, examination of records are made

- Unannounced inspections can also be done and samples can be taken.
- Inspection report regarding compliance of organic standards is submitted to the certification committee.
- Certification is granted.
- Grower releases the stock for sale with certification mark (India Organic Logo).

## Types of certification

**Foreign** Certification: The inspection and certification of export oriented organic products in developing countries are certified by the certification bodies based in importing countries. The main disadvantage is that costs of certification are high due to frequent plane trips and Western salaries.

**Co-Certification:** Local branch officers of Western Certification programmes conduct inspection along with local inspection staff who speaks local languages and familiar with local agriculture production conditions. However, inspection work is supervised by the Head Office. This reduces number of plane trips, thereby its cost.

**Indigenous Certification:** Indigenous certification agencies can usually offer cheaper inspection fees as less traveling is required and only local salaries have to be covered. They support the development of domestic markets. For export to the European Union (EU) market, a certificate from a certification body, which is accepted by EU competent authorities is required. To regulate the export of certified organic products, the Director General of Foreign Trade, Government of India has issued a public notice according to which no certified organic products may be exported unless they are certified by an inspection and certifying agency duly accredited by one of the accreditation agencies designated by Government of India.

**Group certification:** Majority of agriculture practitioners worldwide are small holders and are often located in remote areas with long travel times from one place to another. Furthermore, the overall revenue from their agricultural production is usually too small to allow a viable farm inspection by external inspection body for each farmer. Based on these observations, an idea was generated to develop Grower Group Certification where group certification refers to the certification of a group of producers who are in close proximity to one another, whose farms are uniform in most ways and who are organized under one management and marketing system (IFOAM). According to NPOP, certification of an organized group of producers, processors and exports with similar farming and production systems and which are in geographical proximity.

## Grower group certification for organic agriculture

The Organic Agriculture movement will not be successful until it reaches the small holder farmers. Previously, Organic Standards and Certification procedures were designed considering only individual farmers, the potential for smallholder group's certification being realized latter. Group Certification has now been introduced and is most popular and acceptable amongst the Indian farming community. Small holder group certification brings the following benefits:

1. It reduces certification cost per ha per annum as compared to individual farmer certification.
2. Small holder groups gathering farmers with common interests make it possible and easy for government to implement development schemes.
3. It helps the farmers in collective bargaining with buyers.
4. It saves transportation cost incurred by group members in transporting their produce to market. 5. National and international buyers get bulk volume of particular commodity.
6. It ensures a regular supply of consistent quality to buyers and processors.
7. It provides well opportunity for input providers to market among group members.
8. It helps in knowledge /information dissemination among large number of farmers.
9. It contributes to maintenance of social harmony as groups may contain farmers of distinct caste and religion.
10. The growers organized into groups become an easy market for certification agencies.
11. Members can benefits from group training in Organic Agriculture, Biodynamic agriculture, Organic Certification, Organic produce quality , demonstrations, tours , etc.,
12. All groups members can receive advice on seeds, fertility, pests and diseases, post harvest related issues and marketing.

## Who can be considered as a group?

1. Only small farmers can be members of the group covered by group certification. Large farms can also belong to the group but have to be inspected annually by the external inspection body. Processors and exporters can be part of the structure of the group, but have to be inspected by the external inspection body.

2. The farmers of the group apply similar production system and the farms should be in geographical proximity.
3. A group may be organised on itself i.e., as a cooperative, or as a structural group of producers affiliated to a processor or an exporter.
4. The group must be established formally, based a written agreement with its members. It shall have control management, established decision procedures and leagel capacity.
5. When intended for export, the marketing of the products must be carried as a group.

## Principles

1. A substantial part of the inspection work is carried out by internal inspections in the frame work of the internal control system set up by the group.
2. The external inspection body verifies and evaluates the effectiveness of the internal control system and certifies the group as a whole.

## Internal Quality System (IQS)

Group certification is based on the concept of an Internal Quality System comprising of the following:

- Implementation of the internal control system
- Internal standards
- Risk assessment.

An external inspection and certification body should be identified for conducting annual inspection of the individual group / unit. The external inspection agency shall evaluate by checking the IQS documentation, staff qualifications and re-inspecting some farms.

## How to develop an IQS

The following are the minimum requirements for setting up an IQS for grower groups:

- Development of Internal Control System (ICS)
- Identification of producer groups
- Creation of awareness about group certification
- Identification of qualified personnel for maintaining the internal control system
- Give necessary training in production and IQS development
- Preparation of IQS manual containing policies and procedures

- Implementation of the policies and procedures
- Review and improvement of the IQS document for maintaining a harmonized IQS.

**Internal Quality System Manager (IQS Manager)**

IQS manager shall develop and implement the IQS and would be responsible to organize internal inspections, coordinate between field staffs, approval staff, and the external inspection agency. The IQS manager shall have defined procedures to approve or sanction farmers. The responsibility of IQS manager shall be to ensure that all the standards requirements are fully implemented by the group.

**Internal inspectors** Adequate number of internal inspectors shall be identified from within the group. The inspectors shall be qualified and well versed with the standards to perform internal inspections.

**Approval manager / committee** Qualified person or approval committee shall be designated from within the group to take the approval decision. The approval manager/committee shall be well versed with organic procedures of IQS, internal standards and NPOP standards.

**Field officers** Field officers should be identified among the group, one at each production area. The field officer shall train the farmers by organizing field extension services.

**Purchase officers** Purchase officers shall be identified who would be responsible for correct purchase of produce from the farmers. The purchase officer is required to be well versed with IQS.

**Warehouse manager** If there are separate warehouses, it may be necessary to have a warehouse manager who would be responsible for handling the produce. He/she shall be well versed with the procedures of IQS for proper implementation.

**Processing manager** If a processing unit is operated by the IQS operator, it may be necessary to assign a processing manager. The processing manager is required to be trained in the handling procedures. When the processing of the produce is being organized in a company, the latter needs to be inspected by the certifier and would be responsible for processing according to the internal handling rules. In such case, the processing unit shall have a formal contract with the grower group.

**Internal Control System (ICS)** An internal system to regulate the functioning of any group is known as Internal Control System. Internal Control System is implemented to manage and apply

**Internal Standards** The internal standards shall be prepared in local language by the IQS manager for the region of operations under the framework of NPOP standards. If the farmers are illiterate, the internal standards shall contain illustrations in the text for better understanding. The internal standards would contain:

- Definition of production unit
- How to deal with part conversion
- Conversion period

Farm production norms for the entire production unit (e.g. seeds, nutrient management, pest management, soil management, approved inputs, prevention of drifts, livestock husbandry management).

## Harvest and post harvest procedures

**Conflict of interest** The IQS personnel shall not have any conflict of interest that might hinder the work. All possible conflicts shall be declared in a written statement. In such cases, the IQS shall ensure that alternative solutions are found.

**Scope of certification** The certification shall be granted to the group with reference to the regulations / standards adopted by the group.

**Trade** The group will market the products under a single entity. For trading the products from the group of producers, the IQS shall draw up relevant procedures.

Procedures for Implementation of Internal Control System

For maintaining the internal control system, the following procedures shall be adopted by the grower group.

**Registration of members** All members of the group will be formally (legally) registered under a single entity.

**Provision of documents to the members of the grower group** Each member of the grower group will be supplied with docket in local languages, which will contain the following –

- Copy of IQS manual
- Internal standards document
- NPOP document (Each member / staff shall be communicated when there is a revision in the standards.)
- Definition of the production unit Farm Entrance Form (farm data sheet), including last use of prohibited inputs

- Field records (main cultivation measures, use of inputs, harvested quantities, post harvest procedures): remark: may be included in internal farm checklist.
- Prevailing farming system and package of practices available for the area
- Details and description of the various steps required for the process flow right from cultivation to harvest and sales of the products.
- Written contract (for formal commitment) of each grower within the group x Annual farm inspection checklist
- Information on training programmes and provision of advisory services by the field officers.

## Internal inspections

At least two inspections of the group (one in growing season of each crop) shall be carried out by the internal inspector and will be documented.

The inspection will be carried out in presence of the member or his representative and must include a visit of the whole farm, storage of inputs, harvested products, post harvest handling and animal husbandry.

The internal inspector will also verify if the internal standards have been followed and whether the conditions of the previous internal inspection have been fulfilled.

The visit of the internal inspector will be documented in the farm inspection checklist duly signed by the inspector and counter-signed by the member or his representative.

In case of severe non-compliance, the results will be reported immediately to the IQS manager and all measures will be taken according to the internal sanction procedures.

**External Inspections:** The external Inspection and Certification Agency will re-inspect some of the farms for the evaluation of the grower group for efficient internal control system for compliance with the NPOP Standards. The sampling plan for inspection shall be based on the inspector's perception of risk based on the following factors:

1. Size of holding
2. Number of the members in the group
3. Degree of similarity between the production system and crop system
4. Inter-mingling / contamination
5. Local hazards

**Yield Estimates** Yields will be estimated for each crop for individual farmer in the group. This activity should be carried out especially during harvesting and should be counter-checked with the estimates during buying.

**Internal Approvals:** The IQS manager will have a defined procedure to approve or impose sanction on the farmers in the group. All internal farm checklist are screened by internal approval staff with special focus on the critical control points of risk / difficult cases. x The approval committee for providing internal certification status will check the assessment of the internal inspector. If necessary, conditions will be set out for achieving compliance with the NPOP. x The next competent person or committee must confirm results of the internal inspection in an approval procedure.

## Participatory certification

Organic producers around the world have been developing methods to guarantee the organic integrity of their products for fifty years. Today, what are generally referred to as Third Party Certification systems have become the dominant means of Organic Guarantee for world trade and Indian producers have accredited Third Party Organic Certification agencies to choose from. While Third Party Certification is an essential component to world trade, there are downsides to the system. The inherent expense and paperwork required in a multi level system discourages most small organic producers from being certified at all. This limits local and domestic trade as well as access to organic products. Worse yet, it limits the growth of the Organic Movement as a whole. In an attempt to reduce the inequality, a number of alternative methods to guarantee the Organic integrity of products have been developed for small domestic producers, and they are growing rapidly. These alternative programs are now collectively referred to as Participatory Guarantee Systems (PGS). The term PGS embodies the active participation of producers & other stakeholders in their organic guarantee process.

Participatory Guarantee System (PGS) is a quality assurance initiative that is locally relevant, emphasize the participation of stakeholders, including producers and consumers and operate outside the frame work of third party certification. As per IFOAM (2008) definition "Participatory Guarantee Systems are locally focused quality assurance systems. They certify producers based on active participation of stakeholders and are built on a foundation of trust, social networks and knowledge exchange".

- PGS is a process in which people in similar situations (in this case small holder producers) assess, inspect and verify the production practices of each other and take decision on organic certification (PGS-Green and PGSOrganic).
- PGS system has number of basic elements which embrace a participatory approach, a shared vision, transparency and trust. Participation is an essential and dynamic part of PGS.

- Key stakeholders (producers, consumers, retailers, traders and others such as NGOs, Societies/Gram panchayaths/State/Central Govt. organization/agencies/farmer etc.,) are engaged in the initial design, and then in the operation of the PGS.
- In the operation of a PGS, stakeholders (including producers) are involved in decision making and essential decisions about the operation of the PGS itself.
- In addition to being involved in the mechanics of the PGS, stakeholders, particularly the producers are engaged in a structured ongoing learning process, which helps them improve what they do.
- This process is facilitated by the PGS group itself or in some situations a supportive NGO/Societies, Gram panchayaths, State/Central Govt. organization/agencies etc. The learning process is usually "hands-on and involves field days or workshops.

The idea of participation embodies the principle of collective responsibility for ensuring the organic integrity of the PGS.

## Guiding principles for participatory guarantee system

Guiding Principles for Participatory Guarantee System in tune with the international trends and IFOAM s PGS Guidelines, PGS India system is also based on participatory approach, a shared vision, transparency and trust.

In addition it gives PGS movement a National recognition and institutional structure without affecting the spirit of PGS.

**Participation:** Participation is an essential and dynamic part of PGS.

**Key stakeholders:** Producers, consumers, retailers, traders and others such as NGOs, Societies, Gram panchayaths, State/Central Govt. organization/agencies etc.,) are engaged in the initial design, and then in the operation of the PGS and decision making.

The idea of participation embodies the principle of a collective responsibility for ensuring the organic integrity of the PGS.

## This collective responsibility is reflected through

- Shared ownership of the PGS
- Stakeholder engagement in the development process
- Understanding of how the system works and
- Direct communication between producers and consumers and other stakeholders, producers, consumers, retailers, traders and others such as

NGOs, Societies, Gram panchayaths, State/Central Govt. organization/ agencies etc.

Together these help to shape the integrity based approach and a formula for trust. An important tool for promoting this trust is having operational processes that are transparent. This includes transparency in decision making, easy access to the data base and where possible farms are open to participation and visits of consumers.

Participation of traders/ retailers or consumers in decision making may not be possible under all situations, but their participation in any form will increase the credibility and trustworthiness of the group.

**Shared Vision:** Collective responsibility for implementation and decision making is driven by common shared vision. All the key stakeholders (producers, consumers, retailers, traders and others such as NGOs, Societies, Gram panchayaths, State/Central Govt. organization/agencies etc.) support the guiding principles and goals, PGS is striving to achieve. This can be achieved initially through their participation and support in the design and then by joining it. This may include commitment in writing through signing an application/ document that includes the vision. Each stakeholder organization (or PGS group) can adopt its own vision conforming to the overall vision and standards of PGS- India programme.

**Transparency:** Transparency is created by having all stakeholders, including producers and consumers, aware of exactly how the guarantee system works to include the standards, the organic guarantee process (norms) with clearly defined and documented systems and how decisions are made. Public access will be ensured to documentation and information about the PGS groups, such as lists of certified producers and details about their farms and non-compliance actions. These will be available through a dedicated National database websites (PGS-INDIA webportal). But still it does not mean that entire information on National PGS database will be available to everyone.

At the grass roots level transparency is maintained through the active participation of the producers in the organic guarantee process which can include

- Information sharing at meetings and workshops
- Participation in internal inspections (peer reviews)
- Involvement in decision making.

**Trust:** The integrity base upon which PGS are built is rooted in the idea that producers can be trusted and that the organic guarantee system can be an expression and verification of this trust. The foundation of this trust is built from the idea that the key stakeholders collectively develop their shared vision and then collectively continue to shape and reinforce their vision through the PGS. The ways this trust is reflected may depend entirely on factors that are culturally/ socially specific to the PGS group. The idea of"trust assumes that the individual producer has a commitment to protecting nature, maintaining biodiversity, maintenance of soil health, and consumers health through organic production which are free from chemical residues.

## Mechanism for expressing trust worthiness includes

- Declaration (a producer pledge) via a witnessed signing of a pledge document
- Written collective undertaking by the group to abide by the norms, principles and standards of PGS

**Horizontality PGS:** India is intended to be non-hierarchical at group level. This will reflect in the overall democratic structure and through the collective responsibility of the PGS group with sharing and rotating responsibility, by engaging producers directly in the peer review of each other s farms; and by transparency in decision making process.

**National networking:** PGS India while keeping the spirit of PGS intact also aims to give the entire movement an institutional structure. This is proposed to be achieved by networking the groups under common umbrella through various facilitating agencies, Regional Councils and Zonal Councils. To make the system completely transparent and accessible to traders and consumers entire data will be hosted on a common platform in the form of a website (PGS-INDIA). National Centre of Organic Farming shall be the custodian of data, define policies and guidelines and undertake surveillance through field monitoring and product testing for residues. Regional councils and facilitating agencies will facilitate the local groups in capacity building, training, knowledge/ technology dissemination and data uploading on the PGS website. But at every stage it will be ensured that these agencies including apex body do not interfere in the working and decision making of the group. Even if surveillance is done and reports are made, the same will also be put on website in public domain. What action is to be taken on adverse reports will be left to the group and Regional Council.

## Advantages of PGS over third party certification system

In PGS organic farmers have full control over the certification process and are able to produce far more credible and effective system of quality assurance compared to third party certification. Important benefits of this system over third party certification system are as follows:

a) The procedures are simple; documents are basic and use the local language understandable to farmers.

b) All the members are local and known to each other. Being themselves practicing organic farmers have high degree of understanding on day-to-day knowledge or acquaintance of the farm.

c) Peer appraisers are among the group and live in the same village, therefore have better access to surveillance

d) Peer appraisal instead of third party inspections reduces cost

e) Mutual recognition and support between Regional PGS groups ensures better networking for processing and marketing.

f) Empowers farmers with increased capacity building

g) Bring consumers to the farm without the need of middleman

h) Unlike grower group certification system, PGS offer every farmer with individual certificate and each farmer is free to market its own produce independent of group.

i) Consumers and buyers are often involved in production and verification process

j) Random residue testing at regular intervals ensures the integrity and increases the trust.

## Limitations of PGS

i) PGS certification is only for farmers or communities that can organize and perform as a group within the village or in close-by villages with continuous territory and is applicable on, on-farm activities comprising of crop production, processing and livestock rearing (including bee keeping) and off-farm processing "by PGS farmers of their direct products".

ii) Individual farmers or group of farmers having less than 5 members are not covered under PGS. They either have to opt for third party certification or join the existing PGS local group.

iii) PGS is applicable on on-farm activities comprising of crop production, processing and livestock rearing and off-farm processing "by PGS farmers of their direct products".

iv) Off-farm processing activities such as, storage, transport and value addition activities by persons/agencies other then PGS farmers away from the group are not covered under PGS.

v) Off-farm input approval granted by the group is applicable on the members of the same group and cannot be taken as a basis for universal approval for other groups.

vi) Off-farm inputs need to be approved by each group for their member s use on case to case basis.

vii) PGS ensures traceability only up to end till it is in the custody of PGS group. Once the product leaves the custody of PGS group there is no control of PGS on its integrity.

Therefore PGS is ideal for local direct sales/ direct trade between producer and consumer and direct trade of packed finished product with PGS logo between PGS group and traders/ retailers. But Local Groups and buyers in consultation with RC can devise some mechanism with full traceability records to allow use of PGS logo on products packed by traders/ retailers.

## Operational structure

Schematic operational structure of the PGS India is given below:

- Local groups (Farmers group) (LGs)
- Regional Councils (RCs)
- Zonal Councils (RCOFs)
- National Centre of Organic Farming (NCOF) (PGS-INDIA Secretariat)
- National Advisory Committee (NAC)

## Alternative certification options

The word organic is central to the certification (and organic food marketing) process, and this is also questioned by some. Where organic laws exist, producers cannot use the term legally without certification. To bypass this legal requirement for certification, various alternative certification approaches, using currently undefined terms like "authentic" and "natural", are emerging. In the US, motivated by the cost and legal requirements of certification (as of Oct. 2002), the private farmer-to-farmer association, Certified Naturally Grown, offers a "non-profit alternative eco-labelling program for small farms that grow using USDA Organic methods but are not a part of the USDA Certified Organic program.

In the UK, the interests of smaller-scale growers who use "natural" growing methods are represented by the Wholesome Food Association, which issues a symbol based largely on trust and peer-to-peer inspection.

## Organic certification and the Millennium Development Goals (MDGs)

Organic certification, as well as fair trade certification, has the potential to directly and indirectly contribute to the achievement of some of the Millennium

Development Goals (MDGs), which are the eight international development goals that were established following the Millennium Summit of the United Nations in 2000, with all United Nations member states committed to help achieve the MDGs by 2015. With the growth of ethical consumerism in developed countries, imports of eco-friendly and socially certified produce from the poor in developing countries have increased, which could contribute towards the achievement of the MDGs. A study by Setboonsarng (2008) reveals that organic certification substantially contributes to MDG1 (poverty and hunger) and MDG7 (environmental sustainability) by way of premium prices and better market access, among others. This study concludes that for this market-based development scheme to broaden its poverty impacts, public sector support in harmonizing standards, building up the capacity of certifiers, developing infrastructure development, and innovating alternative certification systems will be required.

### International food standards

The body Codex Alimentarius of the Food and Agriculture Organization of the United Nations was established in November 1961. The Commission's main goals are to protect the health of consumers and ensure fair practices in the international food trade. The Codex Alimentarius is recognized by the World Trade Organization as an international reference point for the resolution of disputes concerning food safety and consumer protection. One of their goals is to provide proper food labelling (general standard, guidelines on nutrition labelling, guidelines on labelling claims).

### Private organic certification

Besides the public organic certification regulation EU-Eco-regulation in 1992, there are various private organic certifications available:

- Demeter International is the largest certification organization for biodynamic agriculture, and is one of three predominant organic certifiers. Demeter Biodynamic Certification is used in over 50 countries to verify that biodynamic products meet international standards in production and processing. The Demeter certification program was established in 1928, and as such was the first ecological label for organically produced foods.
- Bio Suisse established in 1981 is the Swiss organic farmer umbrella organization. International activities are mainly focused on imports towards Switzerland and don't support export activities.
- Global Organic Textile Standard (GOTS) is a private standard for Organic clothing for the entire post-harvest processing (including spinning, knitting, weaving, dyeing and manufacturing) of apparel and home textiles made

with organic fibres (such as organic cotton, organic wool etc.). It includes both environmental and social criteria. Established in 2002, the standard is used in over 68 countries and is endorsed by USDA and IFOAM - Organics International. The material must be at least 95% organic, as certified by "recognized international or national standards." If the material is 70% organic, it can be labeled as "made with organic."

## Czech Republic

Following private bodies certify organic produce: KEZ, o. p. s. (CZ-BIO-001), ABCert, AG (CZ-BIO-002) and BIOCONT CZ, s. r. o. (CZ-BIO-003). These bodies provide controlling of processes tied with issueing of certificate of origin. Controlling of compliancy (to (ES) no 882/2004 directive) is provided by government body ÚKZÚZ (Central Institute for Supervising and Testing in Agriculture). *Source: "Information on organic produce of the Ministry of Agriculture of the Czech Republic"*

## France

In France, organic certification was introduced in 1985. It has established a green-white logo of "AB - agriculture biologique." The certification for the AB label fulfills the EU regulations for organic food. The certification process is overseen a public institute ("Agence française pour le développement et la promotion de l'agriculture biologique" usually shortended to "Agence bio") established in November 2001. The actual certification authorities include a number of different institutes like Aclave, Agrocert, COSMEBIO, Ecocert SA, Qualité France SA, Ulase, SGS ICS.

## Germany

In Germany the national label was introduced in September 2001 following in the footsteps of the political campaign of "Agrarwende" (*agricultural major shift*) led by minister Renate Künast of the Greens party. This campaign was started after the outbreak of mad-cow disease in 2000. The effects on farming are still challenged by other political parties. The national "Bio"-label in its hexagon green-black-white shape has gained wide popularity - in 2007 there were 2431 companies having certified 41708 products. The popularity of the label is extending to neighbouring countries like Austria, Switzerland and France.

In the German-speaking countries there have been older non-government organizations that had issued labels for organic food long before the advent of the EU organic food regulations. Their labels are still used widely as they significantly exceed the requirements of the EU regulations. An organic food label like "demeter" from Demeter International has been in use since 1928 and this label is still regarded as providing the highest standards for organic food in the world.

### Greece

In Greece, organic certification is available from eight (8) organizations approved by EU. Most of them are BIOHELLAS and the DIO (Greek: Ïñãáíéóìüò ÅëÝã÷ïõ êáé Ðéóôïðïßçóçò Âéïëïãéêþí Ðñïúüíôùí - ÄÇÙ)

### Ireland

In Ireland, organic certification is available from the Irish Organic Farmers and Growers Association, Demeter Standards Ltd. and Organic Trust Ltd.

### Switzerland

In Switzerland, products sold as organic must comply at a minimum with the Swiss organic regulation (Regulation 910.18). Higher standards are required before a product can be labelled with the Bio Suisse label.

### Sweden

In Sweden, organic certification is handled by the organisation KRAV (agriculture) with members such as farmers, processors, trade and also consumer, environmental and animal welfare interests.

### United Kingdom

In the United Kingdom, organic certification is handled by a number of organizations, regulated by The Department for Environment, Food and Rural Affairs (DEFRA), of which the largest are the Soil Association and Organic Farmers and Growers. While UK certification bodies are required to meet the EU minimum organic standards for all member states; they may choose to certify to standards that exceed the minimums, as is the case with the Soil Association.

The farmland converted to produce certified organic food has seen a significant evolution in the EU15 countries, rising from 1.8% in 1998 to 4.1% in 2005. For the current EU25 countries however the statistics report an overall percentage of just 1.5% as of 2005. However the statistics showed a larger turnover of organic food in some countries, reaching 10% in France and 14% in Germany. In France 21% of available vegetables, fruits, milk and eggs were certified as organic. Numbers for 2010 show that 5.4% of German farmland has been converted to produce certified organic food, as has 10.4% of Swiss farmland

and 11.7% of Austrian farmland. Non-EU countries have widely adopted the European certification regulations for organic food, to increase export to EU countries.

## Asia and Oceania

### Australia

In Australia, organic certification is performed by several organisations that are accredited by the Biosecurity section of the Department of Agriculture (Australia), formerly the Australian Quarantine and Inspection Service, under the National Standard for Organic and Biodynamic Produce. All claims about the organic status of products sold in Australia are covered under the Competition and Consumer Act 2010.

In Australia, the Organic Federation of Australia is the peak body for the organic industry in Australia and is part of the government's Organic Consultative Committee Legislative Working Group that sets organic standards.

Department of Agriculture accreditation is a legal requirement for all organic products exported from Australia. Export Control (Organic Produce Certification) Orders are used by the Department to assess organic certifying bodies and recognise them as approved certifying organisations. Approved certifying organisations are assessed by the Department for both initial recognition and on an at least annual basis thereafter to verify compliance.

In the absence of domestic regulation, DOA accreditation also serves as a 'de facto' benchmark for certified product sold on the domestic market. Despite its size and growing share of the economy "the organic industry in Australia remains largely self-governed. There is no specific legislation for domestic organic food standardisation and labelling at the state or federal level as there is in the USA and the EU".

### Australian approved certifying organisations

The Department has several approved certifying organisations that manage the certification process of organic and bio-dynamic operators in Australia. These certifying organisations perform a number of functions on the Department's behalf:

- Assess organic and bio-dynamic operators to determine compliance to the National Standard for Organic and Bio-Dynamic Produce and importing country requirements.
- Issue a Quality Management Certificate (QM Certificate) to organic operators to recognise compliance to export requirements.

- Issue Organic Produce Certificates (Export Documentation) for consignments of organic and bio-dynamic produce being exported.

**As of 2015, there are seven approved certifying organisations**

- AUS-QUAL Pty Ltd (AUSQUAL)
- Australian Certified Organic (ACO)
- Bio-Dynamic Research Institute (BDRI)
- NASAA Certified Organic (NCO)
- Organic Food Chain (OFC)
- Safe Food Production Queensland (SFQ)
- Tasmanian Organic-dynamic Producers (TOP)

There are 2567 certified organic businesses reported in Australia in 2014. They include 1707 primary producers, 719 processors and manufacturers, 141 wholesalers and retailers plus other operators.

Australia does not have a logo or seal to identify which products are certified organic, instead the logos of the individual certifying organisations are used.

**China**

In China, the organic certification is administered by a government agency named Certification and Accreditation Administration of the People's Republic of China (CNCA). While the implementation of certification works, including site checking, lab test on soil, water, product qualities are performed by the China Quality Certification Center (CQC) which is an agency of Administration of Quality Supervision, Inspection and Quarantine (AQSIQ).The organic certification procedures in china are performed according to China Organic Standard GB/T 19630.1-4-2011 which was issued in year 2011. This standard has governed standard procedure for Organic certification process performed by CQC, including application, inspection, lab test procedures, certification decision and post certification administration. The certificate issued by CQC are valid for 1 year.

There are 2 logos that are currently used by the CQC for labeling products with Organic Certification, these are the Organic Logo and CQC Logo. No conversion to organic Logo now.

There were more than 19000 valid certificates and 66 organic certification bodies until 2018 in China.

## India

In India, APEDA regulates the certification of organic products as per National Standards for Organic Production. "The NPOP standards for production and accreditation system have been recognized by European Commission and Switzerland as equivalent to their country standards. Similarly, USDA has recognized NPOP conformity assessment procedures of accreditation as equivalent to that of US. With these recognitions, Indian organic products duly certified by the accredited certification bodies of India are accepted by the importing countries." Organic food products manufactured and exported from India are marked with the India Organic certification mark issued by the APEDA. APEDA has recognized 11 inspection certification bodies, some of which are branches of foreign certification bodies, others are local certification bodies.

## Japan

In Japan, the Japanese Agricultural Standard (JAS) was fully implemented as law in April 2001. This was revised in November 2005 and all JAS certifiers were required to be re-accredited by the Ministry of Agriculture.

## Singapore

As of 2014 the Agri-Food & Veterinary Authority of Singapore had no organic certification process, but instead relied on international certification bodies; it does not track local producers who claim to have gotten organic certification.

## Cambodia

In Cambodia, Cambodian Organic Agriculture Association (COrAA) is the only organization that is authorized to give certificate for organic agricultural products. It is a nationwide private organization working for the promotion of organic and sustainable agriculture in Cambodia. COrAA has developed both organic and chemical-free agricultural standards and provides third-party-certification to producers following these standards. In addition, the services that COrAA provides include technical training for the conversion from chemical/conventional to organic farming, marketing support, organic awareness building among the general public, and a platform for dialogue and cooperation among organic stakeholders in Cambodia.

## Issues

Organic certification is not without its critics. Some of the staunchest opponents of chemical-based farming and factory farming practices also oppose formal certification. They see it as a way to drive independent organic farmers out of business, and to undermine the quality of organic food. Other organizations such as the Organic Trade Association work within the organic community to foster awareness of legislative and other related issues, and enable the influence and participation of organic proponents.

### Obstacles to small independent producers

Originally, in the 1960s through the 1980s, the organic food industry was composed of mainly small, independent farmers, selling locally. Organic "certification" was a matter of trust, based on a direct relationship between farmer and consumer. Critics view regulatory certification as a potential barrier to entry for small producers, by burdening them with costly and cumbersome process.

### Manipulative use of regulations

Manipulation of certification regulations as a way to mislead or outright dupe the public is a very real concern. Some examples are creating exceptions (allowing non-organic inputs to be used without loss of certification status) and creative interpretation of standards.

In 2013 the Australian Competition and Consumer Commission said that water can no longer be labelled as organic water because, based on organic standards, water cannot be organic and it is misleading and deceptive to label any water as such.

The label itself can be used to mislead many customers that food labelled as being organic is safer, healthier and more nutritious.

### Organic certification mark

**Organic Logo :** A trademark-"India Organic" will be granted on the basis of compliance with the National Standards for Organic Production (NSOP). Communicating the genuineness as well as the origin of the product, this trademark will be owned by the Government of India. Only such exporters, manufacturers and processors whose products are duly certified by the accredited Certification Bodies, will be granted the licence to use of the logo which would be governed by a set of regulations.

**Concept of Organic Logo:** Symbolizing the rhythm of cosmic and earth forces represented by the blue and brown waves of force and energy, 'India Organic' logo celebrates the essence of nature. These forces work in harmony upon the earth's environment and this rhythm is reinforced and supported by the 184 green plant growth. The colours used have a special significance in the logo concept. The cosmic force in blue symbolizes universal purity. Richness of soil, nourished with natural ingredients in organic farming, is symbolized by the earth forces in golden brown. The plant in green uses the colour of nature and natural products untouched by chemicals. The blue background is symbolic of earth's environment that is congenial for life to thrive in and is also free of pollution and harmful chemicals. India Organic etched over the surface authenticates the carrier as "Organic" and also establishes the Indian connection for all the carriers of the mark. Beautifully synthesizing all the elements of our environment, the logo also communicates total adherence to the National Organic Standards.

## General information on organic standards and certification

Organic standards are sets of requirements that describe what practices can be considered organic. Typically, organic standards address various aspects of organic production, namely:

- General farm production requirements and conversion periods;
- Crop production requirements and requirements for the collection of wild products;
- Animal production requirements (including bee-keeping);
- Processing and handling requirements;
- Social justice requirements, and
- Labeling requirements.

Organic standards that are endorsed by the international organic movement are listed in the IFOAM Family of Standards. A list of existing organic government regulations and their implementation status is published annually in The World of Organic Agriculture.

The German Federal Ministry of Food, Agriculture and Consumer Protection collects and publishes full texts of various international organic standards on the their website.

Information on organic standards with a focus on Europe, as well as analyses of differences between various organic standards in relation to EU organic regulation, can be found in the Organic Rules and Certification database.

## Constraints of organic certification

- Very high fees for certification by certifying agencies.
- Lack of awareness of group certification among small & medium farmers.
- Small and medium-sized Indian organic producers are not able to access international markets because of no-certified produce.
- Need of certification in domestic market.

- Under current government policy and as per the standards, it takes approximately three years for a farm in conversion to be certified as organic, and costs are hefty for the small farmer.
- While internationally, our exports cannot hold ground without certification, uncertified brands in India have had some success. Still the domestic market has not developed for the certified organic produce and demand for safe, pure and healthy food is slowly increasing.
- In developed countries, especially in Europe and the US, every supermarket has a green line where a complete range of certified organic products is available. India is just beginning to manage to hold in some supermarkets.

## Conclusion

Organic certification is an important component of organic agriculture. Different standards and agencies are available in the world for facilitating organic agriculture. Therefore a sound knowledge of regulating mechanism of organic certification in the different countries is required. At national level, a well-defined regulatory mechanism is available to understand about procedure of organic certification. However, with needs of consumer & producers, procedures of certification are also changing and hence it needs regular updation of knowledge about certification agencies & regulatory mechanism.

# 10

# Food Processing and Handling

Any handling and processing of organic products should be optimised to maintain the quality and integrity of the product and directed towards minimising the development of pests and diseases.

**Specific requirements**

i) Processing and handling of organic products should be done separately in time or place from handling and processing of non organic products.

ii) Pollution sources shall be identified and contamination avoided.

iii) Flavouring extracts shall be obtained from food (preferably organic) by means of physical processes.

iv) Organic products shall be protected from co-mingling with non-organic products.

v) All products shall be adequately identified through the whole process.

vi) The certification programme shall set standards to prevent and control pollutants and contaminants.

vii) Organic and non-organic products shall not be stored and transported together except when labelled or physically separated.

viii) Certification programme shall regulate the means and measures to be allowed or recommended for decontamination, cleaning or disinfection of all facilities where organic products are kept, handled, processed or stored.

**Besides storage at ambient temperature, the following special conditions of storage are permitted:**

- Controlled atmosphere
- Cooling
- Freezing
- Drying
- Humidity regulation
- Ethylene gas is permitted for ripening

### Pest and disease control

i) Pests should be avoided by good manufacturing practices. This includes general cleanliness and hygiene.

ii) Treatments with pest regulating agents must thus be regarded as the last resort. Recommended treatments are physical barriers, sound, ultra-sound, light and UV-light, traps (incl. pheromone traps and static bait traps), temperature control, controlled atmosphere and diatomaceous earth.

iii) For pest management and control the following measures shall be used in order of priority:

- Preventive methods such as disruption, elimination of habitat and access to facilities
- Mechanical, physical and biological methods
- Pesticidal substances contained in the Appendices of the national standards
- Other substances used in traps
- Irradiation is prohibited.

iv) There shall never be direct or indirect contact between organic products and prohibited substances. (e.g. pesticides). In case of doubt, it shall be ensured that no residues are present in the organic product.

v) Persistent or carcinogenic pesticides and disinfectants are not permitted.

vi) The certification programme shall set up rules to determine which protection agents and disinfectants may be used.

### Ingredients, additives and processing aids

i) All the ingredients of agriculture origin shall be certified organic.

ii) For the production of enzymes and other micro-biological products the medium shall be composed of organic ingredients.

### The certification programme should take into consideration

i) The maintenance of nutritional value

ii) The existence or possibility of producing similar products.

### Processing methods

i) Processing methods should be based on mechanised, physical and biological processes.

ii) The vital quality of an organic ingredient shall be maintained throughout each step of its processing.

iii) Processing methods shall be chosen to limit the number and quantity of additives and processing aids.

**The following kinds of processes are approved**

- Mechanical and physical
- Biological
- Smoking
- Extraction
- Precipitation
- Filtration
- Extraction shall only take place with water, ethanol, plant and animal oils, vinegar, carbon dioxide, nitrogen or carboxylic acids. These shall be of food grade quality, appropriate for the purpose.
- Irradiation is not allowed.
- Filtration substances shall not be made of asbestos nor may they be permeated with substances which may negatively affect the product.

**Packaging**

i) Ecologically sound materials should be used for the packaging of organic products. Packaging materials that affect the organic nature of the contents should be avoided.

ii) Use of PVC materials is prohibited.

iii) Laminates and aluminum should be avoided.

iv) Recycling and reusable systems shall be used wherever possible.

v) Biodegradable packaging materials shall be used.

The materials used must not affect the organoleptic character of the product or transmit to it any substances in quantities that may be harmful to human health.

**Labelling**

i) Labelling shall convey clear and accurate information on the organic status of the product.

ii) When the full standards requirements are fulfilled, products shall be sold as "produce of organic agriculture" or a similar description.

- Where a minimum of 95% of the ingredients are of certified organic origin, products may be labelled "certified organic" or similar and should carry the logo of the certification programme.
- Where less than 95% but not less than 70% of the ingredients are of certified organic origin, products may not be called "organic". The word "organic" may be used on the principal display in statements like "made with organic ingredients" provided there is a clear statement of the proportion of the organic ingredients. An indication that the product is covered by the certification programme may be used, close to the indication of proportion of organic ingredients.
- Where less than 70% of the ingredients are of certified organic origin, the indication that an ingredient is organic may appear in the ingredients list. Such product may not be called "organic".

  Added water and salt shall not be included in the percentage calculations of organic ingredients.

**Storage & transport**

i) Product integrity should be maintained during storage and transportation of organic products.

ii) Organic Products must be protected at all times from co-mingling with non-organic products.

iii) Organic products must be protected at all times from contact with materials and substances not permitted for use in organic farming and handling.

iv) Where only part of the unit is certified and other products are non-organic, the organic products should be stored and handled separately to maintain their identity.

v) Bulk stores for organic product should be separate from conventional product stores and clearly labeled to that effect.

vi) Storage areas and transport containers for organic product should be cleaned using methods and materials permitted in organic production.

# 11

# Organic Farming Practices Postharvest Handling

Optimal quality organic produce that combines the desired textural properties, sensory shelflife, and nutritional content results from the careful implementation of recommended production inputs and practices, careful handling at harvest, and appropriate postharvest handling and storage. This section will focus on an overview of general postharvest handling considerations unique to the marketing of registered or certified organic produce. An introduction to current permitted and restricted postharvest treatments will be presented.

**Planning for Post harvest Quality**: Achieving an economically rewarding enterprise via the marketing of organic produce must begin well before harvest.

- Seed selection can be a critical decision factor in determining the postharvest performance of any commodity. Individual cultivars have variable inherent potential for firmness retention, uniformity, disease and pest resistance, and sensory shelf–life, to list a few key traits. Cultivars chosen for novelty or heirloom traits may be suitable for small-scale production and local marketing but would be disastrous if shipment to more distant markets was attempted. In addition to genetic traits, environmental factors such as soil type, temperature, wind during fruit set, frost, and rainy weather at harvest can have an adverse effect on storage life, suitability for shipping, and quality.
- Cultural practices may have dramatic impacts on postharvest quality. For example, poor seedbed preparation for carrots may result in sunburned shoulders and green core with many of the specialty carrots favored by consumers at farmers markets.
- Planning for postharvest food safety should be included in any crop management plan. Good Agricultural Practices (GAP) need to be developed and formalized for each crop and specific production field to minimize the risk of chemical (ex. heavy metals carryover), physical (ex. sand and soil, wood, plastic or metal shards), and biological (ex. Salmonella, Listeria, mycotoxins) hazards and contaminants.
- Prior land use, adjacent land use, water source and method of application, fertilizer choice (such as the use of manure), compost management,

equipment maintenance, field sanitation, movement of workers between different operations, personal hygiene, domestic animal and wildlife activities, and other factors have the potential to adversely impact food safety.

- It is worth noting that many elements of a GAP plan are likely to be incorporated into the existing organic crop management program and activities.

Programs in place to ensure produce quality may be directly applicable to food safety with minor modifications. Applying food safety programs, in turn, have been shown to directly benefit postharvest quality. Once pre-requisite production programs are in place, a systematic evaluation and implementation plan of Good Agricultural Practices during harvest operations and any subsequent postharvest handling, minimal or fresh-cut processing, and distribution to consumers must be developed.

## Harvest handling

The inherent quality of produce cannot be improved after harvest, only maintained for the expected window of time characteristic of the commodity. Part of successful postharvest handling is knowing what this window of opportunity is under your specific conditions of production, season, method of handling, and distance to market.

Among the benefits of organic production, it is often more common to harvest and market near or at peak ripeness than in many conventional systems. However, organic production often includes more specialty varieties that have reduced or even inherently poor shelf life and shipping traits.

## As a general approach, the following practices can help to maintain quality

1. Harvest during the coolest time of the day to maintain low product respiration.
2. Avoid unnecessary wounding, bruising, crushing, or damage from humans, equipment, or harvest containers.
3. Shade harvested product in the field to keep it cool. Covering harvest bins or totes with a reflective pad greatly reduces heat gain from the sun and reduces water loss and premature senescence.
4. If possible, move product into a cold storage facility or postharvest cooling treatment as soon as possible. For some commodities, such as berries, tender greens and leafy herbs, one hour in the sun is too long.
5. Don't compromise high quality product by intermingling damaged, decayed, or decay-prone product in a bulk or packed unit.

6. Only use cleaned and, as necessary, sanitized packing or transport containers. These operating principles are important in all operations but carry special importance for many organic producers due to limited postharvest cooling opportunities.

## Postharvest storage

Temperature is the single most important tool to maintain postharvest quality. Other than field cured or durable products, removing field heat as rapidly as possible is highly desirable.

i) Harvesting cuts off a vegetable from its source of water. However, it is still alive and will lose water, and therefore turgor, due to respiration.
ii) Field heat can accelerate the rate of respiration and therefore the rate of quality loss.
iii) Proper cooling protects quality and extends both the sensory (taste) and nutritional shelf life of produce.
iv) The capacity to cool and store produce creates greater market flexibility.
v) It is often critical to reach the desired short-term storage or shipping pulp temperature rapidly to maintain the highest visual quality, flavor, texture, and nutritional content of fresh produce.

## The most common cooling methods are

1. **Room cooling**: An insulated room or mobile container equipped with refrigeration units. Room cooling is slow compared with other options. Depending on the commodity, packing unit, and stacking arrangement the product may cool too slowly to prevent water loss, premature ripening, or decay.
2. **Forced-air cooling**: Fans are used in conjunction with a cooling room to pull cool air through packages of produce. Although the cooling rate depends on the air temperature and the rate of airflow, this method is usually 75–90% faster than room cooling.
3. **Hydrocooling**: Showering produce with chilled water is an efficient way to remove heat, and can serve as a means of cleaning at the same time. Use of a disinfectant in the water is essential and the some of the currently permitted products are discussed in the following section. Hydrocooling is not appropriate for all produce. Waterproof containers or resistant waxed-corrugated cartons are required. Currently waxed corrugated cartons have limited recycling or secondary use outlets and reusable, collapsible plastic containers are gaining popularity. Ensuring that the ice is free of chemical, physical, and biological hazards is essential.

4. **Vacuum cooling**: Under vacuum, water within the plant evaporates and removes heat from the tissues. This system works well for leafy crops, such as lettuce, spinach, and celery, which have a high surface-to-volume ratio. Water may be sprayed on the produce prior to placing it vacuum. As with hydrocooling, proper water disinfection is essential. The cost of the vacuum chamber system restricts its use to larger operations.

### Sanitation and water disinfection

Preventive food safety programs, proper sanitation of equipment and food contact surfaces, and water disinfection should be integrated into every facet of postharvest handling.

i) This responsibility applies to the organic producer, handler, processor or retailer. Organic producers, packers, and handlers are required to keep accurate and specific records of postharvest wash or rinse treatments, identified by brand name and source.

ii) The proper use of a disinfectant in postharvest wash and cooling water can help prevent both postharvest diseases and food-borne illnesses. Since most municipal water supplies are chlorinated and in recognition of the vital role of water disinfection, organic growers, shippers, and processors may use chlorine within specified limits.

iii) All forms of chlorine (ex. liquid sodium hypochlorite, granular calcium hypochlorite, and chlorine dioxide) are restricted materials as defined by existing organic standards.

### Cleaners, Sanitizers and disinfectants

A partial list of allowed cleaners, disinfectants, sanitizers, and postharvest aides follows:

**Acetic acid:** Allowed as a cleanser or sanitizer. Vinegar used as an ingredient must be from an organic source.

**Alcohol, Ethyl**: Allowed as a disinfectant. Alcohol must be from an organic source. Alcohol, Isopropyl may be used as a disinfectant under restricted conditions.

**Ammonium sanitizers:** Quaternary ammonium salts are a general example in this category. Quaternary ammonium may be used on non-food contact surfaces. Its use is prohibited on food contact surfaces, except for specific equipment where alternative sanitizers significantly increase equipment corrosion.

**Bleach.:** Calcium hypochlorite, sodium hypochlorite and chlorine dioxide are allowed as a sanitizer for water and food contact surfaces.

In California, product (fresh produce) wash water treated with chlorine compounds as a disinfectant cannot exceed 10 ppm residual chlorine measured downstream of product contact.

**Detergents:** Allowed as equipment cleaners. Also includes surfactants and wetting agents. All products must be evaluated on a case-by-case basis.

**Hydrogen peroxide:** Allowed as a water and surface disinfectant.

**Ozone:** Considered GRAS (Generally Regarded As Safe) for produce and equipment disinfection. Exposure limits for worker safety applies.

**Peroxyacetic acid:** Water and fruit and vegetable surface disinfectant.

**Other Postharvest Treatments**

**Carbon dioxide:** Permitted for post-harvest use in modified and controlled atmosphere storage & packaging. For crops that tolerate treatment with elevated $CO_2$ (e" 15%), suppression of decay and control of insect pests can be achieved.

**Fumigants:** Naturally occurring forms are allowed (ex. heat vaporized acetic acid). Must be from a natural source.

## Importance of optimal storage and shipping temperature

The stress rapid and adequate cooling as a primary method of postharvest handling, many vegetables are sub-tropical in origin and susceptible to chilling injury. Chilling injury occurs when sensitive crops are exposed to low temperatures that are above the freezing point. Damage is often induced in a very short time of exposure but not apparent for several days or until transfer to warmer display conditions. Some examples of sensitive crops are basil, tomato, eggplant, green beans, okra, and yellow crookneck squash.

For some vegetables different parts have distinct sensitivities, such as eggplant where the cap or calyx is more sensitive and turns black before the fruit itself is affected. The effects of chilling injury are cumulative in some crops. Chilling injury may not be apparent until produce is removed from low temperature storage. Depending on the duration and severity of chilling, after several hours to a few days of warmer temperatures, chilling symptoms become evident in the following ways

- Pitting and localized water loss
- Browning or other skin blemishes
- Internal discoloration
- Increased susceptibility to decay
- Failure to ripen or uneven color development
- Loss of flavor, especially characteristic volatiles
- Development of off-flavors

# 12

# Economic Considerations, Viability Export and Marketing of Organic Products

Organic agriculture systems are usually more profitable than conventional farming systems. This difference results from a combination of yield changes, input cost reductions, and price premiums.

There are solid bases of studies that suggest organic is equal to or more profitable than conventional farming. After a transitional period of 3 to 5 years, organic systems can produce up to 95 per cent of conventional yields. Additionally, organic farming is less dependent on fossil fuels, expensive inputs, and annual loans, making it less vulnerable to financial market fluctuations. Organic is a low-waste system that emphasizes quality over quantity, meaning it uses less land for the same profit.

## Profitability of conventional Vs organic farming

The studies that suggest organic is equal to or more profitable than conventional farming. The part of competitive edge in organic product comes from the premium price which is driven by consumer demand that organic farmers can get for their products even though when profits are adjusted for 50 per cent of the current organic premium.

During the initial years organic systems are incapable of reaching the same yields as conventional systems. After a transitional period of 3 to 5 years, organic systems can produce up to 95 per cent of conventional yields. Additionally, organic farming is less dependent on fossil fuels, expensive inputs, and annual loans, making it less vulnerable to financial market fluctuations as most of inputs are managed from own farm in organic farming. Organic is a low-waste system that emphasizes quality over quantity, meaning it uses less land for the same profit. Conventional crop subsidies exacerbate the problem, incentivizing farmers to grow more than they can sell, which causes excess pollution, overuse of natural resources, and food waste.

## Organic farming & natural capital

Organic farming can stand its ground under traditional economic evaluations, but when we start to incorporate natural capital values, organic becomes the clear winner for long-term profitability. For examples of how organic utilizes and increases various natural capital functions:

Adding organic matter to the soil each year (a foundational organic practice) increases the soil's ability to store carbon dioxide. Whereas the conventional farming has long been the recipient of crop subsidies.

- Organic matter in the soil also increases the soil's water holding capacity, reducing pressure on water resources and making organic farms more resilient to drought. means cheaper food production over time by curtailing cost on pumping underground water.
- Biodiversity on organic farms offers myriad financial benefits. A diverse crop system means a succession of blooms that can feed insect populations (and provide them with habitat) year-round. These beneficial insects help to keep down populations of harmful insects, eliminating the need for pesticides, and providing pollination services to increase harvest yields.
- Genetic diversity on organic vegetable and seed farms acts as a well-endowed gene bank for potential new varieties that will be resilient against future environmental changes, insect populations, and diseases-a service that is essential to global food security, not to mention tasty food!

## Economic considerations

- Labour requirements (quantity and timing of labour);
- Total net return, that is, income (or use) from main crop and other crops and livestock, minus the cost of the inputs used for the production;
- Long-term productivity: the effect of present production on the soil and implications for future yields;
- Marketing possibilities: in times when consumers are willing to pay a premium, improved marketing possibilities should be taken into account when production decisions are made.

From the results obtained, in one study, it is clear that the organic lemon orchard is the investment with higher profitability with respect to conventional one, because the entrepreneur can obtain a higher price in the market, guaranteed by certification. In the conventional lemon, despite a higher yield per hectare compared to the organic lemon (+21%), the sales price of the product on the market is lower, because in most cases the farms address the product to local market and to processing industry. Moreover, consumers have a greater willingness to pay for organic products with respect to conventional ones, driven by a perception that organic farming results in fewer negative environmental impacts and greater benefits for human health than conventional systems, recognizing them a premium price. Production cost value of an organic farm is lower than with respect to a conventional farm. Therefore, considering the results obtained from the longterm research experiments, the profitability of the organic farm is higher respect to conventional one due to premuium price of organic produce and reduction of input cost by using off-farm wastes.

In Europe, most farm comparisons show profits for organic farms between ± 20% of conventional (Nieberg and Oppermann 2003). Four factors usually account for the positive end of these income results.

**First**, operating costs for organic farms may be up to one-third lower, particularly for energy, chemicals, and drugs. Variable input costs are 50 to 60% lower for cereals and grain legumes, 10 to 20% lower for potatoes and horticultural crops, and 20 to 25% lower for dairy cows.

**Second**, where premium prices are available in organic markets, the likelihood of a superior net income situation is even greater.

**Third**, many organic farmers achieve higher net income by making more direct links with consumers, which allows them to capture a greater percentage of the consumer dollar.

**Fourth**, organic farms may be more resilient in the face of poor weather. For example, during 5 yr of droughty conditions in the Rodale Research Centre 22-yr cropping system trials, organic corn yields surpassed conventional ones by an average of 28 to 34% depending on the organic system. In years of "regular" conditions, organic yields have been comparable to that in conventional production (Pimentel et al. 2005).

## Economic viability

Organic agriculture is the best insurance policy that India can have for its population with better performance on productivity, environmental impact, economic viability and social well-being.

Organic Farming of today is not traditional agriculture. The principles governing organic farming are more scientific than even the principles followed in modern agriculture. Thus, turning organic on such a scientific scale requires a full fledged movement and not just piecemeal efforts. The public sector needs to play a public role in working with farmers to evolve farmers' varieties of the desired crops that India grows and needs so that our biodiversity, sustainable agriculture and diverse food systems can be kept alive.

Many farmers are apprehensive about adopting organic farming due to the high production cost and the three-year transition period when farmers have to wait before getting their farms certified. This issue was addressed in the US by food manufacturers offering financial incentives to offset the waiting period. Ardent Mills pays farmers more remuneration for crops grown on land undergoing transition and helps them choose rotational crops they can sell to supplement their income. Kashi has created a logo, "Certified Transitional", to label products made from farms that are undergoing the process of transition.

The courage shown by farmers to convert from conventional to organic is laudable. Though there is lower yield, these farms are more profitable and environmentally friendly, provide several ecosystem services, numerous social benefits and deliver nutritious foods with relatively less pesticide residues compared to conventional farming. Organically managed soils release less carbon dioxide per hectare per year than conventionally managed soils. New studies indicate that using the best management practices in organic systems over a long period of time can produce equal yields, or even outdo those of conventional systems. The Centre's announcement for allocation of ₹ 1 billion for organic market development and ₹ 3 billion for the participatory guarantee scheme is commendable. Indian farmers are using inputs manufactured from energy-intensive processes and, in some cases, from imported sources resulting in a burden on the exchequer. They could follow organic practices and use available bio-wastes to transit towards a circular economy. Consumers should consume responsibly and stakeholders should prevent wastage along the supply chain. Meanwhile, organic agriculture in India will continue to grow and play a larger part in safely feeding 1.5 billion Indian mouths in 2030.

By providing appropriate extension services, projects can support farmers identifying and implementing suitable management practices to achieve these goals. In order to be able to manage the farm in optimum way, farmers not only need to lean organic farming methods, but also to develop an understanding on the interrelations between different components of an agro- ecosystem. In the long term, an organic farm will only be viable if the farmer is aware of and producing what the market wants.

## Why organic farming has not caught up yet in India?

Development of organic agriculture as an alternative tool to address the ill-effects of chemical-based cultivation practices is a recent phenomenon in India. It had achieved dramatic progress in the beginning but could not maintain the pace. The growth of organic agriculture in India has been accomplished by three categories of farmers. The first category is from no input or low input use zones, practising it as a tradition or by default with no organic certification such as the tribes of north-east region. The second and third groups are certified and non-certified farmers, who have recently adopted organic farming realising the ill-effects of modern agriculture and benefits under organic cultivation.

Key problems faced by organic farmers during the transition phase are non-realisation of premium. A number of State Governments have already made significant strides in organic farming such as Sikkim, Mizoram and Uttrakhand to turn the States completely organic. Among all the states, Madhya Pradesh has covered largest area under organic certification followed by Himachal Pradesh and Rajasthan. However, to accomplish the desired dream, importance must be given to have a mechanism compensating farmers' sacrifice during initial year of land conversion. The emerging business opportunity in retailing of organic farm produce has drawn the attention of many private parties. This has led to establishment of direct link between farmers and retailers/exporters.

## International scenario

Demand for organic food products is growing rapidly across the globe and amounted to $64 billion in 2012.Commercial organic agriculture is now practised in more than 186 countries in an area of about 71.5 million hectare representing approximately 2.8 million farmers along with 549 certification bodies and 732 affiliates of International Federation of Organic Agriculture Movement (IFOAM) from 113 countries. The leading producers are Australia, European countries, Argentina and the US. Organic agricultural methods are internationally regulated and legally enforced by many nations, based on the standards set by the IFOAM established in 1972.

## Exports

During 2013-14, India exported 135 products, realisation from which was to the tune of $403 million including $183 million contributed by exports of organic textile. Major destinations for organic products from India are the US, EU, Canada, Switzerland, Australia, New Zealand, South-East Asian countries, West Asia, South Africa, etc.

Soybean (70 per cent) lead among the products exported followed by cereals and millets other than basmati (six per cent), processed food products (five per cent), basmati rice (four per cent), sugar (three per cent), tea (two per cent), pulses and lentils (one per cent), dry fruits (one per cent), spices (one per cent).

## Certification requirements

Total area under organic certification in India in 2013-14 is estimated to be 4.72 million ha with 15 per cent are certified and the rest under forest area. India has the highest number of organic producers in the world (5,97,873), mainly due to small holdings.

The country has internationally acclaimed certification process for export, import and domestic markets which is regulated by National Programme on Organic Production. There are at least 18 accredited certification agencies which are responsible for the certification process. Though Government initiatives such as National Project on Organic Farming, Horticulture Mission for North-East & Himalayan States, National Horticulture Mission, National Project on Management of Soil Health and Fertility, Rashtriya Krishi Vikas Yojana and also Network Project on Organic Farming of Indian Council of Agricultural Research aims at promoting organic agriculture in the country.

However, there is a wide gap in scientific validation and research compared to the progress in the same for general agriculture. Also, there is a need to aid farmers with advisory services (technical and managerial support to form cluster and adopt best management practices).

## Export of organic products

India holds a unique position among 186 countries practising organic agriculture: it has 6,50,000 organic producers, 699 processors, 669 exporters and 7,20,000 hectares under cultivation. India produced around 1.24 million MT of certified organic products which includes all varieties of food products namely Sugarcane, Cotton, Oil Seeds, Basmati rice, Pulses, Spices, Tea, Fruits, Dry fruits, Vegetables, Coffee & their value added products. The production is not limited to the edible sector but also produces organic cotton fiber, functional food products etc.

**Exports:** India plans to export organic food and non-food products three times by 2020. Vegetables exporters in India are expected to make the most of this opportunity. In 2015-16, some of India's top markets were the EU, the US, Canada, Korea and Australia. The Indian government has been undertaking measures to promote organic farming with the aim to improve soil fertility and help to double the farmers' incomes by the year 2022. Export of organic foods has risen by almost 25% between 2015-16 and 2016-17 from Rs 19.76 billion to Rs 24.78 billion at a time when overall farm exports grew by less than 1% from '1,074.31 billion to just Rs 1,084.26 billion, official data showed.

Overall Farm Exports and Organic Exports (in Rs. billion)

| Year | Total Exports | Organic Exports |
|---|---|---|
| 2014-15 | 1313.43 | 20.99 |
| 2015-16 | 1074.31 | 19.76 |
| 2016-17 | 1084.26 | 24.78 |

Oil seeds-Soybean (70%) lead among the products exported followed by Cereals and Millets other than Basmati (6%), Processed food products (5%), Basmati Rice (4%), Sugar (3%), Tea (2%), Pulses and Lentils (1%), Dry fruits (1%), Spices (1%) and others. The overall organic market can hit Rs 10,000 crore to Rs 12,000 crore by 2020 through export as against the current Rs 4000.

While the annual quantitative ceiling for export on lentils and pulses was increased from 10,000 tonnes to 50,000 tonnes those on organic wheat, edible oils, sugar and non-basmati rice have been exempted completely. Most of the organic food is exported to Canada, Europe and West Asia.

## Organic export obstacles

Despite many initiatives, a recent survey-based study covering 418 organic farmers across different states of India suggests that a move to organic farming methods may not be that easy and organic farmers are not getting the expected premium price for their produce. The study highlights five key issues faced by organic farmers that are affecting their livelihood and income.

- First, the supply chain is underdeveloped and small and mid-sized farmers located in hilly regions and tribal belts find it extremely difficult to access the market. There is a shortage of pack houses and refrigerated vehicles, which leads to spoilage.
- Second, while the government is subsidising farmers under the Participatory Guarantee System (PGS) for India, which is a self-certification process supported through the PKVY scheme, these farmers are not allowed to export. In fact, the APEDA has made it mandatory to have a third-party certification for exports.
- Third, as a farmer converts his/her land from conventional chemical-based farming to organic farming, there is a risk of loss in yield due to the withdrawal of chemical inputs and high-yielding varieties of seeds. A number of countries, such as the United Kingdom, have carefully designed subsidies to compensate for the yield loss during the conversion period. However, in India, there is no such subsidy.
- Fourth, there is a serious shortage of good quality organic inputs, which increases the risk of loss of yield.
- The fifth and the biggest challenge faced by organic farmers is the lack of an organic policy for the domestic market and imports. In the absence of regulation on labeling standard for organic production and logo, it is not possible to distinguish an organic product from a conventional product.

## Marketing

In terms of marketing strategy it is important to know the culture and some basic "habits" of the target consumers, for example where they do their shopping and at what times of the day and also how they get their information about food, health issues, etc. This knowledge is not only important in order to identify the main channels to provide information to (potential) customers but also to decide on where to sell the organic products, what to sell, at what time of the day to sell, etc. This in turn has consequences for the purchase and delivery of the organic produce from the producers.

## Methods to communicate their products and organisations to consumers

**Consumers meetings**: IIRD, for example, organises consumers meetings in association with women's groups and social clubs like the Lions, Rotarians, etc. to discuss the importance of organic agriculture, food security and health.

**Invitations letters:** to directly invite individuals or organisations to activities and special events.

**Printing and distribution of leaflets:** Promotional materials are printed in mass quantities and distributed, providing information about the organisation, products, and sometimes activities.

**Mass media:** Articles are published in the printed media such as newspapers and magazines, describing the benefits of organic agriculture.

**Advertisements:** Several organisations regularly advertise in local newspapers that have wide circulation. In addition, stills or moving ads may be broadcasted through local TV network.

**Direct marketing**: The bazaar, market or shop itself presents a forum for person-to-person marketing of organic products. This forum allows consumers to inquire about organic products or agriculture directly from the producer or staff of the marketing organisation.

**Word of mouth**: Publicity may also be generated through personal interactions among consumers. If consumers are satisfied with the organic bazaar, market or shop and their purchases, they may recommend it to their friends and colleagues.

**Field visits**: Many organisations organise regular meetings between consumers and organic producers. By sharing experiences and offering suggestions, these meetings have been found to be useful to both parties.

**Consumer newsletter**: Some organisations send regular newsletter to their members or regular customers which include information about upcoming events, articles about organic agriculture or health issues, recipes, etc.

**Display information and photographs**: Keystone, for example, produced a video on honey collection, which is shown in their shops to sensitise the customers about the people behind the product.

**Participation in trade and food fairs**: Organic producers could display and sell their products at these events, which are regularly organised in most countries at both the national and local level.

**Packaging and logo**: Packaging should be attractive and be in line with the organic identity of the products.

## Marketing constraints

There are many marketing constraints, which make the task for a small organic marketing initiative even more challenging, for example:

1. Lack of supportive policies from the government to encourage organic production.
2. Governments supporting domestic market development for semi-organic products such as "safe" or "IPM".
3. Lack of infrastructure/equipment to produce high quality products. High certification costs.
4. Lack of market information.
5. Producers are not organised and/or do not have a central handling facility.
6. No sufficient economy of scale to reach the market.

## Promotional efforts for marketing of organic produce

Several initiatives taken at government, corporate and individual level for promoting marketing of organic produce.

1. APEDA made a separate cell for organic import and export, regulating certification process and organizes meetings on OA issues under National Programme of Organic Production (NPOP). Export-import bank, NABARD,KVIC are also funding for OA projects. Big programmes e.g National Horticulture Mission, Rasrtriya Krishi Vikas Yojana of Ministry of agriculture provides huge funds to organic inputs and soil health. Almost in every part of country, National Accreditation Board for Testing and Calibration Laboratories (NABL), Gurgaon, gives license and monitor the laboratories working for food quality testing.
2. Corporate associations e.g. ASSOCHEM, FICCI etc. are frequently organize workshops, business meets on organic marketing. ICCOA, Bangalore annually organizes international organic trade fair "Biofach-India".

3. Big corporate houses and many exclusive corporate are entering in contract farming, value addition and export of organic produce. Even, at most of the supermarkets, malls organic products are displayed for sell.
4. The most important is the increasing number of small organic farmers groups, associations (e.g. Maharasthra organic farmers forum, Organic farming association of India, Goa etc. and NGOs e.g Navdhanya, DDS,CSA, Jatan, ICRA etc. are facilitating marketing networks to sell organic produce at farm on profitable price to direct consumers and that is making strong linkages between producer and consumers, the ultimate need of fair marketing. It also saves energy by following "grow seasonally-eat locally" and several other advantages.
5. Media is playing a great role for awareness about pesticide residues in food and organic produce as safe substitute.
6. Many states are coming forward to promote OA .Till date 11 states made their policy for OA. Sikkim has been declared first organic state in 2016 and some other made target of 2020 for fully organic.

### Organic Area in India: State-wise Profile

The area under certified organic farming in India recorded at half a million hectares in 2012-13 increased to nearly four times to 1.93 million hectares in 2018-19 registering a compound annual growth rate (CAGR) of 24.1 percent over the said period (Table 1).

**Table 1.** Area under organic farming in major states of India, 2012-13 to 2018-19

Area in hectares

| States | 2012-13 | 2013-14 | 2014-15 | 2015-16 | 2016-17 | 2017-18 | 2018-19 | CAGR, %*CDVI, %* |
|---|---|---|---|---|---|---|---|---|
| Madhya Pradesh | 144240(28.6) | 232887(32.2) | 397547(33.6) | 461775(31.0) | 464859(32.2) | 613396(34.3) | 674052(34.8) | 27.1****13.8* |
| Maharashtra | 66505(13.2) | 85537(11.8) | 134524(11.4) | 198352(13.3) | 224008(15.5) | 235691(13.2) | 250934(12.9) | 26.2****13.0* |
| Rajasthan | 38289(7.6) | 66020(9.1) | 107523(9.1) | 155020(10.4) | 151610(10.5) | 208571(11.7) | 223991(11.6) | 32.8****14.4* |
| Odisha | 18186(3.6) | 49814(6.9) | 81534(6.9) | 95897(6.4) | 92190(6.4) | 105616(5.9) | 95740(4.9) | 26.6***22.4* |
| Gujarat | 45276(9.0) | 46864(6.5) | 49354(4.2) | 76813(5.2) | 64241(4.5) | 81269(4.5) | 93841(4.8) | 13.5****9.9* |
| Karnataka | 27191(5.4) | 30716(4.2) | 52473(4.4) | 93963(6.3) | 81089(5.6) | 86946(4.9) | 83099(4.3) | 23.3***19.6* |
| Sikkim | 43108(8.5) | 60844(8.4) | 76392(6.5) | 75851(5.1) | 75218(5.2) | 76076(4.3) | 75799(3.9) | 7.8***11.1* |
| Uttar Pradesh | 32890(6.5) | 44670(6.2) | 53955(4.6) | 61082(4.1) | 56249(3.9) | 55198(3.1) | 62838(3.2) | 8.9***10.4* |
| Punjab | 1601(0.3) | 1534(0.2) | 2678(0.2) | 961(0.1) | 1033(0.1) | 1385(0.1) | 8909(0.5) | 15.3$^{NS}$*93.2* |
| Others | 87154(17.3) | 104153(14.4) | 227891(19.2) | 270222(18.1) | 233041(16.1) | 322346(18.0) | 369018(19.0) | 26.6****16.5* |
| All India | 504440 | 723039 | 1183871 | 1489936 | 1443538 | 1786494 | 1938221 | 24.1****12.8* |

*Figures in parenthesis indicate the percent of respective All India figure*, ** and *** indicate significance at 10 percent, 5 percent, and 1 percent respectively.Data Source: National Programme of Organic Production (NPOP).*

The state of Madhya Pradesh has always been positioned at number one in terms of the organic area. The acreage under organic farming in Madhya Pradesh increased from 144240 hectares in 2012-13 to as high as 674052 hectares accounting for 34.8 percent of the country's organically cultivated area in 2018-19. The organically managed area in Madhya Pradesh expanded with a CAGR of 27.1 percent which may be attributed to the fact that the state has an inherent advantage for organic farming as there are many natural grasslands and a large expanse of the area with very low input usage. Maharashtra stood in second place with an organic area of 66505 hectares (13.2 % of country acreage) in the year 2012-13 increasing to 250934 hectares (12.9 %) in 2018-19 with the CAGR of 26.2 percent over the study period. These two states with 47.7 percent of India's total cultivated organic area in 2018-19 were followed by Rajasthan with 11.6 percent, both Odisha and Gujarat close to five percent each. The area under organic cultivation in all the selected states except for Punjab increased, the highest being in the case of Rajasthan (32.8%). The rest of the Indian states accounted for a 19 percent share in the country's organic area in 2018-19, up from 17.3 percent in 2012-13. All the nine states together accounted for nearly 81 percent of the country's entire organic area, which is indicative of a greater degree of awareness about the advantages of organic farming in these states than that in the rest of the Indian states.

Observing the wide state-wise differentials in organic area, it was considered imperative to examine whether the share of different states in the country's organic area is commensurate with their respective shares in the cultivated area. The comparatively higher share in the organic area in comparison to that in the cultivated area in case of a particular state can well be deemed as an indicator of the state's inclination towards going organic. The driving force behind the switch over can either be to attain the sustainability of agriculture or to make the most of the market opportunities thrown open by the increasing consumer demand for organic products may it be on premium price. Table 2 is directed toward examining this issue.

The information presented therein revealed that Madhya Pradesh with one-ninth of the country's cultivated area accounted for the highest share in the country's organic area, which increased from 28.6 percent in 2012-13 to 34.8 percent by 2018-19. It is well documented that the policy initiatives to enhance organic oilseeds cultivation have brought about a turnaround in Madhya Pradesh's organic farming scenario (Mahajan *et al*, 2013). Even though the organic area in Maharashtra more than quadrupled, its share in the country's organic area dropped from 13.2 percent in 2012-13 to 12.9 percent by 2018-19, which was relatively higher than its respective share in the cultivated area (12.4 %). Considering the most recent period of 2018-19, apart from Madhya Pradesh and Maharashtra, other states with the share in the country's organic area being higher than that in the cultivated area are Odisha (4.9 vs. 3.0 %) and Sikkim (3.9 vs. 0.1 %).

**Table 2.** Share of major organic states in India's organic area and cultivated area

Percent@

| State | 2012-13 | | 2015-16 | | 2018-19 | |
|---|---|---|---|---|---|---|
| | Organic Area | Cultivated Area | Organic Area | Cultivated Area | Organic Area | Cultivated Area* |
| Madhya Pradesh | 28.6 | 11.0 | 31.0 | 10.9 | 34.8 | 10.9 |
| Maharashtra | 13.2 | 12.4 | 13.3 | 12.3 | 12.9 | 12.3 |
| Rajasthan | 7.6 | 12.5 | 10.4 | 12.9 | 11.6 | 12.9 |
| Odisha | 3.6 | 3.1 | 6.4 | 3.0 | 4.9 | 3.0 |
| Gujarat | 9.0 | 7.4 | 5.2 | 7.4 | 4.8 | 7.4 |
| Karnataka | 5.4 | 7.0 | 6.3 | 7.2 | 4.3 | 7.2 |
| Sikkim | 8.5 | 0.1 | 5.1 | 0.1 | 3.9 | 0.1 |
| Uttar Pradesh | 6.5 | 11.8 | 4.1 | 11.8 | 3.2 | 11.8 |
| Punjab | 0.3 | 3.0 | 0.1 | 3.0 | 0.5 | 3.0 |

*@percent of the respective figure for India*cultivated area pertains to 2015-16*

As discussed earlier, a point worth a mention here is that Punjab popularly known as grain bowl of India accounting for three percent of the country's GCA has just half a percent share in organic acreage. On the other hand, Sikkim, declared as the organic state of India on Jan 14, 2016 (Meek and Anderson, 2020), with 0.1 percent of India's cultivated area contributes close to four percent to area apportioned to India's organic area.

The importance given to organic farming in a particular state has been ascertained by the proportion of organic area in the respective cultivated area. The pertinent information on this aspect has been portrayed in Table 3.

**Table 3.** Organic area as percent of cultivated area in major organic states of India

| States | 2012-13 | 2013-14 | 2014-15 | 2015-16 | 2016-17* | 2017-18* | 2018-19* |
|---|---|---|---|---|---|---|---|
| Madhya Pradesh | 0.94 | 1.51 | 2.59 | 3.05 | 3.07 | 4.05 | 4.45 |
| Maharashtra | 0.38 | 0.49 | 0.78 | 1.15 | 1.30 | 1.37 | 1.46 |
| Rajasthan | 0.22 | 0.36 | 0.61 | 0.86 | 0.84 | 1.16 | 1.24 |
| Odisha | 0.41 | 1.11 | 1.82 | 2.28 | 2.20 | 2.52 | 2.28 |
| Gujarat | 0.44 | 0.45 | 0.48 | 0.75 | 0.62 | 0.79 | 0.91 |
| Karnataka | 0.28 | 0.31 | 0.52 | 0.94 | 0.81 | 0.87 | 0.83 |
| Sikkim | 55.98 | 79.02 | 99.21 | 98.51 | 97.69@ | 98.80@ | 98.44@ |
| Uttar Pradesh | 0.20 | 0.27 | 0.33 | 0.37 | 0.34 | 0.34 | 0.38 |
| Punjab | 0.04 | 0.04 | 0.07 | 0.02 | 0.02 | 0.03 | 0.22 |
| All India | 0.36 | 0.51 | 0.84 | 1.07 | 1.03 | 1.28 | 1.39 |

**cultivated area pertains to 2015-16@ quoted as 100% in the literature*

From the perusal of the table, It can be readily observed that the proportion of India's cultivated land apportioned to organic cultivation has increased from 0.36 percent in 2012-13 to 1.39 percent in 2018-19. Currently, 1.5 percent of the world's agricultural land is organic and the country having the highest organic shares has been reported as Liechtenstein (38.5 percent) (FiBL 2020). In India, the state of Sikkim is an exceptional one to have attained the status of the first fully organic state in the world. It can be seen that as an outcome of the policy efforts, Sikkim, up from 56 percent in 2012-13, had cent percent of its cultivated area under organic farming since 2016. Some states having the share of organic farming higher than that at the country level are Madhya Pradesh (4.45 %), Odisha (2.28%) and Maharashtra (1.46 %). The states having even less than one percent of farm area under organic farming during the year 2018-19 were Gujarat (0.91%), Karnataka (0.83%), Uttar Pradesh (0.38%) and Punjab (0.22%). Similar trends could be witnessed for the entire period of 2012-13 to 2018-19. The foregoing analysis points towards the potential to increase the organic area. It has been estimated that if each the top four states viz. Rajasthan, Maharashtra, Uttar Pradesh, and Madhya Pradesh in terms of cultivated area, (together accounting for 48 percent of the country's cultivated area) allocate additional one percent to the existing organic area, the absolute addition would be to the tune of 0.67 million hectares, which is equivalent to 8.7 times that in Sikkim alone.

## Organic Production in India: State-wise Profile

The information related to organic production in the selected states of India from the year 2012-13 to 2018-19 has been presented in Table 4.

**Table 4.** Organic production@ in major organic states of India, 2012-13 to 2018-19

MT

| States | 2012-13 | 2013-14 | 2014-15 | 2015-16 | 2016-17 | 2017-18 | 2018-19 | CAGR,%CDVI, % |
|---|---|---|---|---|---|---|---|---|
| Maharashtra | 270744(20.4) | 296847(24.1) | 217112(19.8) | 361671(27.1) | 255935(21.7) | 377308(22.5) | 860976(33.0) | 15.8*38.6 |
| Madhya Pradesh | 425849(32.0) | 397015(32.3) | 321587(29.3) | 365551(27.4) | 391598(33.2) | 575346(34.3) | 738886(28.3) | 9.7*21.3 |
| Karnataka | 322455(24.3) | 258336(21.0) | 254664(23.2) | 282633(21.2) | 164165(13.9) | 154923(9.2) | 370577(14.2) | -3.7$^{NS}$27.2 |
| Uttar Pradesh | 47733(3.6) | 55575(4.5) | 50166(4.6) | 57416(4.3) | 88183(7.5) | 117359(7.0) | 142512(5.5) | 21.0***15.2 |
| Rajasthan | 90021(6.8) | 54119(4.40) | 59450(5.4) | 58534(4.4) | 64245(5.4) | 94029(5.6) | 134613(5.2) | 8.91$^{NS}$28.3 |
| Odisha | 48725(3.7) | 33668(2.7) | 28679(2.6) | 40788(3.0) | 30661 (2.6) | 74642(4.4) | 88949(3.4) | 13.2$^{NS}$35.0 |
| Gujarat | 37942(2.8) | 56217(4.6) | 60621(5.5) | 55466(4.1) | 41611(3.5) | 75305(4.5) | 66106(2.5) | 6.9$^{NS}$17.4 |
| Punjab | 880(0.07) | 2178(0.18) | 507(0.05) | 107(0.01) | 711(0.06) | 113(0.01) | 744(0.03) | -19.5$^{NS}$78.3 |
| Sikkim | 79(0.01) | 91(0.01) | 68(0.01) | 109(0.01) | 185(0.02) | 435(0.03) | 424(0.02) | 38.8***31.7 |
| Others | 84571(6.4) | 75056(6.1) | 102900(9.4) | 113677(8.5) | 142812(12.1) | 206101(12.3) | 203598(7.8) | 19.5***10.7 |
| All India | 1328999 | 1229102 | 1095754 | 1335952 | 1180106 | 1675561 | 2607385 | 10.2*23.0 |

*@ inclusive of production from land in conversionFigures in parenthesis indicate the percent of respective All India figure*, ** and *** indicate significance at 10 percent, 5 percent, and 1 percent respectively Data Source: National Programme of Organic Production (NPOP)*

It needs to be highlighted again that the selected states include seven top-ranking states (Maharashtra, Madhya Pradesh, Karnataka, Uttar Pradesh, Rajasthan, Odisha and Gujarat) on the standpoint of organic production. The states of Punjab and Sikkim had found a place in the list of selected states on their merit; the former being agriculturally the most advanced state in the country and the latter, a well-acclaimed organic state.

India at large produced 1.33 million MT of organic products in 2012-13 which nearly doubled to 2.61 million MT by 2018-19 registering a statistically significant CAGR of 10.18 percent over the above said period. At the beginning i.e. in 2012-13, Madhya Pradesh was positioned at number one contributing 32 percent to India's organic production, followed by Karnataka at 24.3 percent and Maharashtra at 20.4 percent and distantly followed by Rajasthan (6.8 %), Odisha (3.7 %), Uttar Pradesh (3.6 %) and Gujarat (2.8%). Sikkim and Punjab, the much-acclaimed states on various accounts together accounted for merely 0.08 percent of the country's organic produce. Due to the great efforts of increasing awareness about sustainable agriculture and policy initiatives, organic production has got momentum in the selected states but at a varying rate. Due to a comparatively higher rate of growth (CAGR:15.8 %) in the case of Maharashtra in comparison to that in the case of Madhya Pradesh (CAGR: 9.7 %), the states have swapped the places and Maharashtra has come at the top. A special mention needs to be made about Karnataka, the state which has seen quite many ups and downs as endorsed by the Cuddy Della Valle Index (27.2%) has shown a remarkable jump in production, which as per 2018-19 figures stands at 3.7 lakh MT, out of which 87 percent is accounted for by organic sugarcane (Ganur 2017). Among all the states, Maharashtra (as per the latest available figures of 2018-19) records the highest organic production to the tune of 860976 MT followed by Madhya Pradesh (738886 MT), Karnataka (370577 MT), Uttar Pradesh (142512 MT), Rajasthan (134613 MT), Odisha (88949 MT), Gujarat (66106 MT), Punjab (744 MT) and Sikkim (424 MT). The nine selected states together accounted for 92.2 percent of the country's organic production.

It is pertinent to recall here that organic production figures discussed in earlier sections are inclusive of the production from land in conversion, which is not certified. The table that follows presents the current status of the share of certified organic production in total production.

**Table 5.** Share of certified organic production in total organic production, 2018-19

| Category | Total Production* | Certified Organic | |
|---|---|---|---|
| | MT | MT | % of India |
| Sugar crop | 991640 | 984730 | 99.30 |
| Oil Seeds | 727156 | 727148 | 100.00 |
| Fibre Crops | 313833 | 312939 | 99.72 |
| Cereals & Millets | 269734 | 269547 | 99.93 |
| Pulses | 71875 | 71875 | 100.00 |
| Plantation Crops including tea/coffee/coconut | 61544 | 61321 | 99.64 |
| Spices and Condiments | 56253 | 56208 | 99.92 |
| Medicinal/Herbal/Aromatic Plants | 48425 | 48424 | 100.00 |
| Fruits and Vegetables | 42940 | 42937 | 99.99 |
| Flowers | 11016 | 11016 | 100.00 |
| Dry fruits | 8864 | 8834 | 99.66 |
| Fodder seeds/Crops | 1851 | 1851 | 100.00 |
| Tuber Crops | 289 | 289 | 100.00 |
| Miscellaneous | 1964 | 1964 | 100.00 |
| All India | **2607385** | **2599083** | **99.68** |

*@ inclusive of production from land in conversionData Source: National Programme of Organic Production (NPOP).*

As per the 2018-19 figures, 99.68 percent of India's reported organic production is certified organic. It can be observed that oilseeds, pulses, medicinal and herbal plants, ornamental plants, fodder, and timber crops enumerated in organic production are 100 percent certified organic. In the case of the rest of the categories, the proportion of certified production in total production was well above 99 percent.

## Commodity Wise Competitive Strength of Indian States in Organic Production

The analysis was carried out for the most recent period of 2018-19 and all the product groups.

**Table 6.** Commodity wise top organic producer states with respective share, 2018-19

| Commodity | Organic Production, MT | Percent share of top contributing states | | | | | % of India |
|---|---|---|---|---|---|---|---|
| **Sugar Crops** | 984730.1 | Maharashtra 62.6 | Karnataka 32.9 | Madhya Pradesh 2.1 | Uttar Pradesh 1.5 | Uttarakhand 1.0 | 100 |
| **Oil Seeds** | 727148.2 | Madhya Pradesh 67.7 | Maharashtra 20.4 | Rajasthan 8.9 | Odisha 1.0 | Gujarat 0.7 | 98.7 |
| **Fibre Crops** | 312939.4 | Madhya Pradesh 29.4 | Odisha 23.6 | Maharashtra 18.7 | Gujarat 16.3 | Rajasthan 10.9 | 98.9 |
| **Cereals & Millets** | 269546.7 | Uttar Pradesh 32.5 | Madhya Pradesh 27.2 | J&K 12.2 | Assam 8.8 | Uttarakhand 4.6 | 85.4 |
| **Pulses** | 71874.9 | Madhya Pradesh 56.2 | Rajasthan 15.5 | Maharashtra 13.2 | Uttar Pradesh 8.7 | Uttarakhand 3.4 | 97.0 |
| **Plantation** | 61320.5 | West Bengal 32.2 | Assam 23.5 | Kerala 23.1 | Tamil Nadu 12.8 | Karnataka 6.0 | 97.5 |
| **Spices** | 56207.7 | Madhya Pradesh 26.4 | Maharashtra 16.5 | Karnataka 14.7 | Kerala 14.3 | Rajasthan 8.1 | 80.0 |
| **Medicinal, Herbal, Aromatic** | 48423.8 | Uttar Pradesh 65.4 | Rajasthan 7.7 | Madhya Pradesh 7.6 | Karnataka 5.9 | Chhattisgarh 5.5 | 92.1 |
| **Fruits** | 35802.3 | Karnataka 22.8 | Andhra Pradesh 22.1 | Gujarat 20.4 | Odisha 11.6 | Tamil Nadu 10.1 | 87.1 |
| **Flowers** | 11015.8 | Karnataka 91.9 | Madhya Pradesh 7.5 | Gujarat 0.2 | Uttarakhand 0.2 | Rajasthan 0.1 | 99.9 |
| **Dry Fruits** | 8834.2 | Goa 27.8 | Kerala 18.9 | Maharashtra 18.3 | Odisha 13.9 | J&K 10 | 89 |
| **Vegetable** | 7134.7 | Rajasthan 36.3 | Gujarat 13.3 | Madhya Pradesh 11.7 | Karnataka 10.8 | Maharashtra 8.7 | 80.8 |
| **TOTAL** | 2599083 | Maharashtra 33.0 | Madhya Pradesh 28.4 | Karnataka 14.1 | Uttar Pradesh 5.5 | Rajasthan 5.2 | 86.2 |

As regards the sugar crops which accounted for 38 percent of the country's organic production, the top five contributing states are Maharashtra, Karnataka. Madhya Pradesh, Uttar Pradesh and Uttarakhand with Maharashtra alone accounting for 62.6 percent, and another one third was produced by Karnataka. Madhya Pradesh accounting for two-third of organic oilseed production was the front runner followed by Maharashtra, Rajasthan, Odisha, and Gujarat. These five states together produced 98.7 percent of India's organic oilseeds.

Madhya Pradesh again led the rest of the states in case of fiber crops, pulses, and spices; the contenders of the next four slots being different for these crops. The five states namely Madhya Pradesh, Odisha, Maharashtra, Gujarat, and Rajasthan together accounted for 98.9 percent of certified organic fiber crop output. In the case of cereals and millets; Uttar Pradesh and Madhya Pradesh contributed nearly 60 percent to the total production. In the case of organic pulses; nearly three fourth of the total produce was contributed by only two states namely Madhya Pradesh and Rajasthan. In the case of organic spices, the three top-ranking states viz. Madhya Pradesh, Maharashtra, and Karnataka contributed nearly 58 percent, Kerela another 14.3 percent, and Rajasthan 8.1 percent to the country's total production. Out of the study states, Karnataka was the only state contributing to organic plantation production. The five states, West Bengal, Assam, Kerela, Tamilnadu, and Karnataka in the descending order of contribution together accounted for 97.5 percent of the organic production of plantation crops the country.

The state of Sikkim, much acclaimed for being truly organic state was conspicuous by its absence in the organic production scene of the country portrayed in the table above. The same holds good for Punjab and the future may see Punjab foraying into the organic cultivation once the efforts to bring about diversification in Punjab agriculture culminate into reality.

## Conclusion

The foregoing analysis conclusively establishes that the policy initiatives to give a boost to organic production have fructified in terms of increased production and subsequently export. However, given the large land base in the country, there is ample scope for propagating organic agriculture. The position of India in world organic agriculture can be improved by propagating/ incentivizing the already registered farmers to increase the area under organic agriculture. The farmers, particularly in the states which are by default organic, need to be sensitized about the certification process to facilitate certification. Low cost and hassle-free certification need to be put in place. Even group certification can be encouraged through cooperatives and FPOs. Keeping into view the export market trends, the competitive strength of the states in producing a particular commodity may be exploited.

**Way forward :** Although Considerable development has been done in India for enhancing production to marketing of organic produce. The most interesting aspect of leaving CS by farmers despite of high financial and technical support and shift to OA with little support, is showing OA system viability. In 1999 only 40000 ha agriculture area was recorded as certified organic that increased upto 240000 ha in 2011 (six times) within a decade(APEDA,2013) , if the non certified organic areas( mainly drylands) is also be added this will be much high figure. Therefore, it is the need of the hour that a multi-direction action plan has to be prepared for wider adoption and marketing of organic produce. In this financial year (2015-16) govt of India started an exclusive and massive programme namely *Paramparagat Krishi Vikas Yojna* (PKVY) to provide all round support organic agriculture in India. Therefore, organic would be future of Indian agriculture.

# 13

# Role of Organic Farming in Enhancing the Income of Small and Marginal Farmers in Dryland Areas

In India's pre-independence era (before the 1950s), agriculture was a system of harnessing nature for the sustenance of human beings, similar to the presently defined organic farming. That is, Indian farmers relied on the use of crop rotation, crop residues, animal manures, legumes, green manures, off-farm organic wastes, and biological pest controls to maintain soil productivity, supply plant nutrients, and control insects, pests and other weeds on their farms.

Following independence, rapid population growth in India placed great pressure on land and on these traditional farming systems; huge demands for food grains led to increased use of fertilizers and pesticides to boost production. Many of the gains in production during the last 4-5 decades resulted from the "Green Revolution," a campaign of technological interventions in agriculture widely adopted by farmers in developing countries. Expansion of irrigation to cover rainfed areas, popularization of hybrids/transgenic varieties of crops, and use of synthetic chemical fertilizers and pesticides were the major technologies promoted. This paid rich dividends in India, quadrupling food grain production from 50 million metric tons in 1950-51 to 211 million metric tons in 2001-02, and enabling India to become self-sufficient in food grain. Now a second green revolution is also in the offing, to boost agricultural production and meet the estimated requirement of 337 million metric tons by 2011-12.

Unfortunately, these positive developments have been accompanied by gradual and negative side effects such as secondary salinity, decreases in soil fertility, growing insect resistance to pesticides, and increased costs of production, which are challenging the sustainability of conventional agricultural production at high levels (Balak Ram 2003). Consequently, interest in organic agriculture as an ecofriendly system of cultivation is growing at both national and global levels. This shift towards organic production is supported by consumers who are aware of health hazards: demand for food grown organically is increasing by 20-25% in developed countries where awareness is comparatively high.

Organic agriculture is indeed being pursued in India; the National Programme of Organic Products (NPOP) was launched in 2000. Its aim is mainly to create certification facilities; since its inauguration, 2.5 million ha (~6.2 million acres) have been certified as organic, providing 115 238 metric tons of produce by the end of 2004-05 (Gauri 2005). However, this produce is aimed largely at the export market, and largely uses intensive agriculture practices that are not sustainable in drylands, where land degradation is already a serious issue. Furthermore, the drylands also lack local food security and employment, problems that intensive monocrop-oriented systems do not address very well.

Organic agriculture, a holistic system that focuses on improvement of soil health, use of local inputs, and relatively high-intensity use of local labour, is an admirable fit for drylands in many ways, and the drylands offer many benefits that would make it relatively easy to implement. And, in fact, India's National Project on Organic Farming (NPOF), launched in 2004, has given top priority to the drylands (NPOF 2005). The rest of this article will detail, first, the numerous ways in which the Indian drylands and organic agriculture are a good fit for each other; and second, several steps that should be taken to help promote this development.

"Organic agriculture, a holistic system that focuses on improvement of soil health,use of local inputs, and relatively high-intensity use of local labour, is an admirable fit for drylands in many ways, and the drylands offer many benefits that would make it relatively easy to implement."

## Specific benefits of organic farming for the drylands of India

Soil and climate conditions in India's drylands make them particularly well suited to organic agriculture. These marginal lands, with their marginal soils, tend not to respond well to intensive farming practices. They are actually better suited to low-input farming systems that make ample use of biodiversity (Sharma et al., 2000). In turn, organic farming, with its central focus on maintaining and improving soil health, its avoidance of pollutants, and its reliance on local inputs and labour, can materially advance the economic and ecological health both of the drylands and of the people who live there.

To begin with, semi-arid and arid dryland soils typically are poor in water-holding capacity as well as organic matter (Sharma 2000). In some areas depth of soil is another limiting factor for agricultural production. Addition of organic matter, a cornerstone of organic farming practices, will not only improve the physical condition of these dryland soils but also greatly improve their ability to supply balanced plant nutrients.

Furthermore, as stated in the introduction, drylands desertification is increasing at an alarming rate due to over-exploitation of natural resources. This is mainly because of production-enhancing technologies that are inappropriate in drylands

(Dhir 1997). For example, use of tractors increases wind erosion and damages natural regeneration of trees and grasses. Over-use or improper use of canal irrigation can cause waterlogging and salinity (not to mention malaria epidemics). Excessive groundwater pumping has decreased the groundwater table drastically in tube well-irrigated areas. In many locations where intensive-input agriculture systems are followed, soil fertility is decreasing and certain severe crop pests are becoming resistant to synthetic pesticides. These all are indicators of improper land use leading to desertification; adoption of organic farming practices suitable for drylands can help to ameliorate these conditions.

Another serious problem in drylands is the lack of sufficient food security or economic opportunity for the many people who live there. The population in these drylands areas is currently growing at a rate of 2.8% per year. Yet, under current practices, many dryland farmers are unable to earn a year-round livelihood in their own villages. The consequent migration of villagers to cities and nearby states for livelihood during drought further imbalances the economic development of these drylands. Since organic farming is labour-intensive and relies on local inputs, it can provide both improved local food security and ample opportunities for local employment and proper utilization of this precious human resource (Gupta and Sharma 1996).

To be sure, many farmers resist implementing organic systems because they fear a drop in productivity and thus income, during the years while synthetic fertilizer/pesticide use is discontinued and the soil is gradually built up by organic means. However, the longest transition periods occur where pesticide use has previously been greatest. And, because of the lack of a reliable water supply, average fertilizer use in the semi-arid, rainfeddrylands—67% of India's agricultural area—is already very low (36.4 kg/ha) compared to the national average of 76.8 kg/ha. In the actual desert areas, fertilizer use is negligible (FAI 1998). Pesticide use is also very low. Furthermore, large parts of the drylands are still categorized as "virgin," meaning no synthetic inputs have been used there to date. This makes a quick shift to organic agriculture, with no drop in productivity, much easier.

Due to climatic variability, farming systems in drylands traditionally mix crops, trees, animals, grasses etc. Such diversified systems have been found efficient in nutrient recycling and restoration of soil fertility, the basic aims of organic farming; they minimize pest incidence as well. Furthermore, India's traditional farmers possess a rich body of wisdom, based on long observation and practice, concerning soil fertility management and pest control; this can be drawn on to further strengthen organic systems (Sharma and Goyal 2000). These two factors will also aid the quick development of more efficient, more productive organic farming systems in these areas.

**In terms of input supply, the drylands are very rich in local resources suitable for supporting organic farming**

- Some of the best sources of biopesticides, like neem (*Azadirachtaindica*), Karanji (*Pongimiapinnata*), and *Calatropis* spp, are abundantly available in drylands (Rajeshwar Rao 1999).
- Similarly, minerals like rock phosphate, gypsum and lime are available naturally in large quantity in Rajasthan. These minerals help improve soil conditions and supply plant nutrients.
- Finally, properly handled and aged animal and human wastes are among the best organic fertilizers available; traditional dryland farming systems are well supplied with both. Indeed, in such areas animal populations are growing at 1.5 to 2.5% per year (Purohit 1998).

## Strategies for promoting organic farming in drylands

As shown above, there is immense potential for adoption of organic farming in India's drylands (Sharma 2001). Socio-economic and ecological reasons also favour it (Arora and Mohan,1986): many aspects of drylands that would pose constraints for the intensive agriculture system of the Green Revolution can be converted into opportunities for organic farming. The following specific steps would go a long way towards promoting this development.

### 1. Popularize organic farming without compulsion of certification

In drylands, farmers are very poor and are largely unable to afford the cost of certification programs. Promoting certification as a universal requirement of organic farming has thus had a negative impact on its adoption by smallholders in these areas. A more appropriate approach for drylands, at least during a certain initial period until the industry is established, would be to promote non-certified organic farming with a goal of providing local employment and improving local/regional food security rather than developing export crops.

### 2. Promote ley farming

In India's drylands, rotation of grasses and food grains is known as ley farming. This system is traditional; research by the Central Arid Zone Research Institute (CAZRI) has validated it as one of the best options for fertility restoration in these regions (Muthana et al. 1985).

### 3. Integrate efforts of supporting agencies

Several government agencies and NGOs are working individually to promote organic farming. This individual approach, however, may result either in lack of adequate funding or lack of adequate knowledge of organic farming and/or marketing techniques. For example, rural ministries have a scheme of "margin money" in which they subsidize 25-30% of loans made to establish vermicomposting units; but they are unable to ensure a steady market for the produce of such units. Similarly, state agriculture departments and the institutes that make up the Indian Council of Agricultural Research (ICAR) have a wealth of information on organic farming but are unable to provide financial support for initiating such programs. Thus there is need for all related agencies to create integrated programs, linking the storehouses of technologies with financial institutions, in order to effectively promote organic farming. Since funding agencies always have the financial upper hand, they must take the lead to ensure sufficient technology backup and marketing of these products. This could be effected through a coordinating committee with representatives from all concerned organizations.

**4. Encourage decentralized input supply**

Local, decentralized production of all inputs for organic farming should be encouraged not only so that local resources can be utilized but also so that village-level employment can be generated. Locally produced inputs are also much less likely to be adulterated. All of this organic input production should be legally categorized as cottage industry. Subsidies or micro-financing to help set up small-scale input production units that meet local requirements could be provided to village cooperatives and self-help groups (that is, informal groups of people with a joint livelihood system and a common bank account). Banning inter-state transportation of organic inputs could be another effective method for keeping such production local.

**5. Adopt improved methods of composting**

In general, farmers understand the importance of adding organic matter to drylands soils, but the majority apply undecomposed animal and crop wastes. This not only reduces the availability of nutrients to plants but also invites several pests. It would be better to apply these materials after composting them (Durgudeet al. 1996) with any of several suitable methods. Because these composting methods would help reduce disease and improve local sanitary conditions, they could be popularized and financially supported under India's existing "Clean village scheme." This scheme, sponsored by the government's Ministry of Rural Development and executed by state governments and volunteer organizations, focuses on safe disposal and use of human and animal wastes as a means of improving hygiene and health at the village level.

### 6. Increase public awareness and build capacity

Conferences, seminars, and farmers' fairs may be organized to raise awareness and encourage adoption of organic farming. Programs demonstrating how to establish organic systems, and training in how to produce and manage organic inputs, may be started at the village level. Under the NPOF, sufficient provision has been made to train farmers for organic production and internal control and to develop both model organic farms and a nationwide network of organic service providers (to provide guidance, establish farmers' groups and arrange organic inputs). However, most of these programs focus on irrigated areas because of ensured success and export possibilities. A rational allocation policy for drylands is needed.

### 7. Subsidize organic inputs and produce

Subsidies may be provided for organic inputs and produce while the industry is still getting established. In India, subsidies are mainly provided by the national government and channelled through state agriculture departments; the technique is well-tested, having already been used for the synthetic fertilizer and pesticide industry. Indeed, subsidies have been provided for setting up biofertilizer and vermicomposting units under NPOF and for setting up export schemes under NPOP. Additional subsidies could be provided for:

- Setting up organic input production units for composting, biopesticides etc.
- Compensating organic farmers during the period of conversion to organic techniques, to compensate for yield reductions if any.
- Establishing village-level grading and packaging units for organic produce.
- Developing local and regional marketing infrastructure for organic produce in dryland areas, where regional/local food security is more important than crops for export.

### 8. Promote high-value crops

The dryland climate favours quality production of several spices and medicinal plants that are in already in great demand both nationally and internationally. Furthermore, demand for organic versions of these high-value products is increasing. Thus, once the local market is well established and the potential for growing crops for export has increased, they may be promoted to dryland farmers as excellent plants to include in the overall design of their organic system.

### 9. Develop organic farming clusters of villages

Since the drylands are already an area of focus for governmental development programs based on a watershed approach, clusters of villages previously established for such programs (Khan 2002) may be converted into organic clusters of villages by providing technical support. This will be cost-effective and make the eventual certification process of organic produce easier for these villages once the local organic produce market has been well established.

### 10. Develop certification programs and marketing chains

Ultimately, once the drylands organic agriculture industry is strong, cost-effective facilities for sale of certified organic produce in regional and global markets may be created. The cost of certification should be met by the exporters.

### 11. Conduct ongoing research on organic systems

ICAR institutes and State Agriculture Universities have already done much research on various components of organic farming such as vermicompost, biopesticides, biofertilizers, and agroforestry. However, more research is needed to integrate these efforts and assess their effects. Ongoing research is needed on all the following aspects of organic agriculture simultaneously (Sharma 2001):

- Survey and scientific validation of traditional organic farming practices.
- Use of locally available resources for production of manures and biopesticides.
- Assessment of economic and ecological returns from organic vis-à-vis intensive agriculture systems.

### Conclusion

Due to low, erratic rainfall and poor soils in India's drylands, farming systems based on synergism with nature, and thus essentially "organic," have been followed for centuries. This opens the door to immense possibilities for improving the soil health and overall environment, and providing sustainable livelihoods in these areas, by updating these age-old systems with modern research and techniques. The key to achieving this is enhanced collaboration and cooperation among all the agencies and programs that have an interest in supporting these goals. By working together, the goals of promoting organic practices, improving soil health and local food security and, ultimately, creating markets for organic produce both regionally and globally, can be realized for the benefit of all of India.

# 14

# Integrated Organic Farming Systems

Indian farmers were all organic farmers before the advent of synthetic fertilizers, pesticides, mechanization, etc. It was mostly during the mid-sixties and afterwards that farmers started using the synthetically produced inputs (fertilizers, pesticides, etc.), which brought about "Green revolution" in our country. Green revolution not only dramatically increased the productivity of many crops (wheat, rice, etc.) leading to self-sufficiency in food grain production in 1970's but also help in building up buffer stock of food grains for the burgeoning millions of population. Acc to Sangha (2014), intensive agriculture (or modern agriculture) helped achieving dramatic increase in crop yield ensuring food security and commercialization of agriculture, but led to:

- Loss of soil productivity and pollution of soil resources- Uncontrolled and unscientific use of chemical fertilizers without adding adequate quantities of organic manures has resulted in the deterioration of soil health and environment.
- Pollution and depletion of water resources- Pollution of water resources mainly due to excessive use of fertilizers and pesticides and decline in availability of water for human use due to increase in cultivation area under irrigation are the alarming issues that have emerged due to current agricultural practices
- Loss of crop and wild plant diversity- As more and more area has been brought under monoculture cultivation, the diversity of wild plants and other cultivated crops has decreased.
- Increased vulnerability to the impacts of climate change and poor adaptive capability of farmer-Climate change makes these highly intensive agricultural systems very sensitive to increase in temperature and decrease in rainfall.

  As per the agriculture census of 2015-16, the majority of land holdings in India- 86 percent are small (1-2 hectares)and marginal(with holdings of less than one hectare). These households' incomes are already below what they spend on consumption expenditure and having less ability to buy chemical inputs. Thus, only eco-friendly and low cost technologies with minimal dependence on external resources are to be promoted for sustainability.

To sustain and satisfy as many as their needs, the farmers include crop production, livestock, poultry, fisheries, beekeeping etc. in their farms. When farming components are identified considering the agro-climatic conditions, farmers need and market demand, organic farming can provide much higher income than conventional practice. To make the organic agriculture sustainable, it is necessary to utilize all the resources available on- and off the farm effectively by integrating complementary and supplementary enterprises. Thus, an Integrated Organic Farming System (IOFS) model comprising different modules of cereals, pulses, oilseeds, vegetables crops, fruits, dairy unit, fodder crops, water harvesting and nutrient recycling components have been developed to meet the diverse requirement of the farm household while preserving the resource base and maintaining the environment.

**Integrated Organic Farming System (IOFS)** is an innovative concept of on-farm resource management strategy to achieve economic and sustained agricultural production to meet the diverse requirement of the farm household while preserving the resource base and maintaining high environmental quality (Das *et al* 2019).

Components of IOFS when carefully chosen and executed depending upon the suitability of the farm, fulfills the input demand internally and totally avoids purchase of any external inputs. This will not only help to reduce the production costs but also enhances the quality and quantity of produce and improves environmental quality through recycling of organic wastes. IOFS protecting the long-term fertility of soils by maintaining organic matter (OM) levels, fostering soil biological activity, nitrogen (N) self-sufficiency by the inclusion of legumes, effective recycling of OM including crop residues and livestock wastes, and weed, disease and pest control rely primarily on crop rotations, natural predators, diversity, organic manuring, and resistant varieties.

Agriculture is a vital source for economic assistance and all round progress of our country. Present agriculture has taken shape of a business enterprise, which requires a contingent land use planning, and fullest use of available input resources for the highest productivity. Availability of scientific information about the latest technology and their effective transfer from scientist's venue to the farmer's field is of immense utility.

Organic farming is developing rapidly and now practiced in more than 120 countries of the world. Globally 30.4 million ha of the agricultural land are currently managed organically. India ranks as 33$^{rd}$ in total area under organic cultivation. India has now become a leading supplier of organic herbs, organic spices, organic basmati rice etc. to the developed nations. The export realization during 2007-08 was around 78 million US dollars registering a 200% growth over the previous year. Though organic farming is an inevitable phenomenon, it alone cannot feed the billion population of India. To ensure the food and nutritional security, rather than promoting organic farming enmasse and universally, it would be desirable to carefully delineate areas for or no use of fertilizers and other agro-chemicals.

Organic farming is a form of agriculture, which excludes the use of synthetic fertilizers and pesticides, plant growth regulators, livestock feed additives and genetically modified organisms. As far as possible, organic farmers rely on crop rotation, green manure, compost, biological pest control and mechanical cultivation to maintain soil productivity and control pests. Organic farming is often contrasted with conventional or mainstream farming. Organic farming is a production system which favours maximum use of organic materials crop residues, animal excreta, legumes, on and off farm organic wastes, bio-pesticides etc. and discourage use of synthetically produced agro-inputs for maintaining soil health, productivity and pest management under the conditions of sustainable natural resources and healthy environment. Use of organic manures have been found to be promising in arresting the decline in productivity through correction of deficiencies of secondary and micronutrients and its beneficial influence on the physical and biological properties.

Organic Agriculture and the multi-dimensional challenges of future food and farming systems-

- Increased productivity of agriculture in a sustainable manner.
- 70% of the world's poor live in rural areas (< USD 2/day)
- 90% of farms in the world are less than 2 ha covering 60% of the arable land worldwide

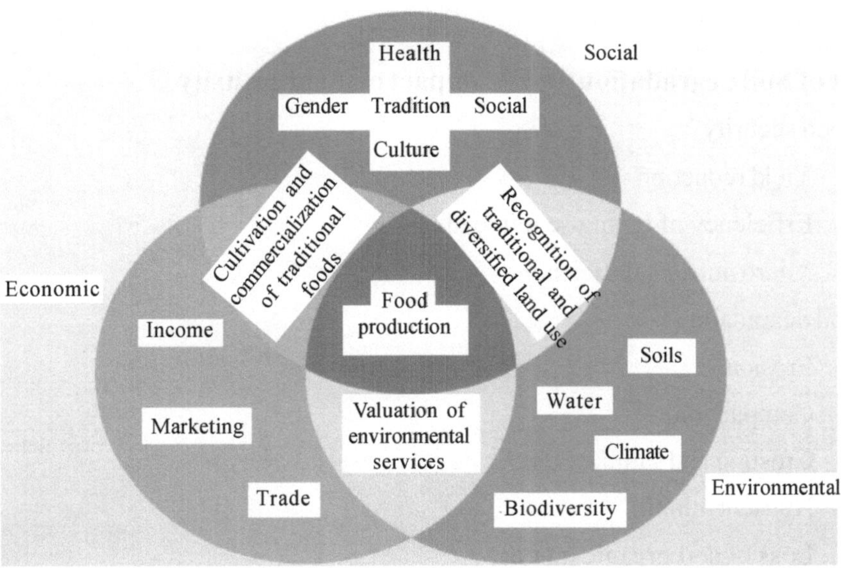

- Widespread subsistence production in isolated and marginal locations with low levels of technology
- Widespread food insecurity in spite of sufficient food being produced at global level food

**Thus**

Organic principles may contribute to a valuable framework for a future sustainable agricultural production-

From focus on increased To holistic integration of productivity alone →

Natural Resource

Management with food and nutritional security

## OA is a viable approach thatcan be suitable for smallholders

- Particularly useful in difficult environments
- Reduces risk by encouraging localized input production Fostering soil and water conservation
- Encouraging the diversification of production

## Organic Agriculture can help in raising the productivity and incomeof low-input agricultural systems

- There seems to be a strong indication that the proliferation of organic agriculture could be a viable strategy to improve livelihoods in Asia's rural areas.

## Effect of Soil degradation and its impact on food security

- Food security
    - Yield reduction
    - Efficiency of input use reduced
    - Micro nutrient deficiency
- Soil degradation
    - Erosion
    - Compaction
    - Crusting and salinization
    - Nutrient mining
    - Loss of soil organic matter

### So this situation

Need for paradigm shift in land husbandry and follow principles and practices of soil management and adopt "Integrated Organic Farming System"

Solutions for soil and food quality improvements-

- Mulching and recycling organic residues
- improve soil structure and quality
- Water conservation and water use efficiency
- Adoption of diversified cropping systems, indigenous foods
- Agro-forestry and mixed farming
- No-till agriculture
- On-farm experimentation including various enterprises and adaptation at farm site
- Use of micronutrient rich bio-fertilisers
- Inoculating soils for improved Biological Nitrogen Fixation
- Biological microbial processes to increase P-uptake

## IOFS perspective under Indian conditions

More than 65% of the country's cultivated area is rainfed, cultivation depends on crop rotation, crop residues, animal manure, legumes and biological pest control. Most of the farms in remote and marginal areas are subsistence type. Organic farming is growing rapidly among Indian farmers and entrepreneurs, especially in low productivity areas, rain-fed zones, hilly areas and the northeastern states where fertilizer consumption is less than 25 kg/ha/year (Mitra and Devi, 2016). The major challenge in 'organic agriculture' is the non-availability of huge quantity of organic inputs for satisfying the farm demand.Use of animal excreta based manure alone is not sufficient for meeting the nutrient needs of the crops. It is therefore, necessary to utilize all the resources available on- and off the farm effectively. Thus, focus should be on integrated organic farming system (IOFS) by integrating complementary and supplementary enterprises such as crop, fruits and vegetables, livestock, poultry, fish, multipurpose tree species, mushroom, etc. along with adequate nutrient recycling strategies (Das et al., 2019).

- The potential of IOFS in India is signified by the fact that the Indian farming sector has abundant resources like livestock, crop residue, water, aquatic weeds and forest litter (Bhattacharya and Chakraborty, 2005) which facilitates IOFS. Inherent advantages such as its varied agro-climatic regions, local self-sustaining agri-systems, sizeable number of progressive farmers and ready availability of inexpensive manpower translate into the potential to cultivate organically a vast basket of products. IOFS will not only promote organic food production but also reduce dependence on external resources through efficient recycling of on-farm biomass and other resources.

- The ministry of agriculture and farmers' welfare developed integrated organic farming systems (IOFS) models for the benefit of farmers. IOFS models were developed in different states of India under the scheme of All India Network Program on Organic Farming (AL-NPOF). This program was developed by the Indian Institute of farming Systems Research.

## Management of Integrated Organic Farming

Management of Integrated organic farming system involves-

1. Organization of crop, livestock and other enterprises (say horticulture, poultry, goatary, bee keeping, vermicomposting, mushroom cultivation etc.) in such a way that it harmonizes rather than conflicts with natural systems.
2. Achievement of closed cycle to the greatest extent possible between soil, plants, animals and people and an avoidance of environmental pollution.
3. Maintenance of soil fertility for optimum production, relying primarily on renewable resources.
4. Reduction of pest and disease incidence through a carefully designed farm rotation and enterprises structure; use of resistant varieties; the encouragement of beneficial pest predators; and the use of other biological pest control techniques.
5. Use of appropriate farm machinery and cultivation techniques which reduces non-renewable resource consumption.
6. Use of forms of animal husbandry which respect the welfare and behavioral needs of farm livestock and other enterprises at the farm
7. Enhancement of the environment in such a way that wildlife flourishes and it is enjoyable for people both working within the system and viewing it from outside.
8. There should be best recycling among the product and byproducts available at the farm.

## These principles will lead to a wider definition of quality

Advantages of Integrated Organic Farming-

1. Organic manures produce optimal conditions in the soil for high yields and good quality crops.

2. They supply all the nutrients required by the crops (including micronutrients).
3. They improve plant growth and physiological activities of plants.
4. They improve the soil physical properties of soil enhancing the source of energy for microbes which help in aggregation.
5. They improve the soil chemical properties such as source supply and retention of soil nutrients and promote favourable chemical reactions.
6. Most of the organic manures are wastes or byproducts which on accumulation may lead to pollution. By way of utilizing them for organic farming, pollution is minimized.
7. Organically grown crops are believed to provide healthier and nutritionally superior food.
8. Organic farming helps to avoid chain reaction in the environment from chemical sprays and dusts.
9. Organic farming helps to prevent environmental degradation and can be used to regenerate degraded areas.

## Integrated Organic Farming Systems

The Integrated Organic Farming Systems can be taken up with the objective to generate maximum income and employment by combining different enterprises namely field crop, multistoried cropping, pomology, olericulture, floriculture, animal husbandry, fishery, duckery, poultry, mushrooms, biogas, agro-forestry and commercial nursery by recycling product and by products.

In one hectare of land an organic farming system can be taken up including field crop, multistoried cropping, pomology, olericulture, floriculture, animal husbandry, poultry, mushrooms, biogas, agro-forestry and commercial nursery. A biogas plant can be installed to meet the energy need of the family. Slurry generated at the plant can be used as manure for crops. Every early morning raw dung collected from the cattle shed of the central farm was fed to the biogas plant. Brinjal, okra, cowpea and maize were grown in kharif and tomato, watermelon, cucumber, bitter gourd, and bottle gourd etc. were grown in the rabi season. Fodder crop was grown to supply greens to the dairy. Mango, ber, lemon and yam had been planted around the farm to utilize all the available space. An ornamental unit was developed in front of the farm house to provide the mental food. The area allotted to different components of the farm is mentioned in Table1.

**Table 1.** Area allocation to different components at Organic Farming Unit

| S.No. | Particular | Area (sq. m) |
|---|---|---|
| 1. | Farm house | 100 |
| 2. | Threshing floor | 300 |
| 3. | Cattle shed (Dairy unit) | 50 |
| 4. | Biogas plant | 20 |
| 5. | Pomology | 1000 |
| 6. | Floriculture | 500 |
| 7. | Multistoried cropping | 1000 |
| 8. | Vegetables | 1200 |
| 9. | Upland crops | 1550 |
| 10. | Mushroom , agro forestry, | 2200 |
| 11. | Poultry unit | 80 |
| 12. | Forage unit, compost pit, | 2000 |
| | **Total** | **10,000** |

Integrated Farming System (IFS) is the answer to the problem of increasing food production and farm income and for improving nutrition of small-scale farmers with limited resources without any adverse effect on environment and agro-ecosystems. In an organic cropping system, the amount of by-products can be as high as or higher than marketable produce. This may go to waste if not utilized in an animal enterprise. Hence, integration of different agricultural and allied enterprises with crop activity as base will provide a way to recycle low cost produces at farm level/no cost waste materials of farm from one enterprise to another and thus reduce the cost of production of economic produce and finally to enhance the net income of the farm as a whole.

The future of Indian agriculture depends on the development of appropriate farming system as applicable to resource poor farm families and as suited to different agro ecological zones. Restricting the use of purchased inputs in organic farming could be achieved through multiple cropping and diversified farming including crops, animal husbandry, vegetables, floriculture, forestry and sericulture etc. The demand must be met from limited resources- land, water, energy and labour. Increasing food supplies with limited natural resources is a daunting challenge to the scientific community. Under this situation, Organic IFS seems to be the answer, considering the current scenario in agriculture in India. Besides facilitating cash income and increased employment opportunities, IFS minimizes the quantum of purchased inputs in farming by effective recycling of products and by-products among the component enterprises; reduce the ill effects of inorganic fertilizers and chemicals (pesticides, herbicides etc.)/ or pollution hazards.

## Integrated Organic Farming and Biodiversity

Integrated Organic Farming increases biodiversity at every level of the food chain- all the way from lowly bacteria to mammals. This is the conclusion of a review of studies from around the world comparing organic and conventional agriculture. According to the researchers, organic farming aids biodiversity by using fewer pesticides and inorganic fertilizers and by adopting wildlife friendly management of habitats where there are no crops, including strategies such as not weeding close to hedge and by mixing arable and livestock farming. Integrated farming particularly benefits some bird species leading to good environment.

Integrated organic farming is emerging as a dynamic alternate farming system for crop as well as other products at the farmers field itself. It is gaining quick acceptability among farmers and consumers. It offers environmental protection, biodiversity enhancement, reduced energy used and $CO_2$ emission, sustainable resource use, increased crop yields without over- reliance on costly external inputs, environmental and biodiversity protection etc. are indirect benefits of Integrated organic farming which further adds to the economy of the country.

There is no doubt that Indian land resources and farming efficiency can result in rising per hectare productivity on all farms provided all the factors of production are fully exploited.

## Details of IOFS models and success stories in different states of India

### 1. Meghalaya

#### IOFS model for valley land at ICAR Complex, Umiam (Das *et al* 2019)

The IOFS model for valley land (950 m above mean sea level) comprising different enterprises such as cereals (rice and maize), pulses (lentil, pea), oilseeds (soybean, rapeseed), vegetable crops (Frenchbean, tomato, carrot, okra, brinjal, cabbage, potato, broccoli, cauliflower, chili, coriander, etc.), fruits (Assam lemon, papaya, peach), dairy unit (a milch cow + calf), fodder crops, central farm pond, farmyard manure pits and vermicomposting unit (Table 2) have been developed (0.43 ha) at ICAR Research Complex for Northern Eastern Hilly Region, Umiam, Meghalaya (950 m altitude) since 2005 under Institute programme and later standardized under Network Project on Organic Farming, funded by ICAR-Indian Institute of Farming System Research since 2012 (Fig. 1& 2).

**Fig. 1.** IOFS model for valley land at ICAR Complex, Umiam.

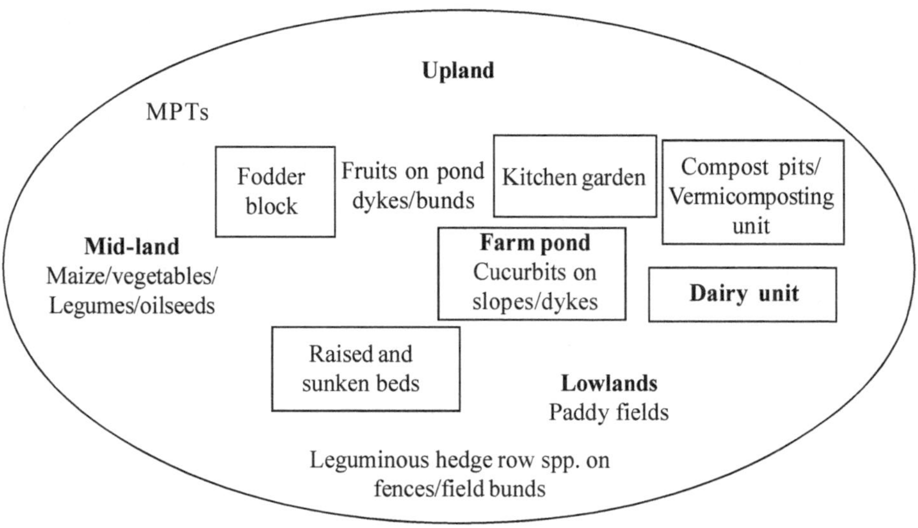


**Fig. 2.** Schematic diagram of IOFS model.

**Table 2.** Economics of the IOFS model (area=0.43 ha)

| S.No. | Components | Area (ha) | Cost of cultivation/ year (Rs) | Gross returns/ year (Rs) | Net returns/ year (Rs) |
|---|---|---|---|---|---|
| 1. | Crops Cereals, pulses, oilseeds, vegetables, fruits and fodder crops | 0.373 | 27,129(48%) | 65,370(55%) | 38,241(61%) |
| 2. | Dairy 1 milch cow + 1 calf | 0.004 | 20,250(36%) | 36,000(30%) | 15,750(25%) |
| 3. | Fishery Composite | 0.05 | 4,760(9%) | 17,000(14%) | 12,240(20%) |
| 4. | Nutrient cycling Vermicompost, FYM, Hedgerow planting, residue recycling, Rock phosphate, liming | 0.01 | 3,700(7%) | 0 | -3,700(-6%) |
| | **Total** | 0.43 | 55,839 | 1,18,370 | 62,531 |
| | Farmers' practice I (Rice-Fallow) | | 10,100 | 25,800 | 15,700 |
| | Farmers' practice II (Rice-vegetables in small scale) | | 18,975 | 43,493 | 24,518 |

REY of 18 t/ha was recorded under the IOFS model compared to 4.0 t and 6.75 t/ha under farmers' practice-I & II, respectively

A farm pond of 460 m$^2$ area with average depth of 1.5 m was part of the IOFS model for life saving irrigation and aquaculture. Climbing vegetables such as bottle gourd, chow-chow, cucumber, ridge gourd etc., were grown on a structure created above water bodies in one side of the pond dyke for vertical intensification. Pumpkin was raised in another side of the pond and allowed to crawl on the ground/ pond dyke slopes. The washings from the dairy unit were diverted to fish pond for promoting growth of zooplankton and phytoplankton for fish growth. The solid waste from cow shed was used for FYM making and vermicomposting. The total cost of cultivation was recorded at Rs. 55,839/- per year under the IOFS model with an area of 0.43 ha. Maximum expenditure was incurred in crop component of the model with 48% of the total cost of cultivation. Dairy unit with one adult cow and one calf registered 36% of the total cost of cultivation, while fishery component recorded 9% of the total cost of cultivation (Table 2). For maintaining vermicomposting unit of 72 m$^2$ area and other important operations like hedgerow planting, residue recycling, rock phosphate application and liming, the expenditure incurredwas Rs. 3700/- which account to 7% of the total cost. Since, the soil of the northeast India and study site is acidic in nature, application of rock phosphate (100-150 kg/ha) and liming (500 kg/ha in furrow in alternate years) is required for phosphorus nutrition as application of organic manure alone is not enough to meet the crop requirement. A total net return of

Rs. 62,531/- per year was achieved under the IOFS model which is much higher than the region's farmer common practices of rice monocropping or improved practice of rice-vegetables cropping system (Table 1). The highest contribution towards the total net return was contributed by crop component of the model (61%) followed by dairy (25%) and fishery component (20%). The fish production was 136 kg. The net return from dairy component was calculated only in terms of milk production since the cowdung produced was recycled back into the model which was used as manure for crop production. The production of vermicompost from model was 1500 kg annually and it was used in the farm itself for nutrient supplement to crops.

For 0.43 ha area, the total nutrient requirement for organic crop production has been estimated at nitrogen (N) 64.7 kg, phosphorus ($P_2O_5$) 23.1 kg and potassium ($K_2O$) 53.8 kg. On farm nutrient recycling in IFOS could produce an amount of 59.7 kg N, 18.9 kg $P_2O_5$ and 51.9 $K_2O$ (Table 3). Hence, 92% of the total N requirement, 82% of the total $P_2O_5$ requirement and 96% of the total $K_2O$ requirement could be met within the model itself and only 8% of the total N requirement, 18% of the total $P_2O_5$ requirement and 4% of the total $K_2O$ requirement is required to be met from the external source to sustain the model. The nutrient requirement of the model from external source would be reduced substantially with the efficient recycling of pond silt, intercropping with legume, use of biofertilizers such as Azotobacter, Rhizobium, phosphorus solubilizing microorganism etc.

**Table 3.** On-farm nutrient supply balance sheet under IOFS model (area=0.43 ha)

| **Components** | **Nutrient requirement (kg)** | | | **On-farm nutrient recycled (kg)** | | | **Nutrient balance (kg)** | | |
|---|---|---|---|---|---|---|---|---|---|
| | **N** | **$P_2O_5$** | **$K_2O$** | **N** | **$P_2O_5$** | **$K_2O$** | **N** | **$P_2O_5$** | **$K_2O$** |
| Cereals (Rice, maize) | 21.1 | 7.5 | 17.5 | 6.6 | 2.3 | 12.2 | -14.5 | -5.3 | -5.3 |
| Horticultural crops (Vegetables, fruits) | 31.4 | 11.2 | 26.2 | 11.8 | 2.6 | 9.0 | -19.6 | -8.6 | -17.2 |
| Dairy | 0 | 0 | 0 | 13.1 | 4.9 | 6.6 | 13.1 | 4.9 | 6.6 |
| Others (Oilseeds, Pulses, Green manuring crops, fodder etc.) | 12.2 | 4.4 | 10.1 | 28.2 | 9.1 | 24.1 | 16.0 | 4.8 | 14.0 |
| **Total** | **64.7** | **23.1** | **53.8** | **59.7** | **18.9** | **51.9** | **-4.9 (-8%)** | **-4.2 (-18%)** | **-1.9 (-4%)** |

Considering the benefits from the IOFS model with a net return of Rs. 62,531 per year (Rs. 5,211/month) from 0.43 hectare area, a net income of Rs. 5,100 per month or Rs. 174 per day was achieved which is a modest amount for living by a four-member family (2 adults and 2 children). Assuming food requirement and other expenditure per day for a four-member family (2 adults and 2 children) for rice (1,300 g, Rs. 33), dal (150 g, Rs. 15), oil (300 g, Rs.15), vegetables (1000 g, Rs. 25), fruits (400 g, Rs. 20,), fish (110 g, Rs.17), meat (100 g, Rs.15), others (milk, egg etc. Rs. 15), a total of Rs. 155 is required per day towards food. However, there is further need to enhance the income to meet the other requirement of the family i.e. healthcare, schooling, clothing etc.

## IOFS models demonstrated at village levels in cluster approach (Layek *et al* 2020)

For disseminating organic production technology developed in the ICAR Research Complex for NEH Region, Umiam, in 2013, a model village concept under Network Project on Organic Farming-Tribal Sub Plan (NPOF-TSP) was conducted in the village of Mynsain in Ri-Bhoi district of Meghalaya with financial assistance from ICAR-Indian Institute of Farming Systems Research, Modipuram.Several farmers in village started practicing organic farming in IOFS mode. They integrated crops (rice, maize), vegetables (tomato, French bean, potato, lettuce, and carrot), livestock (dairy/ piggery), water harvesting (jalkund) etc (Fig 3). Water from micro water harvesting structure like Jalkund used for live saving irrigation in winter months. It increasedthe crop productivity and diversify their homestead farming to growing remunerable crops and rearing cattle, pigs, poultry, etc.

## Flow chart

Identification of suitable land for developing IOFS near the household

↓

Creation of rain water harvesting structure jalkund/farm ponds

↓

Establishment of animal rearing unit dairy/piggery/poultry

↓

Construction of compost pit /vermicomposting unit for efficient recycling of animal excreta and other farm biomass like kitchen waste, weeds etc.

↓

Development of kitchen garden and cultivation of year round vegetables having high market demand (okra, brinjal, tomato, chilli, cabbage, cauliflower, broccoli, lettuce, leafy mustard, carrot, coriander, spinach etc.)

↓

Cultivation of cash crops like ginger or turmeric for income generation
Cultivation of Maize followed by French bean

↓

Fruit crops like pineapple, Assam lemon, papaya, banana, guava etc, at corners of the field

↓

Construction of bamboo machan over jalkund and cultivation of pumpkin, bottle gourd

↓

Promotion of fodder or other multipurpose trees Marketing of produce

**Fig. 3.** Different components of IOFS model in adopted organic villages of Meghalaya

Two individual farmers, Mr. JrillMakroh and Mrs. SkolaKurbah obtained net returns of Rs. 46,695 and Rs. 31,100 from their respective 0.27 and 0.21 ha IOFS models which is equivalent to Rs. 1,73,702/ha/year and Rs. 1,48,946/ha/year, respectively. The net return obtained from the IOFS models were significantly higher as compared to the farmers' practice of maize-fallow or cultivation of maize (rainy season) followed by vegetables in about 30% of the areas (winter season). The IOFS models established in these two farmers field could meet 76 - 95.1% of N, 68.6 - 82.0% of $P_2O_5$ and 85.5 -96.0 % of total $K_2O$ requirement of the systems.

## 2. Tamil Nadu

### IOFS model in TNAU, Coimbatore (https://tnau.ac.in)

A one acre Integrated Organic System Model has been established for irrigated upland conditions at Eastern Block Farm, TNAU, Coimbatore for effective utilization of on farm resources, which enables doubling of small and marginal organic farmers income and employment. The major components of the IOFS model includes crops (system based), dairy, manure pit, perennial tree crops, pest repellent cafeteria, fodder crops etc. The details of the IOFS model are given below:

### Crop component (3400 m$^2$)

### Cropping Systems

1. Green manure – Bhindi – Maize (0.12 ha)
2. Green manure – Cotton – Redgram (0.12 ha)
3. Fodder grass (Cumbu-Napier variety CoCN4 and Desmanthus (0.10 ha)

### Agro forestry (250 m$^2$)

Neem *(Azardiractaindica),* Pungam *(Pongamia pinnata),* Gmelina *(Gmelina arborea),* Ailanthus *(Ailanthus excelsia),* Wood Apple (*Feronia elephantum*), Bamboo (*Bambusa vulgaris*), Orchid tree (*Bauhinia sp.*), Mahua (*Madhuca longifolia*), Red bead tree (*Adenantherapavonina*), Malabar neem (*Melia dubia*), Jamun (*Syzygiumcumini*), Portia tree (*Thespesia populnea*),Paradise tree (*Simarouba glauca*)andFig tree (*Ficus glomerata*)

### Pest Repellent Cafeteria (250 m$^2$)

Jatropha (*Jatropha curcas*), *Adhatoda* (*Adhatodavasica*), Yellow oleander (*Cascabelathevetia*), Notchi (*Vitex negundo*),Indian laurel (*Calophyll-uminophyllum*), Netted custard apple (*Annona reticulata*), Bael (*Aegle marmelos*), Tanner's cassia (*Cassia auriculata*), Karanda (*Carissa carandas*), Paper flower (*Bougainvillea glabra*), Lemon grass (*Cymbopogan citrate*)and Crepe jasmine (*Tabernaemontanadivaricata*).

### Dairy (20 m$^2$)

Milch animal: 2 cows with one calf

### Manure pit (30 m$^2$)

Residue from the crops and manure from Dairy unit will be converted into compost and used as manure for crops

### Supporting activities (50 m$^2$)

Threshing floor, cattle shed, cow dung pit, cow urine collection unit composting units, nutrition garden etc.

### Border plants

Banana, *Glyricidia*, coconut, *Desmanthus*, annual moringa, teak and curry leaf

### Impacts

The IOFS model resulted in net return of Rs. 1,04,924/- per year (87% from crops, 11% from livestock, 2% from others) and Benefit-cost ratio of 2.30. This model generated an employment of 464-employment man days per year and contributes 143.2 kg of Nitrogen, 33.8 kg of Phosphorous and 56.3 kg of Potassium/ha/year as nutrient contribution through green manuring, FYM, vermicomposting, mulching and weed and crop residue compost additions.

## 3. Kerala

### IOFS model in Kerala (https://www.indiafilings.com)

The IOFS model comprising of crops (Turmeric, coconut, fodder, yam, banana, tapioca and vegetable cowpea) + livestock (2 cows) has been developed at

Calicut, Kerala. Model could generate net income of Rs 1,23,000/acre and 89% of seeds/planting material/ nutrients requirements are fulfilled within the system.

## 4. Sikkim

### IOFS model in east Sikkim (https://www.indiafilings.com)

A 1.25 acre IOFS model comprising of cropping systems (rice-vegetable pea, rice-potato-dhaincha, maize-soybean-buckwheat, rice-toria-dhaincha, rice-cabbage-dhaincha, coriander-radish-broccoli-fenugreek and cauliflower-pea-beet root-spinach) + 2 cows+ 50 poultry birds in Timpyem and Nandek in east Sikkim. Net income of Rs 1,37,000/year can be obtained from this model.

### IOFS modelsin ICAR-NOFRI, Tadong, Gangtok (Avasthe *et al* 2020)

In order to overcome the problems of irregular economic return from agriculture, a holistic, resource based, client-oriented and interactive approach i.e., integrated organic farming system (IOFS) model was designed and developed at Institute level and in farmers' fields in participatory mode for marginal and small farmers having land holding of d" 1.2 acre (Table 3). The performance of each model consisting of agriculture, horticulture, livestock (dairy/poultry) and fisheries components was also evaluated at Research Farm and replicated at farmers' field. Maize-Black gram/rajmash-buckwheat/pea-based cropping system along with fruits (Sikkim mandarin, guava etc.), vegetables (year-round seasonal vegetables under low cost poly tunnel and rain shelters), livestock (dairy/poultry/ pig), and vermicompost were integrated in d" 1.2 acre land which increased the cropping intensity up to 300% with 481% - 518% higher net return and employment for 270 to 295 days as compared to only 60±10 man-days in maize/rice-based mono-cropping system (Table 4).

**Table 4.** IOFS model developed by ICAR-NOFRI

| S.No. | Model component | Model area (ha) | Cost of production (Rs) | Gross return (Rs.) | Net return (Rs.) | B:C ratio | Total system employment |
|---|---|---|---|---|---|---|---|
| 1. | Agriculture + horticulture + dairy + fishery + poultry + vermicomposting | 0.6 | 114625 | 244550 | 129925 | 2.13 | 295 |
| 2. | Agriculture + horticulture + dairy + vermicomposting | 0.6 | 116695 | 238685 | 121990 | 2.05 | 270 |
| 3. | Agriculture + horticulture + dairy + pig + vermicomposting | 0.6 | 119600 | 248560 | 128960 | 2.08 | 285 |

## Demonstration of IOFS in Sikkim (Avasthe *et al* 2020)

The majority of activities were initiated at the villages adopted by the Institute in East Sikkim district and the same were gradually replicated in the other three districts of Sikkim.Individual farmers with aptitude for adopting innovative technologies to enhance farm income through integrated organic farming system (IOFS) approach were brought under the project activities. The farmers/ entrepreneurs interested to begin a medium to large-scale cash crop and poultry farming were facilitated by linking them with funding agencies, research institutions and marketing agencies (Table 5).

**Table 5.** Details of farmer's field IOFS models

| Name of district with altitude | Model component | Total no. of villages | Area (ha) |
|---|---|---|---|
| East Sikkim:Altitude -900 to 1300 m AMSL | Agriculture + horticulture + dairy + fishery/poultry/pig + vermicomposting + cardamom based agroforestry model | 15 | 27.6 |
| West Sikkim:Altitude -900 to 2700 m AMSL | Agriculture + horticulture + dairy + fishery/poultry/pig + vermicomposting + cardamom based agroforestry model | 05 | 9.5 |
| North Sikkim:Altitude -900 to 4300 m AMSL | Agriculture + horticulture + poultry/pig + vermicomposting + cardamom based agroforestry model | 09 | 24.6 |
| South Sikkim:Altitude -900 to 1800 m AMSL | Agriculture + horticulture + poultry/pig + vermicomposting | 12 | 7.5 |
| **Total** | | 41 | 69.2 |

## Socio-economic profile after adoption of IOFS

### 1. Employment generation

People living below the poverty line were mostly dependent on agriculture and livestock component for their livelihood and had minimum scope of employment for 3 to 4 months during the cropping season. However, the introduction of second crop after maize, rice, integrated organic farming system, year-round vegetable production, kiwifruit production, semi-intensive poultry farming and pig farming created year-round employment avenues for farm women and rural unemployed youth.

An integrated organic farming system with agri + horti + dairy + fishery + vermicompost components generated employment for 270 to 295 days as compared to only 60±10 man-days in maize/rice-based monocropping system. The Vanaraja poultry having the capacity of 50 to 100 nos. of birds generated employment for farm women/youth round the year and generated income of at least of Rs. 2500 to 3000/-per month. Similarly, pig farming with 2 sows has the potential to generate employment for a person for about 100-140 days with monthly income > Rs. 2,000/- per month. Enhancing the number of sows from 2 to 10 with proper training in animal health care and vaccination may have tremendous opportunity to earn at least of Rs. 25,000 to Rs. 30,000/- per month. Overall, innovative scientific interventions significantly increased the financial status of tribal farmers over their existing farming conditions.

## 2. Improvement in purchasing power

After fulfilling the nutritional requirement at household level, IOFS model provides an additional amount of Rs. 70,000 ±10,000 per annum to the farmers. Hence, IOFS supported the poor farmers to spend more towards the purchase of essential commodities for household. The poor farmers who otherwise were not able to send the kids to school became capable to spend some money for education of their children and also could take better care of their health. Additionally, from the IOFS model, many landless tribal farmers obtained new means to sustain their livelihood by initiating small poultry units of 50 to 100 nos. of Vanaraja backyard poultry birds. After training at ICAR Research Complex, Sikkim Centre improved variety of poultry was reared to earn Rs. 2000 to 5000/- per month. Organic Kiwifruit cultivation also provided new avenues of income generation at high altitude areas to the farmers with land unsuitable for the cultivation of annual crops. Many self-help groups, NGOs in Sikkim have adopted IOFS for livelihood improvement at different villages and obtained tremendous success. The Govt. of Sikkim is also now promoting the IOFS models among the farming community of Sikkim. The performance of IOFS run by few successful beneficiaries is presented below (Table 6).

**Table 6.** Impact of IOFS in increasing purchasing power(Avasthe *et al* 2020)

| S.No. | Name and address of beneficiaries | IOFS module | Economic implication | Social implication |
|---|---|---|---|---|
| 1. | Mrs. Pemkit Lepcha, Timpyem, East Sikkim | Agriculture + horticulture + dairy + poultry + pig + vermicomposting | Net income Rs 1.094 lakhs per year | Managing family with her additional earning. Received Mahindra Krishi Samridhi Award under Youth Category in 2017 |
| 2. | Shri Nim Tshering Lepcha, Nandok, East Sikkim | Agriculture + horticulture + cattle + fishery/ poultry + pig + vermicomposting + large cardamom | Net income Rs 3.25 lakhs per year | Managing family with his additional earning and constructed pucca house, and is planning to develop agro-tourism. Felicitated by Sh. Khorlo Bhutia, |

| | | | | |
|---|---|---|---|---|
| | | | | Secretary, Food Security and Agriculture Development Department, Government of Sikkim as progressive farmer in 2015. |
| 3. | Mrs. Doma Bhutia Thanka-Lingtam | Agriculture + horticulture + dairy + poultry + pig + vermicomposting | Net income Rs 1.12 lakhs per year | Managing family with her additional earning. |

## Other benefits of integrated organic farming system

- ***Productivity:*** The demand for food, feed, fodder and fiber is increasing with an increase in population, but the availability of land for agriculture is decreasing. IOFS provides an opportunity to increase economic yield per unit area per unit time by virtue of intensification of crop and allied enterprises.
- ***Profitability***: Use waste material of one component at the least cost. Thus, reduction of cost of production and form the linkage of utilization of waste material and elimination of middleman interference in most inputs used. Working out net profit/ BC ratio is increased (Table 1&3).
- ***Potentiality or Sustainability***: Organic supplementation through effective utilization of by-products of linked component is done thus providing an opportunity to sustain the potentiality of production base for much longer periods.
- ***Risk minimization:*** A single commodity-based agriculture is more vulnerable to climatic, biotic (pests and diseases) and economic (relative prices of input and output) changes compared to IOFS. IOFS are often less risky than single enterprises based production system.
- ***Balanced Food*:** Components of varied nature are linked to produce different sources of nutrition.
- ***Environmental Safety:*** In IOFS waste materials are effectively recycled by linking appropriate components, thus minimize environment pollution.
- ***Recycling*:** Effective recycling of waste material (crop residues and livestock wastes) in IOFS. Therefore, there is no reliance to outside inputs – fertilizers, agrochemicals, feeds, energy, etc.
- ***Income Rounds the year*:** Due to interaction of enterprises with crops, eggs, milk, mushroom, honey, cocoons silkworm, it provides flow of money to the farmer round the year. There is higher net return to land and labour resources of the farming family.

- ***Adoption of New Technology***: Resourceful farmers (big farmer) fully utilize technology. IOFS farmers, linkage of dairy/mushroom / sericulture / vegetable. Money flow round the year gives an inducement to the small/ marginal farmers to go for the adoption of technologies.
- ***Saving Energy***: To identify an alternative source to reduce our dependence on fossil energy source within short time. Effective recycling technique the organic wastes available in the system can be utilized to generate biogas. Energy crisis can be postponed to the later period.
- ***Meeting Fodder crisis***: Every piece of land area is effectively utilized. Plantation of perennial legume fodder trees on field borders and also fixing the atmospheric nitrogen. These practices will greatly relieve the problem of non – availability of quality fodder to the animal component linked.
- ***Solving Fuel and Timber Crisis***: Linking agro-forestry appropriately the production level of fuel and industrial wood can be enhanced without determining effect on crop. This will also greatly reduce deforestation, preserving our natural ecosystem.
- ***Agro – industries***: When one of produce linked in IOFS are increased to commercial level there is surplus value adoption leading to development of allied agro-industries.
- ***Increasing Input Efficiency***: IOFS provide good scope to use inputs in different component greater efficiency and benefit cost ratio.
- ***Climate change mitigation and adaptation:*** The IFS has an important role in reducing atmospheric CO2 concentration because (a) integration of woody perennials (trees and shrub) enhances C storage, (b) integration of livestock provides manure, and application of manure enhances C storage in the soil, and (c) integration of woody perennials and livestock with arable crops provides nutrients to crops via effective recycling of resources which reduces need of fertilizers and hence, indirectly saves fossil fuels.

## Constraints in IOFS

- ***Low amount of nutrients and high C: N ratio of different organic residues***: Organic nutrient sources like farmyard manure, compost, poultry manure, straw, etc. contain very low amount of nutrients, not sufficient to meet the nutrient requirements of crops thereby forming a large gap exists between the available potential and their utilization. Slow release especially when temperature remains low for most period of crop growth makes the synchronization of nutrient release and their uptake by crop plants a difficult task. Although many organic residues like paddy straw, wheat straw, sugarcane trash, can be used to make up the plant nutrients need yet they have a wider C: N ratio. If they are not fully decomposed and added as such, will cause immobilization of nutrients. Therefore, it is not possible to meet the nutrient requirements of crops entirely from organic sources, if 100% cultivable land is converted to organic farming.

- ***Low yield of crops:*** With the use of organic sources of nutrients, the yield of the crop is very low, especially during initial stages, although it may stabilize later, yet complete dependence on pure organic farming will not be sustainable in the long run especially in Indian conditions due to huge food demand of ever-increasing population.
- ***Small holding and poor farmers:*** They are not directly connected to markets to sell farm products (Kaur and Toor 2015) main reason for lower profit of the farmer.
- ***Market and Infrastructural problems:*** Market development, especially domestic markets, continues to be one of the biggest challenges. Going to major markets for the selling of small quantities of additional marketable farm product produced from small farms, aren't profitable. Closed corner location, remoteness and isolation from major markets are the prime constraints. High consumer prices may also limit consumer. There is tremendous lack of infrastructural facilities for processing, packing, storage etc. to meet the organic standards (Kaur and Toor 2015) and to avoid contamination of organically produced food. On one hand, target groups of organic food products such as big hotels, restaurants, airlines, cafes, etc. which can afford to pay premium prices for high quality organic foods are completely lacking. On the other hand, common people cannot afford to pay higher prices for organically produced food indicating low domestic consumption.
- ***Control of weeds:*** Organic sources of nutrients promote profuse proliferation of weeds that compete with the plants for different nutrients, space, light, water, etc. On the other hand, weedicides cannot be used in organic farming, and thus economic weed control remains a challenge in organic farming.
- ***Scarcity of biomass:*** Readily available and abundant organic sources are not enough to meet the requirements of composts, vermicomposts etc. A lot of organic wastes and plant residues get lost in the fields either because of difficulty in collection or to handle or use properly. Moreover, people in rural areas use plant residues like straw, stubbles, pruned material and cow dung for producing charcoal as an energy source, because dependence on other energy sources like LPG, diesel, petroleum products etc, proves costly (Sharma and Pradhan 2011).
- ***High input costs:*** Local or farm renewable organic resources like neem cakes, groundnut cakes, cow dung, earthworms, etc are becoming costlier

day by day than the conventional or industrially produced chemical fertilizers & pesticides. Chemical fertilizers are easier to purchase given the farmer has purchasing power.

- ***Lack of awareness:*** There is lack of awareness and knowledge about modern methods or techniques of composting, vermicomposting etc., among the farmers from the preparation as well as application point of view and thus both quality and efficacy are poor at the end (Kaur and Toor 2015). Selection and management of appropriate enterprises in IOFS also required sufficient awareness and knowledge, which farmers generally lack.
- ***Certifying oriented problems:*** Before producing marketable products an organic farm has to have a transition period of 1 to 3 years depending upon the certifying agency's requirements and during this period the farmers have to grow the crops as per standards set for organic farming and thus produce about $3/4^{th}$ of the normal yield. Yet they have to market the produce in the open general market. Small and marginal farmers can hardly afford to do so. Moreover, the farmers won't opt for certification because of the costs involved as well as the extensive documentation that is required by certifiers.

**Niche areas for IOFS in India**

- ***Northern east hilly region***: Advantages to go for IOFS production (Das *et al* 2020).
    - o Firstly, the use of inorganic fertilizers and chemicals is meager in the region as farmers are having apathy towards use of agro-chemicals.
    - o Secondly,the difference in production gap due to adoption of organic agriculture is expected to be negligible in NEH region; rather there is scope for enhancing productivity with good organic management and the organic premiums would boost earning of the hill farmers.
    - o Thirdly, it is an added advantage that all the households are maintaining livestock (pig, poultry, cattle, goats, etc.) and the region is having about 26 million livestock population producing sufficient quantity of on-farm manures, which could be effectively used for organic agriculture.
    - o Some other strengths of NER for organic farming are
    - o The region is home to some of the high value niche crops like Assam lemon, Joha rice, medicinal rice, sticky rice (Manipur), pineapple, aromatic rice of Tripura (harinarayana&kalikhasa), Khasi mandarin, ginger, turmeric (Meghalaya), French bean, King chili, bird eye chili (Nagaland) Assam tea, large cardamom, ginger, pine apple and passion fruits which has high market demands.

- o Less pressure on food demand due to thin population density as compared to vast natural resources available in the region.
- o Dependence of mid and high altitude farmers on within farm renewable resources.
- o Time tested indigenous farming systems and use of indigenous technical knowledges in agriculture are mostly prevailing in the region.

- ***Arid and semi-arid regions*** (Rathore *et al.,* 2019)
  - o IOFS involving integration of different enterprises (crop, livestock, horticulture, forestry, poultry and fish) and recycling of wastes reduced cost of production and enhanced productivity, profitability, resource use efficiency, generated more employment and minimized resources degradation and risks due to drought as cultivation is rainfed in these areas.

Integrated organic farming system (IOFS) is the best option for enhancing sustainability, consumer safety, market profitability and livelihood security in particular regions of India. The research and extension in the IOFS area is necessary which majorly emphasize on the local needs and conditions. Dissemination and adoption of the IOFS practices among the existing farmers is very essential to know the importance of judicious implementation of organic farming which enhances the adaptability of the crops under changing climatic scenario. This will lead to economic improvement and empowerment of farmers under organic conditions.

## Strategies needed to promote organic farming based farming system in India

In Indian context, to promote and make organic farming based farming system economically viable following issues are needed to be addressed.

- Adequate research and extension support needs to be provided in order to improve region specific farming techniques and disseminate findings for conversion and management of organic farms in farming system mode.
- Researchers should study and quantify the role of organic agriculture in mitigating the climate change and ill effects of modern agriculture and improve resource sustainability.
- Central and State government should acknowledge organic farming as an emission mechanism to reduce green house gases and sequester carbon. They should help farmers by promoting organic agriculture through research and extension services.

- Government should recognize organic agriculture in Kyoto Protocol carbon credit mechanism.
- Organic market development sector needs major thrust on developing supply chains and related infrastructure to ensure competitive price of organic produce to the grower in domestic and international markets.
- Mission – Mode programs for on-farm demonstrations, training for capacity building of institutions, organic farmers, service providers, NGOs and processing / packing industry, with full research backup are needed. Model organic farms are need to be established in Public-Private-Partnership mode.
- Government support is required for cheaper access to organic certification of farms.
- There is a need to create the data base on farming system in relation to type of farming system, infrastructure, economics, sustainability etc. under different farming situation.
- Need to develop research modules of farming system under different holding size with varying economically viable and socially acceptable systems.
- The assessment and refinement of the technologies developed at research station at cultivators' field. * Need to prepare a policy a contingent planning to counteract the weather vagaries/ climate threats under different farming situations.
- Need to prepare a policy draft for the consideration of planners for its promotion at large scale with nominal financial assistance either through short/medium /long term loans and other promotional advantage.

# 15

# Good Agricultural Practices (GAP) Concept & Strategies

## Concept

The concept of Good Agricultural Practices (GAP) has evolved in recent years in the context of a rapidly changing and globalizing food economy and as a result of the concerns and commitments of a wide range of stakeholders about food production and security, food safety and quality, and the environmental sustainability of agriculture. These stakeholders include governments, food processing and retailing industries, farmers, and consumers, who seek to meet specific objectives of food security, food quality, production efficiency, livelihoods and environmental benefits in both the medium and long term.

Broadly defined, GAP applies available knowledge to addressing environmental, economic and social sustainability for on-farm production and post-production processes resulting in safe and healthy food and non-food agricultural products. Many farmers in developed and developing countries already apply GAP through sustainable agricultural methods such as integrated pest management, integrated nutrient management and conservation agriculture. These methods are applied in a range of farming systems and scales of production units, including as a contribution to food security, facilitated by supportive government policies and programmes.

## Good agricultural practices for crop production

### Selection of crops / varieties

Crop production involves the selection of annual and perennial crops, their cultivars and varieties, to meet local consumer and market needs according to their suitability to the site and their role within the crop rotation for the management of soil fertility, pests and diseases, and their response to available inputs.

### Good agricultural practices related to crop production will include

- Selecting the most appropriate cultivars and varieties on an understanding of their characteristics, including:

    - Response to sowing or planting time
    - Productivity
    - Quality
    - Market acceptability
    - Nutritional value
    - Disease and stress resistance
    - Climatic adaptability
    - Response to fertilizers and agrochemicals
- Devising crop sequences to optimize use of labour and equipment
- Maximizing the biological benefits of weed control by:
    - Competition
    - Mechanical
    - Biological
    - Herbicide options
- Providing non-host crops to minimize disease

**Nutrient management**

The physical and chemical properties and functions, organic matter and biological activity of the soil are fundamental to sustaining agricultural production and determine, in their complexity, soil fertility and productivity. Appropriate soil management aims to maintain and improve soil productivity by improving the availability and plant uptake of water and nutrients through enhancing soil biological activity, replenishing soil organic matter and soil moisture, and minimizing losses of soil, nutrients, through erosion, runoff and leaching into surface or ground water.Though soil management is generally undertaken at field/farm level, it affects the surrounding area or catchment due to off-site impacts of runoff, sediments, nutrients movement, and mobility of livestock and associated species including predators, pests and bio-control agents.

**Good agricultural practices related to nutrient management include**

- Maintaining and improving soil organic matter through the use of compost, manure and other soil ameliorants (material that you add to the soil to improve it)
- Rotating crops-to enable the soil to build up and replace the nutrients extracted by the previous crop. On occasion the land may be left fallow or converted to pasture

- Maintaining soil cover to provide a welcoming environment for the microorganisms and earthworms living in the soil. These organisms help breakdown soil and make it more fertile
- Rational mechanical and/or conservation tillage – by not over ploughing or rotovating the soil. Ensuring hard pans or dust bowls are not created
- Rotating pastures by moving the animals from paddock to paddock
- Adding chemical fertilizers only after soil analysis to reduce wastage and toxic build-up in the soil or pollution of aquifers and surface water
- Applying chemical fertilizers as recommended by the manufacturers and the Ministry of Agriculture. Application methods should match the formulation of the fertilizer, e.g. foliar fertilizers should be in solution and applied directly to leaves, whereas a granular formulation would be applied in the drip zone
- Establishing beds or rows along the contours on sloping land
- Maintaining a protective cover crop on sloping land to reduce erosion and inevitable soil loss

### Water management

Agriculture carries a high responsibility for the management of water resources in quantitative and qualitative terms. Careful management of water resources and efficient use of water for irrigation where applicable, and for livestock, are criteria for GAP. Efficient irrigation technologies and management will minimize waste and will avoid excessive leaching and salinization.

### Good agricultural practices related to water management will include

- Improving soil structure and increasing soil organic matter content; that will help the soil hold more water and reduces the frequency with which water has to be applied
- Applying only those production inputs (including waste or recycled products of organic, inorganic and synthetic nature) that will not contaminate of water resources
- Monitoring crop and soil water status through the use of simple techniques like sample rolling or through the use of electronic probes
- Accurately scheduling irrigation
- Routinely checking the quality of water from on-farm wells and managing water tables to prevent excessive extraction
- Preventing soil salinization by adopting water-saving measures and re-cycling where possible

- Providing adequate, safe, clean watering points for livestock in extensive, open range systems
- Using drip irrigation techniques as opposed to sprinkler systems
- Managing ground and soil water by proper use, or avoidance of drainage where possible. Thus using the pitch of drains to slow water movement out of the plot
- Directing guttered water from buildings into storage tanks or ponds
- Adopting techniques to enhance the functioning of the water cycle by establishing permanent cover and maintaining watersheds
- Maintaining or restoring wetlands as necessary

### Crop protection

Maintenance of crop health is essential for successful farming for both yield and quality of produce. This requires long-term strategies to manage risks by the use of disease and pest-resistant crops, crop and pasture rotations, disease breaks for susceptible crops, and the judicious use of agrochemicals to control weeds, pests, and diseases following the principles of Integrated Pest Management.

### Good agricultural practices related to crop protection will include

- Identifying the known pests an diseases of the crop
- Listing and being guided by the conditions under which the known pests and diseases strike
- Using resistant cultivars and varieties
- Using crop sequences, associations, and cultural practices that maximize biological prevention of pests and diseases
- Learning how to assess the balance status between pests and diseases and beneficial organisms of the crops you are planting
- Utilize organic control practices where and when applicable
- Deciding on interventions following consideration of:
  - Farm productivity
  - Environmental implications of agrochemicals
  - Integrated pest management (IPM)
- Maintaining accurate records of agrochemical use.
- Storing and using agrochemicals according to legal requirements of registration for individual crops, rates, timings, and pre-harvest intervals

- Ensuring that agrochemicals are only applied by trained and knowledgeable persons
- Ensuring that equipment used for the handling and application of agrochemicals complies with established safety and maintenance standards

## Harvest and on-farm processing and storage

Product quality also depends upon implementation of acceptable protocols for harvesting, storage, and where appropriate, processing of farm products.Harvesting must conform to regulations relating to pre-harvest intervals for agrochemicals and withholding periods for veterinary medicines. Food produce should be stored under appropriate conditions of temperature and humidity in space designed and reserved for that purpose. Operations involving animals, such as shearing and slaughter, must adhere to animal health and welfare standards.

Good practices related to harvest and on-farm processing and storage will include those that harvest food products following relevant pre-harvest intervals and withholding periods, provide for clean and safe handling for on-farmprocessing of products. For washing, use recommended detergents and clean water; store food productsunder hygienic and appropriate environmental conditions; pack food produce for transport from the farm in clean and appropriatecontainers; and use methods of pre-slaughter handling and slaughter that re humane and appropriate for each species, with attention to supervision, training of staff and proper maintenance of equipment.

## Potential benefits of GAP

- Appropriate adoption and monitoring of GAP helps improve the safety and quality of food and other agricultural products
- It may help reduce the risk of non-compliance with national and international regulations, standards and guidelines, regarding permitted pesticides, maximum levels of contaminants in food and non-food agricultural products, as well as other chemical, microbiological and physical contamination hazards
- Adoption of GAP helps promotes sustainable agriculture and contributes to meeting national and international environment and social development objectives.

# 16

# Lessons Learnt From Organic Farming

Exhaustion of the natural resources, destruction of the agrosphere and environmental deterioration resulting from increased man-made environmental modification expressed as climate change, has posed serious threat to existence of humankind. Growing awareness of environmental and health issues associated with the intensive use of chemical inputs has led to interest in alternate forms of agriculture in the world. The best solution to former problem is organic agriculture. Organic farming unites all agricultural systems that maintain ecologically, economically and socially advisable agricultural production. The adverse effects of synthetic chemicals used in agriculture have changed the mindset of some consumers of different countries who are now buying organic with high premium for health. Policy makers are also promoting organic farming for restoration of soil health and generation of rural economy apart from making efforts for creating better environment. Organic farming is being practiced in more than 100 countries of the world. Organic farming through sustainable agriculture meets not only the food requirements of present generation in an eco-friendly way but also the requirements of future generations and maintains our environment. In India, organic farming, in spite of the reduction in crop productivity by 9.2 per cent, provided higher net profit to farmers by 22.0 per cent compared to conventional farming. This was mainly due to the availability of premium price (20–40%) for the certified organic produce and reduction in the cost of cultivation by 11.7 per cent. In view of rising population and demand for food production, conventional farming systems cannot be neglected but organic farming should be certified where it already exists and promoted to the newer areas to the extent possible because it seems safer bet for sustainable agriculture at a time when advanced technologies are still costly and have to be proved safe for long-term development.

Agriculture is the backbone of the Indian economy. About 67% of our population and 55% of the total work force is dependent on agriculture and other allied activities. Agriculture meets the basic needs of India's ever growing population. It has been estimated that the agricultural growth of around 4% is required for India in order to achieve a double digit growth in Gross Domestic Product (GDP). Despite this, agriculture is confronting various limitations such as fragmentation of agricultural landholding, conversion of agricultural land to non-agricultural uses and low productivity. Still due to the rising demand for quality and value-added products, there are increasing investment opportunities in the sector. Since, the population of world is increasing at an alarming rate and there would be more requirements of the food production to feed the world but in a sustainable manner.

Generally, man is dependent on plants for basic necessities such as food, fibre, medicine, clothing, shelter etc. The plant growth and yield in all ecosystems depend on the cycling and recycling of nutrients between the plant biomass and the organic and inorganic soil stores. Agriculture, over the years developed as a result of man's quest to feed himself, family and his animals. Through time man discovered that agriculture attracts various input from science, art and also a business of producing plants and animal products for the advantageous use of mankind. Plants are primary producers in agriculture. Plants by the process of photosynthesis, take in carbon dioxide from the air, moisture and nutrients from the soil and traps the energy from sun light thereby converting these simple compounds into complex food materials. Animals are secondary agricultural producers. They eat plants or parts of plants and convert the complex compounds present in plants into animal products such as eggs, meat, milk, hides and wool. Traditional/primitive agriculture relies on the soil, rainfall and the local species of plants and animals. Any plants and organisms which affect the productivity of useful plants and animals were avoided or controlled. The productivity in such system was quite low and capable of feeding the farmers and his family alone. Due to population pressure and increased urbanization, agricultural practices improved with use of high external inputs such as agro chemicals for pest and weeds, inorganic fertilizers which in return helped to enhance productivity but with serious adverse effect on environment and making the entire system very unsustainable.

Agricultural production and yields during the last decades have been increasing in the industrialized countries along with global fertilizer and pesticide consumption. The rising global trade with agricultural products and the improved access to fertilizers and pesticides have changed agricultural systems. The rapid development in transportation and communication has enabled farmers to buy their inputs and sell their products further away in larger quantities. These developments resulted in increased food security, whereas a greater variety of food has been offered and diets have changed towards a greater share of meat and dairy products. However, such trend has led to a growing discrepancy among populations and agricultural systems, especially developing countries in Africa have seen very few improvements in production and food security. At the same time, the application of unsuitable farming techniques and the rapid increase of farming inputs use have contributed to the rise of environmental problems due to agricultural activities such as radical decline in biodiversity, pollution of surface and groundwater with pesticides and nitrates, soil degradation and to a some extent, global warming (Tilman *et al.*, 2002). For further repositioning of agriculture for sustainable consumption & income, the knowledge of what is presently going on is needed to shape what will happen in future.

In the more industrialized countries, the increasing concern from the consumer community and from part of the farmers on the negative environmental consequences of intensive agricultural activity, also coupled with the increased demand for healthy food, have both contributed to develop agricultural systems based on sustainable farming practices, primarily focused to preserve the natural resources while ensuring reliable productivity of food. But unfortunately, global warming seems to be the major threat for food security, especially in tropical countries. It is assumed that global warming will worsen the irregularity of rainfall and the drought intensity in many countries. Meanwhile, intensive agriculture which is dependent on agrochemicals and non-renewable fossil fuels is being responsible for over 20 % of global anthropogenic greenhouse gas emissions (Scialabba, 2003). In Asia, this figure could slightly vary chiefly it may be higher as most of the fertilizers used here are nitrogen-based (Stoll, 2002) and the industrial process of manufacturing nitrogen fertilizer emits nitrogen dioxide into the atmosphere which is a strong greenhouse gas. Since green revolution technologies is considered as the major food production system in the world but still there is growing evidence that the green revolution has, at its worst, increased inequality, worsened absolute poverty, and resulted in environmental degradation (IFPRI, 2002).

In developing countries, agriculture is the vital sector for the socio economic development of masses. The trend in agricultural production system from traditional/indigenous to modern agriculture has made the agriculture dynamic and improved overtime. Presently, there is another strategy known as organic agriculture, which is a production system that sustains and maintains the soil health, the ecosystem and the people living in that system and beyond. It is completely dependent on natural processes, biodiversity and cycles adapted that are understood by local agricultural farmers in their conditions but without making the use of external inputs having negative effects. Organic agricultural practices is a combination of innovations, art, science as well as business to promote the environmental gains and ensure fair relationships along with the good quality of life for all involved in it (Ibeawuchi *et al.*, 2015). Organic agricultural practices enhance productivity by the slow release of essential nutrients to the plants as well as maintains soil biological stand. This brings about increase in productivity without causing any adverse effect on the environment. Therefore, ensuring food security, mitigating poverty and conserving the essential natural resources is significantly important (Rothschild, 1998) and is achieved through organic farming and various other means without deteriorating natural resources.

The major challenge for scientists is that it very difficult to feed the ever increasing population with organic food (Moghtader *et al*, 2011). Safe production and secure food supply is one of the major need of low income countries to restore their reservoirs (Arshad and Shafqat, 2012). The concept of food security revolves around the components of agriculture, environment, marketing, employment income, health & nutrition and public policy (Johan, 1999). Organic farming with limited or no use of synthetic chemicals has huge potential to minimize these negative impacts on ecosystem. Therefore organic farming has been considered as a major thrust area for achievement of Millennium Development Goals (MDGs) and to ensure sustainable development. Conversion of chemical based farming systems to organic farming contributes to diminish the contribution of agriculture to global warming. It therefore, contributes to the stability of the food supply which is threatened by the climate change (Aubert, 2007).

Organic and biodynamic agriculture represent some of the various approaches to sustainable agriculture, and several techniques used in such methods (e.g. rotation of crops, inter-cropping, minimum tillage, integration of crops and livestock, mulching) are in fact practiced under various agricultural systems. However, various facts that makes organic and biodynamic agriculture unique, as regulated under various laws and certification programs is that: (i) almost all chemical and synthetic inputs are forbidden, (ii) genetically modified organisms (GMOs) are prohibited and (iii) soil binding crop rotations are mandatory (Anonymous, 1999). Organic agriculture unites all agricultural systems that maintain ecologically, economically and socially advisable agricultural production (IFOAM, 2002). These agricultural systems utilize the natural potential of plants, animals and landscapes and are aimed at the agricultural practice's harmonization with the environment.

Organic farming is an integrated production management system which promotes and enhances agro-ecosystem health including biodiversity, biological cycles and soil biological activity. It focus on the usage of management practices such as the use of off–farm inputs, taking into account that regional conditions require locally adapted systems. This is achieved by using agronomic, mechanical and biological methods where possible, without using any synthetic materials, to accomplish any specific function within the system (FAO, 2002). Organic farming predominantly restricts the usage of external production factors (resources) by putting a limitation to the use of chemically-obtained fertilizers, pesticides and pharmaceutical preparations. Instead, in order to enhance the yield and protect the crops, organic farming involves other agro technical methods and several natural factors. Organic farming is a type of agriculture that rely on various techniques such as green manuring, crop rotation, vermicomposting, biofertilizers, animal husbandry and biological pest control. Thus, organic farming involves cultivation of crop/plant by using organic manures that are environment friendly and supports the life of soil and other beneficial organisms in the soil. Organic farming is becoming popular in developing countries, where farmers make use of animals for tilling the land and manures prepared from dung and other waste material of animals. Thereby, organic farming supports the quality of production and crop yield (Auobamiri and Ahmad, 2014). Most of the developed countries utilize synthetic chemicals to protect plants/crops from insects, pests attack and other plant diseases. The usage of synthetic chemicals may give high yield for some period of time and protects the plants but the continuous usage of such chemical in farming may result in inactivation of some useful organisms in soil; in addition to this, certain insects or pests and other diseases develop resistance to these chemicals and later the chemical may not show its effect on the insects, pests and other harmful organisms to the crops. Organic farming in some developing countries like India is not new approach and is being followed from ancient days. Bio fertilizers are prepared with beneficial microbes which release essential nutrients to soil and promote the crop growth and product yield without any environmental degradation (Rao *et al*, 2014). Organic agriculture developed as a response to the industrialization of agriculture and its associated environmental and social problems. Since organic agriculture actually delivers overall advantages over conventional agriculture is however controversial. Some claim that organic agriculture are more profitable and eco-friendly (Reganold and Wachter, 2016), while others question the role of organic agriculture in future sustainable food systems (Connor and Minguez, 2012).

Organic agriculture adheres to the principles that have been developed by the specific local, climatic, economic, social, cultural and historical features. Organic agriculture is mainly based on four basic principles namely health, ecology, fairness and care (Juma, 2007) and are key for sustainability in agriculture. Organic production systems can provide major contributions to farmer livelihoods and food supply stability by improving fertility of the soil, providing diversity and therefore, ensuring flexibility to food production systems in accordance to many causalities of climate change. Usually, a local marketing initiative's set up in organic farming and operating strategy should match its scope. The appropriate option thus, chosen will depend on the country, the location, nature and availability of product range, scale of planned operation and existing sector/market conditions (Wai, 2000). The impact of organic agriculture in comparison to conventional agriculture has not been extensively studied including various economic and social aspects. Therefore, it is vital that every country should practice organic farming and should minimize the usage of synthetic chemicals, but the green revolution had brought enormous change in food production systems and now it requires organic form of green revolution by meeting the food requirement and to safeguard the soil life and the ecosystem (Sudadi and Sumarno, 2014). The fertility of soil is declining with increase use of synthetic chemicals and the harmful organisms are developing resistance towards synthetic chemicals and these chemicals are dangerous to the ecosystem and the animals. Many countries export and import different food products like crops, fruits, seeds etc. but from these, some of the food products are banned to import due to high chemical content which would be detrimental for human health, if consumed (Ul-Haq *et al.*, 2014).

In India, the organic movement has its origin in the work of Howard (Howard, 1940) who devised and conceptualized most of the views which were later acknowledged by those people who became active in this movement. The objectives of environmental and socio-economic sustainability are the fundamentals of organic farming (Stockdale *et al*, 2001). The key features comprises of safeguarding the long-term soil fertility by maintaining sufficiency through the use of legumes and biological nitrogen fixation, efficient recycling of organic materials inclusive of livestock wastes and crop residues and weed, diseases and pest control focusing primarily on crop rotations, natural predators, organic manuring and resistant varieties. The immense importance is laid on maintaining the fertility of the soil by returning all the wastes back to the soil primarily through compost to reduce the gap between NPK addition and removal from the soil (Chhonkar, 2002). Today, the rapidly increasing population pressure has compelled many countries to use synthetic fertilizers and chemicals to enhance the productivity of farm for meeting their growing food requirements. The continuous and over usage of synthetic chemicals has, however, resulted in deteriorating the soil and human health along with environmental pollution. Therefore, in developed countries farmers are encouraged to convert their existing farms into organic farm.

The key factors influencing consumer demand for organic food is the health awareness and the willingness of the masses to pay for the organic high-priced produce. Generally, consumers of organic products are affluent, health conscious and educated group pushed by strong consumer demand, liberal price premium and concerns about the environment. Because of these unseen advantages, conventional growers are switching to organic farming. Agricultural practices of India date back to more than 4000 years and organic farming is very much native to this country. As mentioned in Arthashastra, farmers in the Vedic period had a fair knowledge regarding seed selection, soil fertility, sowing seasons, plant protection and crop sustainability in different lands (Sofia *et al*, 2006). The farmers of ancient India followed the natural laws and this helped in maintaining the soil fertility over a considerably longer period of time (Chandra and Chauhan, 2004). The modern studies concluded that instead of integration of the two forms (organic & inorganic inputs) so as to achieve better crop yields, the application of only organic inputs alone can fulfill the nutritional requirements of the crop. The interaction between organic and inorganic matter may result in either increase or decrease soil nutrients, depending on the planting and nutrient material in question (Frankenberger and Abdelmagid, 1985).

Organic agriculture is growing speedily and today around 170 countries produces organic food commercially. Organic agriculture is practiced in almost all countries of the world and its share in agricultural land and farms are increasing (FAO, 2002). The market for organic products is increasing at a rapid rate, not only in the major markets like Europe, Japan and North America but also in many other countries including developing countries (Foster and Lampkin 2000; Lin 2003; FAO/ITC/CTA, 2001).

There were 43.1 million hectares of organic agricultural land in India including conversion areas and with 2 million producers. The world's organic producers are highest in Asia (36%) followed by Africa (29%) and Europe (17%). Organic farming is growing rapidly since last decade on sustainable with annual increase of 20% (Avery, 2007; Lotter 2003). The development of organic agriculture worldwide has showed a strong growth since the beginning of the 21st century. Almost 23 million hectares are managed organically worldwide in 2002. Compared to 2000 this is an increase of 31.4%. The total world retail sales of organic products reached US$25 billion in 2002, which increased by 42.9% compared to 2000 (Yussefi and Willer, 2003).

The consumer demand for organic products has increased more than three-fold (although from low levels) in the recent past in global sales (Reganold and Wachter, 2016). Some countries in Northern Europe are recently observing a rapid increase in the sales of organic foods. In Sweden, sales enhanced by 18% in 2016 compared with the preceding year, with organic products now constituting 8.7% of total food sales (Ekoweb, 2017).

## Historical background and emergence of organic farming

Although the term 'organic farming' is gaining popularity in current era, but it was initiated in 10000 years back when ancient farmers started cultivation depending on natural sources only. There is short reference of several organic inputs in our ancient literatures like Arthashastra, Kautilya, Mahabharata, Ramayana, Rigveda etc. In fact, organic agriculture has descended from traditional agricultural practices that were being practiced in countless villages and farming communities. The major milestones in the area of organic farming are presented in Tables 1 and 2.

**Table 1.** Historical perspective of organic farming

| **Ancient period** | |
|---|---|
| Oldest practice | 10000 years old, dating back to Neolithic age, practiced by ancient civilization like Mesopotamia, Hwang Ho basin etc. |
| Ramayana | All dead things - rotting corpse or stinking garbage returned to earth are transformed into wholesome things that nourish life. Such is the alchemy of mother earth – as interpreted by C. Rajagopalachari. |
| Mahabharata (5500 BC) | Mention of Kamadhenu, the celestial cow and its role on human life and soil fertility. |
| Kautilya Arthashastra (300 BC) | Mentioned several manures like oil cake, excreta of animals. |
| Brihad- Sanhita (by Varahmihir) | Described how to choose manures for different crops and the methods of manuring. |
| Rig Veda (2500- 1500 BC) | Mention of organic manure in Ria Veda 1, 161, 10,2500-1500 BC, is Green Manure in Atharva Veda II 8.3, (1000 BC). In Sukra (IV, V, 94, 107 112) it is stated that to cause healthy growth the plant should be nourished by dungs of goat, sheep, cow, water as well as meat. A reference of manure is also made in Vrksayurveda by surpala (manuscript, oxford, No 324 B, Six, 107-164). |
| Holy Quran (590 AD) | At least one third of what you take out from soils must be returned to it implying recycling or post harvest residue. |

*Source:* Bhattacharyya and Chakraborty, 2005

**Table 2:** Key milestones on organic farming

| | |
|---|---|
| Sir Albert Howard (1900-1947) | Father of modern organic Agriculture, developed organic composting process (mycorrhizal fungi) at Pusa, Samastipur, India and published document "An Agriculture Testament". |
| Rudolph Steiner (1922) | A German spiritual Philosopher built biodynamic farm in Germany. |
| J.I. Rodel (1950), USA | Popularized the term sustainable agriculture and method of organic growing. |
| IFOAM | Establishment of "International Federation of Organic Agriculture Movement", in 1972 |
| One Straw Revolution | Release of the book by Masanobu Fukoka (1975), an eminent microbiologist in Japan. |
| EU Regulation | EU Regulation on Organic Food, 1991 |
| Codex | Codex guideline on organic standard, 1999. |

*Source:* Bhattacharyya and Chakraborty, 2005

The development of the organic farming worldwide had gone through three stages namely emergence, expansion, and growth in chronological sequence.

## Stage of emergence (1924-1970)

The commencement of organic farming could mark out back to 1924 in Germany with Rudolf Steiner's course on Social scientific basis of agricultural development and in his theory he considered the human being as part and parcel of a cosmic equilibrium that she/he must understand in order to live in harmony with the environment. Therefore, a balance must be maintained between the materialistic and spiritual phase of life (Herrmann and Plakolm, 1991). These theories were applied to agriculture and laid the foundation of biodynamic agriculture (Kahnt 1986). It was developed at the end of the 1920s in Denmark, England, Germany, the Netherlands and Switzerland (Herrmann and Plakolm 1991; Kahnt 1986; Diercks 1986).

In 1930, politician Hans Mueler gave momentum to organic-biological agriculture in Switzerland. His goals were at once economic, political and social as they anticipated on the ability of the farmer and a much more direct and less cluttered association between the production and consumption stages (Herrmann and Plakolm 1991; Niggli and Lockeretz 1996). Maria Mueler applied these theories to orchard production (Niggli and Lockeretz 1996). Hans Peter Rush, the Austrian doctor adapted these ideas and included them in developing a method for the greatest utilization of renewable resources (Gliessman 1990). Hans Peter Rush and Hans Mueler laid the hypothetical foundation for the organic-biological agriculture and its expansion in the Germanic speaking regions and countries (Niggli and Lockeretz 1996; Rigby *et al.* 2001). Sir Albert Howard was the creator of the organic farming movement. His book 'An Agricultural Testament' summarized his research works at Indore in India of 25 years, where he developed the famed Indore composting process, which utilize the primordial art of composting on the scientific basis and explained the relationship between the health of the soil, plants and the animals (Du and Wang 2001). J.I. Rodale started his research and practice on organic farming in the United States of America. His main objective was to build up and revealed the practical methods of rebuilding natural soil fertility. By 1942, he published the magazine 'Organic gardening'(Coleman 1989). Lady Eve Balfour started her first study know as the Haughley experiment in which she compared the natural and conventional farming methods. Her ideas encouraged the development of the soil association that was founded in England in 1946. The main objective of the soil association was to return humus and soil fertility to their basic place in the biological balance. It was set up on the basis of theories given by Sir Albert Howard in his agricultural testament of 1940 (Soil association 2001). During 1950- 1960s, organic farming (lemaire-boucher) began to embrace in France due to the awareness of doctors and consumers with regard to food and its effect on health (SOEL 2002). Later, Nature and progress association was founded. Mokichi Okada started natural agriculture in Japan in 1935. His major goals were to respect and focus on the function of soil and nature in the agricultural production, and to harmonize the relationship between nature and humans by enhancing soil humus so to have higher yields without the use of fertilizers and agricultural chemicals. The health and environmental issues became a concern in the 1950s-1960s of the last centuries in Japan and thus, it facilitated the development of natural agriculture. The fundamentals of natural agriculture became the essential contents of Japanese agricultural, standard of organic agricultural products (Sheng *et al.* 1995; Yu and Dai 1995).

**Stage of expansion (1970- 1990)**

The research and development programs for organic agriculture extended worldwide after the 1960s, especially, the expansion and dual polarity of organic agriculture started in 1973 with the oil crisis and the growing environmental issues. So, this was the right time for the new ideas, protest movements, appropriate sociological transformations and the proliferation of alternative life styles. The new thoughts comprises of rational use of natural resources, safeguarding the environment, realizing low input and high efficiency, ensuring food security and sustainable development of agriculture such as organic, organic-biological, bio-dynamic, ecological and natural agriculture were remarkably developed in their concepts, research and practical activities (Herrmann and Plakolin 1991; Rigby *et al.* 2001; Du and Wang 2001; May 2001; Pacini *et al.* 2002; Conacher J and Conacher A 1998). In 1970, William Albrecht defines the ecological agriculture as the agriculture system in which the ecological principles were introduced to the production system of organic agriculture (Coleman 1989). In England, the soil association developed the logo in addition to the legally formulated specifications and quality controls that ensures legal guarantee for the consumers (Yussefi and Willer 2003; Soil Association 2001). The largest non- governmental organization of organic agriculture in the world is IFOAM (International Federation of Organic Agriculture Movements) was founded in 1972 (Niggli and Lockeretz 1996). The main organic agriculture associations and research institutions in the world such as FNAB (Federation Nationale d' Agriculteurs Biologiques), FiBL (Forschungsinstitut fuer Biologischen Landbau), which is currently the largest organic research institute worldwide, were founded in 1970s-1980s (FAO 2002; Greene 2001). These organizations played the major role in standardization of the produce and market of organic products and support research and consumer's awareness. The governmental action on organic farming gradually began in the different regions and countries as the guidelines for organic farming. In the United States, the regulation on organic farming was implemented in 1974 in Oregon and in California in 1979 respectively (Greene 2001). The United States Department of Agriculture (USDA) initiated the research program on organic farming on the large scale on 69 organic farms of 23 states and published the report and recommendations on organic farming explaining the development status and potential remained as issues and the research directions. In this report the guidelines and definition for the organic farming was specified, and an action plan for the development of organic farming was called for. The publication of the report and development of organic farming in the United States was a milestone in legislation (USDA 1980). In France, the organic farming regulation was put into practice in 1985 (Graf and Willer 2001; Dai 1999).

## Stage of growth (since 1990)

The organic farming altogether goes through a new stage of growth worldwide in the 1990s. The trade organizations for organic products were founded, organic farming regulations were implemented, and organic farming movement was promoted by both the governmental and non-governmental organizations. The first BioFach Fair was emerged in Germany in 1990 which is currently the largest fair for organic products worldwide (ITC 1999). The federal government of the United States published the regulation for organic food products in 1990 (Greene 2001). The European Commission adopted EU regulation 2092/91 on organic agriculture in 1991. Since 1994, this regulation became a law and was granted in almost all European Union countries (IFOAM and FA0 2002). Simultaneously, the North America, Australia and Japan, the major markets for organic products, published and implemented organic regulations in succession. (Yussefi and Willer 2003; Niggli and Lockeretz 1996). Both, the International Federation of Organic Agriculture Movements (IFOAM) and the Food and Agriculture Organization of the United Nations (FAO) set out guidelines for the production, processing, labeling and marketing of organic produced foods in 1999. These guidelines were of prime importance for international harmonization of the organic farming standards (FAO and WHO 2001). Organic farming had rapidly developed worldwide during this stage. The main drivers of steady market and production growth were the commitment of many retail chains as well as favorable policy conditions. Together these conditions have provided an opportunity for harmonious increase in supply and demand. The legal framework and state support for organic farming research and development was relatively gaining importance since the end of the 1990s.

Therefore, organic agriculture is the holistic production management systems which promotes and enhances agro-ecosystem health including biodiversity, biological cycles and soil biological activity. It laid emphasizes on the use of management practices particularly the usage of off farm inputs, taking into account that regional conditions require locally adapted systems. This is achieved by using, where possible, cultural, biological and mechanical methods, as opposed to using synthetic materials, to fulfill any specific function within the system (FAO and WHO 2001). Terms such as organic, biological, biodynamic and ecological are recognized as organic farming in the EU regulations (Yussefi and Willer 2003; FAO 2002; FAO and WHO 2001).

## Organic agriculture worldwide

Approximately, organic farming is being practiced in 179 countries of the world and the area under organic agriculture is continuously growing (IFOAM, 2004). The total organically managed area is 50.9 million hectares (including conversion areas) in 2016 which is around 6.5 million hectares more than in 2015. About 1.1 percent of the agricultural land is organic. In 11 countries, 10 percent or more of the farmland is organic (FiBL, 2017). The major part of the global organic land area is located in Oceania, Europe and Latin America (Yussefi and Willer, 2003; FiBL, 2003).

## Europe

In the year 2017, 12.7 million hectares were managed organically by more than 280,000 farms in Europe. In the European Union, 7.5 million hectares were under organic management with almost 200,000 organic farms. 2.1 percent of the European agricultural area and 5.1 percent of the agricultural area in the European Union is organic. 27 percent of the world's organic land is in Europe. The countries with the largest organic agricultural area are Germany, Italy and Spain. Currently, there are six countries in Europe with more than 10 percent of organic agricultural land namely Austria, Czech Republic, Estonia, Liechtenstein, Sweden and Switzerland (FiBL, 2017).

The legal protection, grants under rural development programs, and a European as well as several national action plans were the major support for the development of organic farming in the European Union and neighboring countries. One of the chief instruments of the European Action Plan on organic food and farming was an information campaign which was launched during 2008 with the objective of enhancing awareness of organic farming throughout the European Union.

## North America

In North America, almost 3.0 million hectares are managed organically, representing approximately 0.6 percent of the total agricultural area and the 7 percent of the world's organic agricultural land. The major part of the organic land is regulated through the National Organic Programme (NOP) in the U.S. that was issued after the Organic Foods Production Act (OFPA) passed by the Congress on 1990. Interestingly, as clearly stated by the USDA, neither NOP nor OFPA address food safety nor nutrition, rather they are deemed regulations to norm the marketing of organic products in USA. The year 2009 was an important year for the organic agriculturein Canada. On June 30, 2009, the Canada organic regime was established. It includes a new national logo, mandatory national standards and consistent labeling rules (FiBL, 2017).

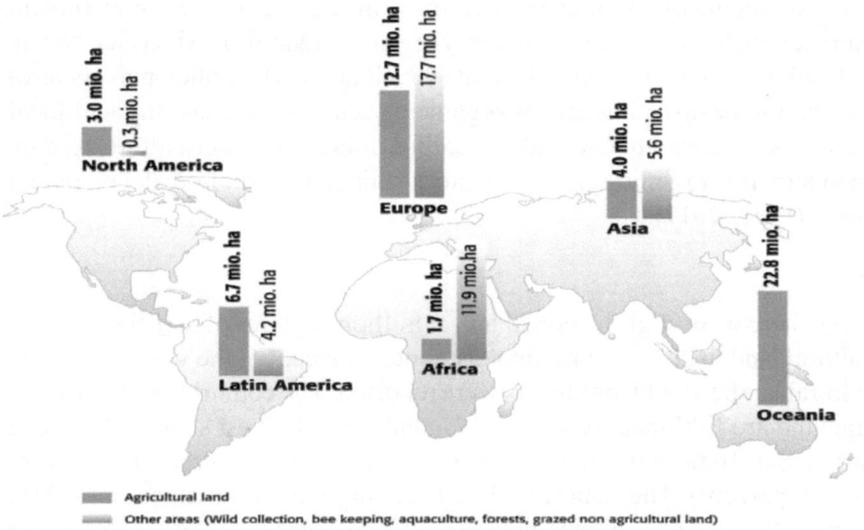


*Source:* (FiBL Survey, 2017)

## Latin America

In Latin America, about 6.7 million hectares of agricultural land is managed organically in 2017 which is approximately 23 percent of the world's organic land. The leading countries are Argentina, Brazil and Uruguay. The highest shares of organic agricultural land are in the Falkland Islands (37 percent), French Guiana, and the Dominican Republic. Most of the organic products from Latin American countries are marketed and sold in the European, Japanese or North American markets. The main crops are grains and cereals, tropical fruits, cocoa, coffee, sugar and meats. 18 countries have legislation on organic farming and three additional countries are currently developing organic regulations. The type of support in Latin American countries comprises of organic agriculture promotion programs and market access support by export agencies. In few countries, limited financial support is also being provided to pay certification costs during the conversion period (FiBL, 2017).

## Asia

In Asia, nearly 4.0 million hectares area is managed organically in 2017. It constitutes about seven percent of the world's organic agricultural land. 500,000 organic producers were reported in India. The leading countries by area are China (1.4 million ha) and India (0.8 million ha). Organic wild collection areas play a vital role in China and India, while aquaculture is significant in Bangladesh, China and Thailand. Even though most of the production is for export purposes but still markets continue to provide domestic growth support in the region. Mixtures of regulatory frameworks co-exist in the region. The government standard-setting bodies have set voluntary organic standards in Malaysia, Nepal, Laos, Thailand, the United Arab Emirates &Vietnam. The policy makers have recognized the positive impacts of organic agriculture on economy and local communities, climate change and the carbon footprint of agriculture so they begun to integrate organic agriculture into sustainable agriculture development initiatives (FiBL, 2017).

## Africa

In Africa, there were slightly more than 1.7 million hectares of certified organic agricultural land in 2017, approximately about 3 percent of the world's organic agricultural land. 540,000 producers were reported. The countries with the most organic land are Ethiopia, Tunisia and Uganda. The highest shares of organic land are in Sao Tome and Prince (7.9 percent), Tunisia (1.8 percent) and Sierra Leone (1.9 percent). The majority of certified organic production is in Africa and is mainly meant for export markets. The European Union, as the major recipient of these exports, is Africa's largest market for agricultural produce. In 2011, considerable achievements were reached especially the African Union's (AU) decision to support organic farming and their subsequent leadership in promoting and further developing strategies for organic farming policies such as the African Ecological Organic Agriculture Initiative and the IFOAM-African Union Conference that took place in November 2011 in Nairobi (FiBL, 2017).

**Table 3:** Percentage of Area under Organic farming among different countries in the world

| Country | Percentage area under organic farming |
|---|---|
| USA | 0.23 |
| UK | 4.22 |
| Germany | 4.10 |
| Argentina | 1.70 |
| Austria | 8.40 |
| Australia | 2.20 |
| Japan | 0.10 |
| Switzerland | 7.94 |
| South Africa | 0.05 |
| Italy | 3.70 |
| India | 0.03 |
| Pakistan | 0.08 |
| Sri Lanka | 0.05 |

*Source:* Roychowdhury *et al.*, 2015)

## Meaning of Organic Farming

The United States Department of Agriculture (USDA) defines the organic farming as a system which avoids or mainly eliminate the use of synthetic inputs (such as pesticides, fertilizers, feed additives, hormones etc) and to the greater extent rely upon crop rotations, crop residues, off-farm organic waste, animal manures, mineral grade rock additives and biological system of nutrient mobilization and plant protection. In another definition, Food and Agriculture Organization (FAO) explained that the organic agriculture is a unique production management system which promotes and enhances agro-ecosystem health including biological cycles, soil biological activity and biodiversity and this is achieved by using on-farm agronomic, biological and mechanical methods without the usage of any synthetic off-farm inputs. National Program on Organic Production (NPOP), India described organic agriculture as a system of farm design and management to create an eco-system which can ensure sustainable productivity without the use of artificial external inputs such as chemical fertilizers and pesticides. Similarly, The International Federation of Organic Agriculture Movements (IFOAM) described the organic agriculture as a production system that maintains the health of soil, ecosystem and people. It is dependent on ecological processes, biodiversity and cycles adapted to local conditions rather than the use of chemical inputs with unpleasant effects. Organic agriculture is a combination of tradition, innovation and science for the well being of the environment, encourage fair relationships and a high-quality of life for all involved.

Thus from the above definitions, it is clear that organic agriculture is an alternative to regular farming system. It makes use of manure, green manure, compost, bone meal instead of using chemical inputs such as fertilizers and pesticides (Amadou and Barbier, 2015). This system utilizes the organic wastes and crops are grown in such a way that it keeps the soil alive and healthy. Microbes are utilized as bio-fertilizers to enhance production without deteriorating the environment (Bagchi, 2015; Kumar, 1999; Kundu and Gaur, 1984; Pagnanelli *et al*, 2010). Organic farming promotes environment friendly agricultural practices without making use of synthetic inputs and is primarily dependent on the organic wastes to raise crops.

However, the modern organic movement is fundamentally different from its original form as it now ensures environmental sustainability at its core in addition to protecting soil health, food nutrition and the health of masses. Hence, the modern definition of organic farming says that it is a method of farming system which particularly aims at cultivating the land and raising crops in such a manner so as to keep the soil healthy and alive by the use of organic wastes (crop, farm and animal wastes, aquatic wastes) and other biological materials along with natural microbes (bio-fertilizers) to provide crop nutrients for enhancing sustainable production in an eco-friendly manner.

## Features of organic farming

1. It utilizes microbes such as bio-fertilizers to release nutrients to the crops (Sane and Mehta, 2015; Alkhader, 2015).
2. It makes use of organic wastes to protect the quality of soil.
3. Nitrogen content required for the process of nitrogen fixation is maintained by using legumes (Ofoegbu *et al.*, 2015; Chandra, 2014; Gharaibeh, 1998).
4. It makes use of natural techniques like crop rotation, organic manures and natural predators instead of using pesticides to control weed and pests.
5. It aims to conserve wildlife and protect the environment.

## Resource use in organic farming

The philosophy of organic farming is that it forms the basis of a production system that is economically, socially and environmentally sustainable. One of the principles of organic farming states that it will avoid pollution and waste by using renewable resources in the production and processing systems (IFOAM, 2005). In addition to this, the environmental benefits of organic agriculture are one of the key drivers for government support (Shepherd *et al.*, 2003; Stolze *et al.*, 2000; Condron *et al.*, 2000; Hansen *et al.*, 2001; Stockdale *et al.*, 2002). Hence, from both the policymakers' and the farmers' viewpoint, there is a actual need to understand how resources are used within farming systems and their impact on the environment.

The concept of resource use can be used to define short-term productivity and long-term sustainability. It is important to maintain a balance between inputs and outputs in all farming systems but is critical where external resources are costly and/or non-renewable. Agreement on how sustainability should be interpreted or precisely defined, let alone measured, is problematic. Increasing awareness of the environmental impact of conventional farming systems has led to a move towards alternatives which explicitly acknowledge the externalities of farming systems (e.g. the impact of pesticides on wildlife or fertilizers on water supplies) and seek to minimize them by changes within the farming system. Alternative approaches to agricultural systems that include sustainability as an explicit goal may often be distinguished from more conventional approaches to agriculture. They emphasize on system instead of single-crop management, enhanced links between crop and livestock enterprises, increased recognition of the importance of biological processes, and increased recycling of nutrients within the farming system.

Currently, the organic (biological/ecological) approach is the only alternative to conventional production systems. Both low-input usage and organic systems are covered under the 'sustainable' and 'alternative' headings (National Research Council, 1989; Edwards *et al*, 1990; Francis *et al*, 1990).

It has been recommended that justifiable environmentally sustainable solutions can only be established by considering a life-cycle analysis approach to environmental systems analysis. Life-cycle analysis is a holistic approach to assess the environmental impact of the production, use and disposal of a product (Payraudeau and HMG, 2005). While this might be a definable concept in relation to a manufactured product (ISO/DIS, 2000), its application to agriculture and especially to multi-enterprise integrated rotational crop and livestock systems is best by extreme complications in relation to the appropriate definition of the system with which any agricultural product is associated. Attempts have been made to use life-cycle analysis to evaluate the environmental impacts of plant and animal production (Payraudeau and HMG, 2005; Mattsson, 1999; Brentrup *et al*., 2001; Williams *et al*., 2006; Cederberg, 2002; Haas *et al*., 2001). In temperate climates, resource budgets are continuously been used to determine the sustainability of organic farming systems. After the restriction on the use of imported materials to build/maintain soil fertility, many studies have focused on nutrient use efficiency (Watson *et al*., 2002; Berry *et al*., 2003).

## Principles of organic agriculture

IFOAM explains four principles of organic farming on which its standards are based. Each of the principles plays a vital role to improve or enhance the long term health of the farm, the surrounding environments and the farming participants which in turn are responsible for the long term success of the farm. The four basic principles include:

1. *Health:* Organic agriculture should improve/enhance the health of soil, plants, animals, human and planet as one and indivisible.
2. *Ecology:* Organic agriculture should be based on living ecological systems and cycles, work with them, emulate them and help sustain them.
3. *Fairness:* Organic agriculture should ensure fair relationships with regards to common environment and life opportunities.
4. *Care:* Organic agriculture should be handled in a precautionary and responsible manner to safeguard the environment and protect the health and well being of current and future generations.

## In addition to this, other three principles of organic farming include

1. To work as much as possible within a close system and draw upon local resources.

2. To avoid/minimize all forms of pollution that may result from agricultural techniques.
3. To enhance the potential of agricultural producers to earn a living through their work.

### Aim/goals of organic agriculture

According to the official standpoint of IFOAM Basic Standards (2002) production and processing of organic produce is based on many goals, among which it is difficult to separate out the priority ones, because all of them are equally important. They are as following:

1. Production of high-quality raw materials, foodstuffs, and other goods in sufficient quantities.
2. Synchronization the production system with the soil's natural cycles, live systems, plant and animal worlds.
3. Acknowledgement of a wide social and ecological impact beyond and within the organic production and processing system.
4. Conservation and improvement of soil fertility and biological activity in support of cultural, local, biological and mechanical methods rather than the employment of external production factors (resources).
5. Preservation of biological and agricultural diversity in and around farms through introduction of a sustainable production system and wildlife protection.
6. Optimum use and preservation of water resources with all their live organisms.
7. Ensuring maximum usage of renewable resources in the production and processing systems and prevent their losses and pollution.
8. Facilitation of local and regional production and movement of goods to consumers.
9. Accomplishment of a harmonious balance between animal production and crop production.
10. Ensuring the conditions under which domestic animals reveal their natural behavior.
11. Making use of packaging materials that can be recycled or can be decomposed in a biological way.
12. Ensuring that each and every person involved in organic farming and processing enjoys the living conditions that satisfy the requirements of a healthy and safe environment.

13. Development of a socially-oriented chain 'production-processing-sale' which would act in accordance with the ecological requirements.
14. Identification of the reality that it is necessary and vital to study the local experience and traditional farming systems.
15. Independence of agriculture from synthetic chemicals, minimal energy consumption in production processes, significantly reduced production expenditures and dependence on external financing.
16. Development of local, national and international markets of organic produce and fair international trade.
17. To produce food free from toxic chemicals and to maintain the soil fertility for long time (Sujatha *et al.*, 2014). This also provides employment opportunity to the agriculture labors in various forms like composting, strip farming and non-chemical weeding etc.
18. The aims of organic farming go beyond food production to include caring for and protecting the environment (landscapes, habitats, climate, air, water and biodiversity) and the wellbeing of people and animals. It is thus appropriate to enquire how a focus on increased yields will affect reaching these wider goals.
19. Seufert and Ramankutty (2017) provide the latest comprehensive review on the costs and benefits of organic agriculture in its current form and revealed that the organic agriculture ensures higher biodiversity and safeguard soil and water quality per unit area, increased profitability and improved food nutritional value.

## Need for organic agriculture

In current scenario, the European Union has banned Indian contingent of mangoes on account of high pesticidal residues. The Indian chillies undergo the same treatment in Saudi Arabia, which happens to be the fifth-largest importer of vegetables from India. Even in India, the most commonly and intensively used pesticide such as endosulphan has been completely banned from its usage in many crops including fruits and vegetables. All these are signals that the concern regarding food safety is getting prominent in both developing and developed economies. Since extensive use of synthetic inputs resulted in various issues such as over-exploitation of the groundwater resources, depletion of fertility of soil and higher susceptibility of crops to various insect-pests attack and diseases, (Sidhu, 2002) organic agriculture is being seen as an alternative in some quarters. But it has to be well understood that organic agriculture aims at optimization rather than maximization. It has been estimated that ushering in organic cultivation in about 10 million hectares by 2020 could save Rs. 10,000 crores of government exchequer with the reduction of fertilizer subsidies (Pratap, 2006).

## Scope and mode to promote organic agriculture

India is one of the leading fruit producing countries in the world, producing about 10% of the world's fruit production (Indian Horticulture Database, 2014). Agriculture is the backbone of the Indian economy as nearly 67% of population and 55% of the total work force is dependent on agriculture and other allied activities. Thereby, agriculture sector has the remarkable potential to grow and satisfies the ever increasing demands Indian population. It has been assumed that Indian agricultural should have the growth rate of 4% or more so as to achieve a double digit GDP growth rate (Chandrashekar, 2010). The scope and prospects of potential organic agriculture in India is evident by the fact that the farm sector has plentiful organic resources like crop residue, water, livestock, aquatic weeds, forest litter, rural & urban solid wastes and agro- industries, bio-products (Bhattacharya and Chakraborty, 2005). The adherence of huge population of the farmer to the natural law in ancient India has played a vital role in maintaining the fertility of soil over a longer period of time (Chandra and Chauhan, 2004). The inbuilt advantages such as its diverse agro-climatic regions, local self-sustaining agri-systems, substantial number of progressive farmers and ready availability of inexpensive manpower translate into the potential to cultivate organically a vast basket of products (Munda, 2006). More than 65% of the country's cultivated area is rainfed on crop rotation, crop residues, legumes, animal manure and biological pest control. Majority of the farms are of subsistence type in remote and marginal areas. Organic farming is gaining much importance among Indian farmers and entrepreneurs, especially in rain-fed zones, low productivity areas, hilly areas and the north-eastern states where fertilizer consumption is less than 25 kg/ha/year (Mitra and Devi, 2016). In fact, North Eastern Region (NER) is considered as home to some niche crops such as Assam lemon, Joha rice, medicinal rice and passion fruits which has high market demands and accounts for 45 per cent of total pineapple production in India (Munda *et al.*, 2007). Viewing the benefits of organic agriculture, Uttaranchal and Northeast states have declared themselves as organic-farming states while Mizoram and Sikkim have declared their intentions to shift to total organic farming (Mitra and Devi, 2016). Sikkim, by practicing organic farming on approximately 75,000 ha of agricultural land has become India's first fully organic state (Sikkim, 2016). Nagaland has 3000 ha area under organic farming whereas Meghalaya has committed to certifying 200,000 hectares of land as organic by 2020 (Hill, 2016). These regions receive very high rainfall (2000 mm to 11000 mm per annum) resulting to the massive production of biomass including weeds, herbs and shrubs (Munda, 2006) when a large part of the land falls under forests, pastures, wastelands etc. Lotter *et al.* (2003) estimated that organic farming performed better in areas having extreme rainfall because of the less run-off and higher absorption of water in the field. The encouragement for organic agriculture initiated first in the rain-fed areas especially in the hilly areas where there is little or no use of chemical fertilizers and other agro-chemicals due to poor resources with small holder farmers (nearly 60% of farms in India are less than one ha). Jammu & Kashmir mostly have a hilly topography. Thereby, majority of its area suffers from run-off and soil erosion losses. Besides, more than 70 per cent of the arable land is un-irrigated where cultivation of crops alone is very risky (Gupta *et al.*, 2005 and Chandra, 2014). The varied agro-climatic zones with potential of diversification makes possible to convert at least these areas into organic farming easily (Wani *et al.*, 2013). With the success of organic farming, some awakening has initiated in the western and southern states like Karnataka and Maharashtra (Khanna, 2012).

## Organic farming V/S conventional/modern farming

The organic agriculture differs from conventional agriculture not only gradually but fundamentally. The continuous practice of organic methods thus, seems to provide a new quality in how the agro-ecosystem works. This functioning cannot be explained by summing up single ecological measures. Organic farming tends to improve fertility of the soil in a way and to an extent which cannot be achieved by conventional farming even if the later constantly respects some ecologically principles.

Organic agriculture is one of several to sustainable agriculture and many of the methods used (e.g. rotation of crops, inter-cropping, mulching, double digging, integration of crops and livestock) are practiced under various agricultural systems. In organic agriculture almost all synthetic inputs are prohibited and soil building crop rotations are mandated, it is this fact that makes organic agriculture unique. The fundamental rules of organic production are that natural inputs are accepted and synthetic inputs are prohibited, but there are exceptions in both cases.

On other hand, today's modern chemical farms have little use for the skilled husbandry which was once the guiding principle of working the land. The main focus today is primarily on productivity - high input in exchange for high returns and productivity (mostly diminishing now however for farmers worldwide). Four important considerations namely what happens to the land, the food it produces, the people who eat it and the communities which lose out are mostly overlooked. The various drawbacks/hazards of modern chemical farming are summarized in table 4.

**Table 4.** Effects of modern farming system

| | |
|---|---|
| Land exhaustion | The constant use of artificial fertilizer, together with a lack of crop rotation, reduces the soil's fertility year by year. |
| Fertilizers | High yield levels are produced by applying large quantities of artificial fertilizers, instead of by maintaining the natural fertility of the soil. |
| Nitrate run-off | About half of the nitrate in the artificial fertilizer used on crops is dissolved by rain. The dissolved nitrate runs off the fields to contaminate water courses. |
| Soil erosion | Where repeated deep ploughing is used to turn over the ground, heavy rains can carry away the topsoil and leave the ground useless for cultivation. |
| Soil compaction | Damage to the structure of soil by compression is a serious problem in areas that are intensively farmed. Conventional tillage may involve a tractor passing over the land six or seven times, and the wheelings can cover up to 90 per cent of a field. Even a single tractor pass can compress the surface enough to reduce the porosity of the soil by 70 per cent, increasing surface run-off and, therefore, water erosion. In the worst cases, the surface run-off may approach 100 percent -none of the water penetrates the surface. |
| Agricultural fuel | As crop yields grow, so does the amount of fuel needed to produce them. European farmers now use an average of 12 tons of fuel to farm a square kilometre of land; American farmers use about 5 tons (1987 figures). |
| Biocide sprays | The only controls used against weeds and pests are chemical ones. Most crops receive many doses of different chemicals before they are harvested. |
| Cruelty to animals | On most modern farms, all animals are crowded together indoors. Complex systems of machinery are needed to feed them, while constant medication is needed to prevent disease. The cruelty involved in managing, breeding, growing and slaughtering farm animals today is unimaginably repulsive and horrifying. |
| Animal Slurry | With so many animals packed together in indoor pens, their manure accumulates at great speed. It is often poured into lagoons which leak into local watercourses, contaminating them with disease-causing organisms and contributing to algae blooms. |
| Imported animal feed | Many farms are not self-sufficient in animal feed; instead they rely on feed brought into the farm. This often comes from countries which can ill afford to part with it. |
| Stubble burning | In countries where stubble is burned, large amounts of potentially useful organic matter disappear into the sky in clouds of polluting smoke. |
| Loss of cultivated biodiversity | Large and other chemical farms tend to be monocultures growing the same crop and crop variety |
| Threat to indigenous seeds and animal breeds and species | Native cultivars and animal breeds lose out to exotic species and hybrids. Many native animal breeds are today threatened with extinction. The same holds true for many indigenous plant varieties which have disappeared within the space of one generation. |

| | |
|---|---|
| Habitat destruction | Agribusiness farming demands that anything which stands in the way of crop production is uprooted and destroyed. The wild animals and plants which were once a common sight around farms are deprived of their natural habitat and die out. |
| Contaminated food | Food, both plant and animal products, leaves the farm contaminated with the chemicals that were used to produce it. |
| Destruction of traditional knowledge systems and traditions | Rural indigenous knowledge and traditions, both agricultural and non-agricultural, is invariably connected to agriculture and agricultural systems. |
| Control of agricultural inputs and food distribution channel | The supply and trading in agricultural inputs and produce is in the hands of a few large corporations. This threatens food security, reducing the leverage and importance of the first and the last part of the supply chain - the farmer and the consumer. |
| Threat to individual farmers | Chemical agriculture is a threat to their livelihoods and changes their lifestyles, unfortunately not for the better. |

*Source:* www.localfoodworks.org

## Components of organic farming

The main components of organic farming are biological nitrogen fixation, biopesticides, biogas slurry crop rotation, residues of crops, etc. Vermicomposting has viewed as a chief component in organic farming which plays a vital role in improving soil fertility and crop growth in a sustainable way. The various components of organic farming are as follows.

1. ***Crop rotation:*** For ensuring sustainable agriculture, the crop rotation should be encouraged on the same land over a period of two years or more for improving soil fertility and control of weeds, insects and diseases. For example, use of legumes in rotation improves fertility of soil.

2. ***Crop Residues:*** Organic farming has great potential of using crop residues and straw of cereals and pulses in recycling of nutrients. Crop residues when inoculated with fungal species improve physico-chemical properties of soil and enhance crop yields.

3. ***Organic manures:*** The organic manure is prepared from biological sources such as plant, animal and human residues. Organic manure helps in enhancing crop growth directly by improving the uptake of humic substances and indirectly by promoting productivity of soil by increasing availability of major and minor plant nutrients through soil microorganisms. The four main types of organic manures are given below:

   a) *Bulky organic manure:* Bulky organic manure comprises of compost, FYM and green manure having lesser nutrients in comparison to concentrated organic manure.

b) *Farm Yard Manure (FYM):* FYM is a combination of well decomposed dung, urine, farm litter and leftover materials (roughages or fodder).

c) *Compost:* Large quantities of waste materials such as weeds, stubble, bhusa, sugarcane trash, vegetable refuse, sewage, sludge, livestock waste, human and industrial refuse can be converted into compost manure by anaerobic decomposition. Compost is used in the same way as FYM and is good for application to different type of soils and crops.

**Table 5.** Nutrient content of some organic manure

| Sources | Nutrient content | | |
|---|---|---|---|
| | N | $P_2O_5$ | $K_2O$ |
| FYM/Compost | 0.5-1.0 | 0.15-0.20 | 0.5-0.6 |
| Poultry manure | 2.87 | 2.90 | 2.35 |
| Cattle dung | 0.3-0.4 | 0.10-0.15 | 0.15-0.2 |
| Sheep and goat manure | 0.65 | 0.05 | 0.03 |

*Source:* Nazir *et al.*, 2016

d) *Green Manuring:* Green manuring refers to the process of adding organic matter to the soil by ploughing and adding undecomposed green plant tissues into the soil for improving physical structure and fertility of the soil. The green manure crop (legume crop) provides organic matter and additional nitrogen. He most commonly used green manure crops are berseem (*Trifolium alexandrium*), cowpea, cluster bean, dhaincha (*Sesbania aculeata*), sunhemp (*Crotalaria juncea*), senji (*Melilotus parviflor*, *Vigna sinensis*), etc.

4. ***Waste:*** The waste is primarily of two type namely industrial waste and municipal & sewage waste. The municipal and sewage waste is an important component of organic waste. The industrial by products such as spent wash & coir waste can be used as manure

5. ***Biofertilizers:*** Biofertilizers are microorganisms that have the ability of enhancing the fertility of soil for example, by fixing atmospheric nitrogen and through mycorrhizal fungi and phosphate solubilisers. These are environment friendly and sustainable way of improving soil fertility. Biofertilizers have biological nitrogen fixing organism which help them in establishment and growth of crop plants and trees, enhance biomass production and grain yields. There are basically two types of bio-fertilizers.

   a) Symbiotic Nitrogen-fixation which includes rhizobium.

   b) Asymbiotic Nitogen-fixation which includes azolla, azotobacter, azospirillium, blue green algae and mycorrhizae.

6. ***Bio-pesticide:*** Biopesticides are commonly derived from plants. They include plant products like alkaloids, phenolics, terpenoids and some secondary chemicals. They are biologically active against fungi, insects, nematodes affecting their behaviour and physiology. Commonly known insecticides are margosa, nicotine, neem, pyrethrum, rotenone etc.
7. ***Vermicompost:*** Vermicompost is organic manure or compost prepared by the use of earthworms that generally live in soil, eat organic matter and excrete it in digested form. These are rich in macro and micronutrients, vitamins, growth hormones and immobilized microflora essential for plant growth (Nagavallemma *et al.*, 2004).

## Methods of organic farming

1. ***Soil management/ Maintenance of soil fertility:*** Soil is the foundation of terrestrial life. Specific soil management practices are required to safeguard and conserve the soil resources (Ghorab and Khalil, 2016). Improvement and preservation of soil fertility is the major issue during introduction of organic farming technology. Studies reveals that organic farming conserves fertility of the soil and improves system stability better than conventional farming (Stolze *et al.*, 2000; Shepherd *et al.*, 2003). A study of cotton production under organic conditions in India showed similar yield levels as in modern cultivation techniques, but the quality of the soil showed advantages for the organic system as indicated by soil organic matter, water stable aggregates and mean weight diameter (Blaise, 2006). The fertility of the soil is maintained by several ways under organic farming (Partap, 2010) such as optimization of agricultural crop within each farm; optimum and efficient use of available organic fertilizing resources (humus, turf, turf-and-humus composts, putrid mud, organic wastes of agricultural production and others); taking the advantages of diversification by increasing areas under perennial grasses, application of bacterial substances and expanding the areas under green fodder; recommencement of planned chemical land reclamation, which would make use of local deposits of limestone and chalk; use of local raw material resources for enhancement of soil fertility (defecate, glauconites, phosphorites, phosphate slag, sapropels, zeolites and others); limiting the expansion of areas under harmful crops, which can further worsen the phytosanitary state of the soil; introduction of minimum tillage & wide cut tillage methods and direct sowing technology; application of contour structural arrangement of the territory in land use, which presumes optimization of the state of natural environment within the territory of watersheds or ravine and gully systems; developing field-protecting and water-protecting measures.

2. ***Ecological benefits:*** The ever rising concern about environmental degradation, declining natural resources and urgency to meet the food requirements of the increasing population are forcing farm scientist and policy makers to critically examine and provide an alternative to chemical agriculture. Nemecek *et al.* (2005) found greenhouse warming potential in organic systems 29 to 32 percent lower on per ha basis than in a mineral fertilizer system and 35 to 37 percent lower than in the conventional manure-based system. The economic returns from the ecological farming system are really significant, if one gets a premium price for the organically grown rice. The declines in productivity or no effect on yields is over looked by the higher prices that farmers get for their organic produce (Rao *et al.*, 2006). Various studies have confirmed that organic agriculture is productive and sustainable (Mader *et al.*, 2002).
3. ***Efficient energy utilization*:** The requirement of energy for organic farms in the production process as measured per rupees of produce is only one third of what it is required for their conventional counterparts. This is because N fertilizer and pesticides are not used by organic farmers and the comparison of total energy input/ha with total energy output favors organic farming systems. In one study it was revealed that the soil organic matter was relatively higher in organically managed soil in comparison to conventional soil despite of similar contents of organic carbon (Marinari *et al.*, 2007). The nitrate leaching can only be reduced by carefully managing crop rotations including catch and cover crops (Thorup-Kristensens *et al.*, 2003). Meanwhile, a study conducted by Hoeppner *et al.* (2006) on the impact of organic versus conventional management on energy use, energy output and energy-use efficiency revealed that a) the energy consumption was 50% lower with organic as compared to conventional management b) energy output was 30% lower with organic in comparison to conventional management c) the efficiency of energy (output energy/input energy) was highest in the organic management.
4. ***Superior food quality*:** Nowadays, food quality is one of the main concerns of scientists as well as consumers. Nitrates in water and farm produce, pesticides residues, desirable components, keeping quality and physiological imbalances are some of the important characteristics of food quality. It is most often ignored fact that by applying integrated methods, differences resulting from cultivation methods are mostly in favor of the organic variants (Velimirov, 2006). The synthetic toxicants are considerably reduced in organic foods, if not completely avoided (Baker *et al.*, 2002; Mahnke-Plesker, 2005). Population belonging to higher and middle economic status are more concerned about the food quality and therefore are willing to pay higher prices for organic food products.

5. ***Waste utilization*:** The ever growing prices of chemical fertilizers have facilitated the usage of organic wastes in the fertilizer practices on the farm. Good manure management implies improved fertilizers value of manure and slurry and fewer nutrient losses. Composting of all organic wastes in general and of Farm Yard Manure (FYM) or feedlot manure in particular is important in organic farming. The self-sufficiency of nitrogen in organic systems is a major advantage in times of shortage of fossil energy (Cormack, 2000). Sewage and sludge use for crop production can form an important component of organic farming, if treatment and application methods are improved further. Lotter *et al.* (2003) found that water holding capacity in organic plots was approximately 100 percent higher than in conventional plots during torrential rains.
6. ***Use of traditional knowledge:*** Most countries have utilized different kinds of traditional organic materials to improve or enhance the productivity and fertility of their agricultural soils. However, several years ago, organic recycling practices in some countries were mainly replaced with chemical fertilizer which were applied to high yielding cereal crops that responded to a high level of fertility and adequate moisture including irrigation. Consequently, the importance of organic matter to crop production and its proper use in soil management is often overlooked or sometimes forgotten. Due to these changes and the failure to implement effective soil conservation practices, the agricultural soils in a number of developing and developed countries have suffered from serious degradation and decrease in productivity because of nutrient run-off, excessive soil erosion and decreased soil organic matter levels. Therefore, in order to achieve self sufficiency in utilization of organic materials such as usage of agricultural residues and urban wastes to protect agricultural soils from water and wind erosion and to prevent nutrient losses through run-off and leaching, organic farming has been looked as the most reliable method of agricultural production.
7. ***Integrated Intensive Farming System (IIFS) or Intensive use of resources*:** IIFS comprises the intensive utilization of farm resources. In order to ensure environment sustainability, such intensification should be based on techniques which are knowledge intensive and which replaces market purchased chemical inputs with farm grown organic/biological inputs. It is concluded that organic agriculture encourage more species and have more abundance of organism groups than conventional farming that ensures better utilization of wide variety of resources (Bengtsson *et al.*, 2005). Pretty *et al.* (2006) suggested several ways to improve water use efficiency in organic agriculture such as reducing evaporation through minimum tillage, using more water-efficient varieties and inducing microclimatic changes to reduce crop water requirements.

8. ***Economic Benefits***: Partap, 2006 in his recent study of Indian market observed that nowadays there are adequate numbers of people who are willing to pay more for safe food supplies from organic market. Some main findings from research on yields suggested that yield equivalent to or better than conventional agriculture can be achieved under organic systems, although often they are not and yields decrease during conversion period but then improve afterwards (Wynen, 1994; Stonehouse *et al*., 2001; Mendoza, 2002). The estimates made by the International Fund for Agricultural Development (IFAD) reported that the income of associated small farmers in organic farming can increase substantially (Giovannucci, 2005) and Partap and Vaidya (2009).
9. ***Biodiversity conservation***: The number of species is more in organic grasslands as compared to conventional grasslands, as a result of which the plant communities are richer in species and structure (Frieben, 1997). In European organic farming systems, many researchers have found greater diversity and abundance of soil and surface-living arthropods such as beetles, parasitic flies, spiders and wasps as well as non-pest butterflies and many other invertebrate species in comparison to conventional farming systems (Feber *et al*., 1997; Stolze *et al*., 2000; Tybrik *et al*., 2004). Organic farming systems are the most convenient tool used by the planners to balance production and conservation (Hole *et al*., 2005). Therefore, many projects and schemes worldwide working to conserve indigenous varieties and seed banks are linked to organic agriculture projects (Stolton, 2002).

## Role of organic farming in sustainable agriculture

In India, about 1.5 million hectare area is under organic farming (this includes certified and area under organic conversion) with 835,000 number of certified organic producers. This accounts for about 2.59% of total agricultural land (Ramesh *et al*., 2010). The majority of small farmers utilize locally available resources in India. As such in many marginal areas of India, organic farming is present not by choice but by default.

In organic animal husbandry, integration of livestock with cropping or integrating more than one livestock species can be the basis of a sustainable and balanced farming system, allowing effective resource use and nutrient recycling (Subrahmanyeswari and Chander, 2008). The future of organic livestock farming appears to be bright as compared to intensive livestock production especially in dryland areas of India (Chander *et al*., 2007). In rainfed areas, livestock is the major source of occupation for the millions of Indian small scale farmers because crops may fail but livestock sustain the life of poor farmers. The maintenance cost of these livestock is very low in comparison to the crossbreds. This type of mixed farming system is common in India. The farmers have little knowledge about the systematic and scientific approach of organic agriculture management of farm (Tewari and Tewari, 2007). However, it is mandatory to provide resources/inputs to the organic farmers in the form of technical knowledge so as to enable their livestock systems to be modified and certified which is essential in organic production systems (Subrahmanyeswari and Chander, 2008).

## Indigenous practices in organic agriculture

Development and practice of organic agriculture began in the tropics when most of the world's land was unused but arable lands were farmed (Evans, 1992; Kellogg and Orvedai, 1996). In Nigeria and India more than 80% of the population works on land (Evans, 1992). Due to availability of arable land, people practiced the agriculture that was primarily free from the use of synthetic chemicals/inputs.

## Indigenous organic agriculture based on the farming system involved

In indigenous organic agriculture, shifting cultivation or slash and burn, mixed cropping or multiple cropping and crop rotation were practiced.

1. ***Shifting cultivation or slash and burn method*:** This type of indigenous farming system is dependent on natural cycles. Therefore, the farmer continued to practice their farming on a piece of land until the fertility of that particular land does not declined before shifting to another fertile plot. In the beginning, the farmer had no intention to go back to the already exhausted lands. Production was mainly based on naturally accumulated nutrients resulting from fallowing. Crude implement such as axe, hoe, matchets etc were used while the farmer and his family met the labour requirements. The system productivity was quite low. But eventually due to increased population and urbanization, the problem of scarcity of land persists and as result of which the fallow periods were reduced. The situation ushered in other modified forms of farming systems. Shifting cultivation was mainly practiced by the ancient people that are now considered a type of organic agriculture that is in practice today because the use of synthetic inputs such as insecticides, pesticides, fertilizers, plant growth hormones etc were not involved. Rather the use of organic manures such as compost, farmyard manure, green manure and poultry droppings were emphasized for improved soil condition and crop production (Asiegbu and Oikeh, 1993).

2. ***Multiple Cropping or Mixed Cropping*:** This method involves the growing of two or more crops on the same piece of land. The benefits of such systems include maintenance of soil conditions, reduction of diseases and pests, facilitates vertical and horizontal variations thus, allowing cultivation

of crops adapted to light and shade. In addition to this, it also permit phased harvesting which occurs in crops with different maturity period and the harvesting can be phase by harvesting the early maturing varieties before the late maturing ones. The phased harvesting serves as insurance to the farmers because when a farmer grows two or three crops on a piece of land in an intercropping or mixed cropping system, the failure of one crop does not indicate the failure of all crops, the remaining one or two crops out of the two or three planted that remained serves as an insurance to the farmer. At organic level multiple cropping received number of farm input of fertilizers while land preparation is mechanically done.

3. ***Crop Rotation*:** In crop rotation, a given combination of crops is grown in a particular sequence for several years on the same piece of land without the loss of soil fertility and reduced yield. This is done to prevent build-up of pest, weeds and soil borne diseases, control soil erosion, maintain and improve soil fertility, reduce the period of peak requirement of irrigation water and conserve soil moisture from one season to the next. The crops in this system were selected on the basis of their complementary or supplementary relationship to one another rather than competitive. Again, the practice is primarily dependent on organically produced input such as compost, farmyard and poultry manures.

4. ***Current practices*:** The concept of organic agriculture is not a new one for Indian economy. It has been practiced by our forefathers long ago in its elementary form before the introduction of inorganic fertilizers. Our forefathers practiced shifting cultivation which is totally dependent on natural fertility. However, engaging in shifting cultivation, crop rotation and mixed cropping systems act as means of maintaining soil fertility and sustainable pest and disease control without the use of synthetic chemicals (Atungwu *et al*., 2009). However, the current practice of organic agriculture is the modification of the already existing practices with more emphasis on creation of integrated, environmentally and economically viable agricultural systems.

5. ***Agroforestry*:** This is an integrated approach introduced by International Council for Research in Agroforestry and involves combining trees and shrubs with crops and/or livestock (Gholz, 1987). The chief products include in this system are fodder, fuel wood and timber (Agromisa, 2003). Fodder, fuel wood and timber reported the practical applications of agroforestry as live fences to control animal movement in the farmstead, hedgerow barriers, windbreaks and shelter belt, alley cropping.

Furthermore, agroforestry systems can be classified based on the basic components of woody perennials, herbaceous plants and animal components as: (a) agrosilvicultural system (production of woody plants and seasonal plants)

(b) silvopastoral system (production of woody plant species and livestock) (c) agrosilvopastoral system (production of woody plant species, seasonal plants and livestock).

6. ***Intercropping/Multiple Cropping/Mixed Cropping*:** It involves planting of two or more crops on the same piece of land in such a way that compatibility is maintained within the crop communities. In the system, soil fertility is maintained by use of leguminous crops along with the non-leguminous ones (Ibeawuchi and Ofoh, 2000; Tom and Asiegbu, 2002).
7. ***Bush Fallow System*:** This system is still practiced in areas where land for agriculture is available. However, production is at a subsistence level even in the current practices. The key feature of this system includes complete exclusion of synthetic input especially in the rural farming communities. But it is not possible to completely prohibit the use of synthetic input with large farms owned by cooperative societies under fallow systems.

## Current scenario of organic farming in India

The 'Sevagram Declaration' played a vital role for promotion of organic agriculture in India since January 1994. Organic farming has grown many folds with the passage of time due to number of initiatives at government and non-government level. The National Programme on Organic Production (NPOP) defined its regulatory framework while the National Project on Organic Farming (NPOF) has represented the promotion strategy and provided necessary support for area expansion under certified organic farming. Before the implementation of NPOP during 2001 and introduction of accreditation process for certification agencies, there was no institutional arrangement for estimation of organically certified area. Initial estimates during 2003-04 suggested that approximately 42,000 ha of cultivated land were certified organic. By 2009 India had brought more than 9.2 million ha of land under certification and it has been increased further.

According to the International Fund for Agriculture and Development (IFAD) about 2.5 million hectares of land was under organic farming in India in 2004. Further, there are over 15,000 certified organic farms in India. India therefore is one of the chief suppliers of organic food to the developed countries. No doubt, the organic movement has again started in India. Organic food and farming systems are a promising and innovative means of handling the challenges faced by the world in the area of agriculture and food production. Organic production has encouraged dynamic market growth, contributed to farm incomes and created employment for more than three decades now. At the same time it delivers public goods in terms of animal welfare, rural development and environmental protection. In addition to this, the development generated by the organic sector have played an vital role in pushing agriculture and food productions generally towards sustainability, quality and low risk technologies.

The various advantages of organic farming for small farmers all over the world include low capital investment, high income, ability to achieve higher premium in the market and the ability to use traditional knowledge. According to a research conducted by the Office of Evaluation and Studies (OE), International Fund for Agriculture Development (IFAD), small farmers in China, Latin America and India can benefit significantly from organic farming and will help in mitigating poverty in these countries. Organic farming tends to minimize modern farmers struggles which involves various debts and farm land mortgages. Organic farming refers to means of farming that does not involve usage synthetic inputs such as chemical fertilizers and chemical pesticides. Numerous small farmers have been practicing organic farming, since they are unaware of the market opportunities they are not able to reap the benefits of organic farming.

**Table 6.** Indian states and agricultural products involved in organic farming

| Indian states involved in organic farming | Main organic agricultural products of India |
|---|---|
| Gujarat | Bajra-mustard-wheat |
| Kerala | Chilly |
| Karnataka | Cereals-cereals |
| Uttaranchal | Cereals-pulses |
| Sikkim | Kholar |
| Rajasthan | Maize Ginger |
| Maharashtra | Soybean |
| Tamil Nadu | Large cardamom |
| Madhya Pradesh | Passion fruit |
| Himachal Pradesh | Bhilwara Urd |
| | Bharatpur Bajra and wheat |
| | Alwar Wheat and bajra |
| | Cotton grass |
| | Dungarpur pulses-cereals |
| | Bajra |
| | Mustard |
| | Til |
| | Wheat |
| | Nagour Guar-cumin |
| | Moong |
| | Ganganagar cotton |
| | Jaisalmer bajra |
| | Jhunjhunu pulses and wheat |
| | Banswara maize |
| | Jaipur Guar |

*Source:* Chandrashekar, 2010

## Strategies to increase yields in organic farming

Recent meta-analyses revealed that globally the yield of organic crops are on average 80% (de Ponti *et al.* 2012), 66–95% (Seufert *et al.* 2012) or 81% (Ponisio *et al.* 2015) of conventional yields. Yield differences vary considerably with crop types (with legumes showing a considerably smaller yield gap than cereals or tubers) growing conditions and management practices. The two most important yield limiting factors in organic crop production are supply of nitrogen (N) and control of perennial (Askegaard *et al.* 2011). These two factors are closely linked as sufficient N is available for rapid early establishment and growth of crops which play a strong role on reducing weed infestation by inducing greater weed suppression ability of the crop (Olesen *et al.* 2007). The most commonly used organic fertilizers such as compost, manure, green manure and organic wastes are quite low in plant available N and this in combination with slow N mineralization in the spring due to low temperatures limits yield in organic crops, especially in the Nordic countries (Dahlin *et al.* 2005).

Yield losses due to the insect-pest attack and diseases also affect the organic-conventional yield gap. The number of crop protection products permitted for organic farming is very limited (EU 2014) although they constitute an important input for decreasing crop losses, especially in some horticultural crops (Letourneau and van Bruggen 2006). The limitation of crop protection products or other effective crop protection measures restricts organic yields. It should be here noticed that copper-based products are most commonly used in European organic farming for crop protection products especially for controlling fungus attacks in vines, fruit crops and potatoes (Niggli *et al.* 2016) but copper fungicides are banned in Scandinavian countries by national legislation. Therefore, organic farmers often has to rely on plant varieties bred for high-input conventional systems i.e. high-yielding varieties with poor weed competitive abilities and shallower rooting depth (Lammerts van Bueren *et al.* 2011) whereas in conventional production systems, these deficits are corrected by the use of herbicides and inorganic nutrients.

To overcome these limiting factors, a wide range of strategies are available. Niggli *et al.* (2016) describe such strategies as summarized in Table 7 for arable crops that are applicable to Northern Europe. Some of the strategies comprise of execution of well-known best practices e.g. the use of appropriate crop rotation design to control weed infestation, disease and pest outbreaks particularly for horticultural crops, which are susceptible to many pests and pathogens (Letourneau and van Brugge, 2006). The new crop protection strategies in combination with the development and increased use of a variety of biological control agents (e.g. bacteria, fungi and predatory arthropods) (van Lenteren, 2012) will be chiefly important to minimize the yield gap. Increased use of resistant varieties is also critical, but these varieties are however not fully resistant, implying that direct crop protection measures will be especially important to secure high yields and product quality in high value crops (Speiser *et al.* 2006).

**Table 7.** Strategies to increase yields in organic arable crops

| Area of intervention | Important for | Strategies to increase yields |
|---|---|---|
| Soil fertility | All crops, but especially on stockless farms | • Crop rotation design and management including optimal management of legume pre-crop effects and green manure crops<br>• Increased crop diversity<br>• Intercropping<br>• New technologies for reduced tillage<br>• Increased cooperation between livestock farms and stockless farms<br>• Adding/promoting supportive microorganisms and fungi in soil |
| Plant nutrients | All crops (except nitrogen for legumes) | • Optimal use of legumes in rotations<br>• Effective use of manures<br>• Increased recycling and use of nutrients from society<br>• Novel treatments of organic food wastes to produce high quality composts<br>• Technological solutions for safe sewage sludge treatments & recycling |
| Crop-weed competition | All crops, but especially in stockless systems without perennial leys | • Crop rotation design and management<br>• New physical weed control strategies & techniques including cover crop management<br>• Use of the false seedbed technique<br>• Precision farming and robots |
| Control of diseases | All crops, but especially potatoes and legumes | • Use of tolerant or resistant crop varieties<br>• Crop rotation design and management<br>• Preventative strategies like intercropping, deep ploughing, optimal planting date etc.<br>• New techniques and products for preventing fungal infections, physical methods and bio-control organisms<br>• Replace copper that is currently used<br>• Use of certified and dressed seeds |
| Control of pests | All crops, several pests in oilseed rape and potatoes | • Crop rotation design<br>• Habitat manipulation (hedgerows, wild flower strips etc.) to strengthen functional biodiversity (e.g. natural enemies)<br>• Physical/biological methods like nets, traps and repellents<br>• Selective pest control products with low negative side-effects |

*Source:* Niggli *et al.*, 2016

## Quality of produce in organic farming

Nowadays, the chief concern for the agricultural scientists and policy makers is environmental hazards and to develop farming methods which are useful to produce quality food. In order to accomplish this goal, the first step is to search for alternate fertilizers which can replace the inorganic source of fertilizers. Several organic materials have been used as fertilizers in organic crop production. But, use of these materials (animal waste and urban sewage waste) in modern agriculture invites pollutions such as heavy metals, chemical residues and parasites (McCalla *et al.*, 1986). To overcome this problem, farmers have adopted a kind of biological fertilizer called Bokashi in Japanese, anaerobically fermented using oil seed sludge, rice bran and fish processing byproduct as materials. A microbial inoculate, including lactic acid bacteria, yeast and actinomycetes, is usually inoculated to materials before fermented. This kind of organic fertilizer is easily stored for long time without bite smell. However, the major problem observed is the low nutrient availability at the early stage of crop growth even if the nutrient availability is hold longer than chemical fertilizers (Xu, 2000). However, the researchers counter the same problem that it is difficult to obtain same yield in organic farming as in conventional farming. Two leafy vegetables were grown under greenhouse conditions and the dynamics of both the plant growth and the organic fertilizer were examined to enlighten the plant limiting factors and benefits of this organic fertilizer. Leaf-picking or leaf-peeling harvest method was adopted to prolong the growth and harvestable period. It has been proved that quality of crop particularly the vegetables and fruits is improved by the application of organic fertilizers (Larson *et al.*, 2000). The nutritional profile of leafy vegetables was studied including sugar contents, vitamin C and the ionic-nitrate and for the analysis, the organic and inorganic grown vegetables were compared. From results, it was observed that the concentration of sugars and vitamin C were significantly higher but nitrate was lower in organically grown vegetables than chemically fertilized vegetables (Maynard and Barker, 1979). In case of chillies, it was noticed that the chillies grown on vermin composted vegetable waste showed higher carbohydrate and protein concentration (Yadav and Vijayakumari, 2004). Organic potato tubers farming may be expected to have significantly higher dry matter contents (19%) that are considered good for processing into French fries without deteriorating the texture of the fries when concentrations exceed 23 per cent (Haase *et al.*, 2007). Similarly, application of FYM at 10 t / ha alone increased the economic yield and quality parameters like protein and amylose content, hulling percentage and milling percentage of rice (Dixit and Gupta, 2000). It is observed that organically grown potato showed 66% higher yield than the conventional crop (Mourao *et al.*, 2008).

## Export pattern of organic products from India

The export of organic food from India is increasing as more farmers are shifting to organic farming. Since the domestic consumption of organic food is quite low in India, so the main market for Indian organic food industry lies in the Europe and US. India has now become a principal supplier of organic basmati rice, organic herbs, organic spices etc. About 53% of the organic food produced is exported from India. This is significantly higher in comparison to the percentage of agricultural products exported. In 2003, only 6-7% of the total agricultural produce was exported from India (Food Processing Market in India, 2005).

The upliftment in the demand for organic food products in the developed countries and extensive support provided by the Indian government so as to enhance the agri-exports, are the major reasons for growth of the Indian organic food industry. The prices of organic food products in India are relatively 20 -30% higher than non-organic food products. This is a very high premium for most of the Indian population where the per capita income is merely USD 800. Though the per capita income in India is rising swiftly but the domestic market is not adequate to consume the entire organic food produced in the country. As a result, exports of organic food are the principal aim of organic farmers as well as the government. The Indian government is dedicated towards promoting organic food production in the country. During the Tenth Five Year Plan, the Indian government has allocated Rs.100 crore or USD 22.2 million for promoting sustainable agriculture. APEDA (Agricultural and Processed Food Export Development Authority) manages the export of organic food (and other food products) in India. The National Programme for Organic Production in India was initiated by the ministry of commerce. The programme provides standard for the organic food industry in the country. Since these standards have been developed taking into consideration international organic production standards such as IFOAM and CODEX, as a result, Indian organic food products are being accepted in the US and European markets. APEDA also provides a list of organic food exporters in India and is presented in table 8. The country wise export pattern for organic products from India is given in table 9.

Since organic farming is labour intensive and labour is quite expensive in developed countries so organic food production costs are higher in these countries. However, in a country like India, where availability of labour is quite high and relatively cheap, organic agriculture is view as a good cost effective solution to the increasing costs involved in chemical farming. Currently most of the organic farmers in India are still in the evolution phase and hence their costs are still high. As these farmers are expected to adopt the organic farming, the production costs are likely to reduce, thus, making India as one of the significant producers of organic food.

**Table 8.** Organic food products exported from India

| Category | Organic Products |
|---|---|
| Organic cereals | Wheat, rice, maize or corn |
| Organic pulses | Red gram, black gram |
| Organic fruits | Banana, mango, orange, pineapple, passion fruit, cashew nut, walnut |
| Organic oil seeds and oils | Soybean, sunflower, mustard, cotton seed, groundnut, castor |
| Organic vegetables | Brinjal, garlic, potato, tomato, onion |
| Organic herbs and species | Chilly, peppermint, cardamom, turmeric, black pepper, white pepper, amla, tamarind, ginger, vanilla, clove, cinnamon, nutmeg, mace |
| Others | Jaggery, sugar, tea, coffee, cotton, textiles |

*Source:* Chandrashekar, 2010

**Table 9.** Country wise export of organic food products from India (2013-14)

| Country | Export volume (Metric tons) | Percentage share | Export value (Rs. crore) | Percentage share |
|---|---|---|---|---|
| Australia | 749.95 | 0.42 | 14.58 | 1.10 |
| Canada | 38545.57 | 21.68 | 182.41 | 13.73 |
| China | 76.35 | 0.04 | 1.57 | 0.12 |
| European Union | 56946.72 | 32.03 | 553.85 | 41.69 |
| Iran | 38.00 | 0.02 | 1.21 | 0.09 |
| Israel | 312.93 | 0.18 | 3.72 | 0.28 |
| Japan | 309.07 | 0.17 | 16.12 | 1.21 |
| Korea Republic | 143.48 | 0.08 | 2.33 | 0.18 |
| Malaysia | 43.44 | 0.02 | 0.91 | 0.07 |
| New Zealand | 599.79 | 0.34 | 4.23 | 0.32 |
| Philippines | 110.11 | 0.06 | 1.88 | 0.14 |
| Singapore | 73.02 | 0.04 | 0.97 | 0.07 |
| Sri Lanka | 78.51 | 0.04 | 2.45 | 0.18 |
| Switzerland | 4306.56 | 2.45 | 33.89 | 2.55 |
| USA | 74942.72 | 42.16 | 498.83 | 37.55 |
| Others | 489.04 | 0.28 | 9.65 | 0.73 |
| Total | 177765.26 | 100.00 | 1328.60 | 100.00 |

*Source:* Deshmukh and Babar, 2015

## Benefits of organic farming

1. ***Better Nutrition*:** Organic food is rich in nutrients and is free from toxic chemicals. It also increases the nutrients in the soil so that grown crop is healthier to consume (Sreerag and Jayaprakas, 2015).
2. ***Higher income*:** The price of organic food is generally 20 - 30% higher than conventional food. This premium is very essential for a small farmer whose income is just sufficient to feed his/her family with one meal.
3. ***Free from toxic chemicals*:** In organic farming, synthetic chemicals are not used to control pests and other harmful plant diseases, which causes cancer and other diseases to the consumers. The organic farming is free from toxic chemicals (Kumar, 2015). In addition to this, the use of chemicals

pollutes the environment as the chemicals go deep inside the soil and water and contaminate them. But organic farming restricts the usage of chemicals and safeguards the environment from contaminants.

4. ***Quality of food***: The organic food have superior quality in terms of nutrients and it tastes better than the food grown by using synthetic chemicals and quality of food is determined by its taste. Brix analysis is used to determine the quality of fruits and vegetables (Sarma and Brar, 2015). Fruits and vegetables grown by organic farming have higher nutrient content and are superior in taste as they are provided with the appropriate time to grow and no chemicals are used in their growth (Sims, 2014; Deivasigamani, 2013; Kumar and Singh, 2013).

5. ***Enhancing the storage period***: Organic food can be stored for relatively longer period due to metabolic and structural integrity in their cellular structure than the other crops grown by using synthetic chemicals (Pavan Kumar and Satyanarayana, 2014).

6. ***Low input cost***: The expenditure on organic farming is low because it need animals to till the land, manures which are easily available and they can prepare their own, and the bio fertilizers are prepared with low investment cost (Namasivayam *et al*., 2013; Densilin *et al*., 2011; Pandit *et al*., 2012). Unlike conventional farming systems, organic farming does not use expensive chemicals and fertilizers. Therefore, it cut down the production costs for farmers (Chriki-Adeeb and Chriki, 2015).

7. ***Less environment pollution***: Organic farming utilizes the fossil fuel by avoiding the use of synthetic fertilizers and helps to slow down global warming as compared to conventional farming systems (Teshome and Kibert, 2015). Organic farming however, functions with minimum environment pollution (Lampkin, 1992; Znaor, 1997). Thereby, organic farming plays a vital role to safeguard the environment and permitting rational use of natural resources (Pretty, 1995; VanMansvelt and Mulder, 1993).

8. ***Improved water holding capacity of the soil***: Organic farming increases the water holding capacity of the soil. The chemical fertilizers required huge quantities of water to activate its molecule whereas organic fertilizers do not demand for such conditions.

9. ***Improves soil fertility***: Organic farming maintains C:N ratio in the soil and increases the fertility and productivity of the soil.

10. ***Profitability of organic agriculture***: Most available studies report better or similar economic performance in organic agriculture than conventional farm, even without taking into consideration the external costs of production (Lampkin, 1992; LEI, 1990; VanMansvelt and Mulder, 1993). This is mainly due to lower production cost and 20-50% higher prices for organic products than conventional products.

11. ***Promising for less developed countries***: Several authors suggested that organic farming is a promising solution for Eastern and Central Europe both from the economic and environmental point of view (Buys, 1993; Znaor, 1994; Znaor, 1997; Znaor, 2002). Options and implication of converting to organic agriculture in the less developed countries have been positively assessed.

## Demerits of organic farming

1. Organic food is more expensive than the food grown with conventional farming practices. Therefore, the consumer has to pay more for organic food (Kumar and Singh, 2014; Kotikal and Math, 2016; Singh and Sharma, 2016).
2. Organic farmer has to work harder as this farming practice requires more interaction of farmer with his crop (Shenbagavalli and Mahimairaja, 2012).
3. The practice of organic farming provides appropriate time for the crop to grow without the use of artificial growth injectors. But this practice would not be able to meet the world's demand for food as the organic food takes more time to grow as compared to the conventional system of growing crops (Brar and Sarma, 2013).
4. Organic farming is a time taking process in getting the result, which makes the farmers to neglect this kind of farming.
5. It requires more labor force and should have regular observation compared to conventional farming.
6. Organic farming is a skill based work and farmers should be trained time to time according to the seasons and the condition of the crops.
7. The productivity is quite low in organic farming as compared to conventional farming, but in conventional farming, the fertility of soil is decreasing time to time with excess use of chemicals.

However, organic farming surely has some disadvantages but still it is a useful form of doing agriculture, which is advantageous both for the ecosystem and the consumers. The soil gains the nutrients & maintains the fertility for longer time.

## Risks involved in organic farming

However, there are both risks and opportunities associated with strategies to increase yields in organic production. The final outcome largely depends on management, i.e. how strategies are managed & worked out to increase the yield of crop. The various risks involved in organic farming are presented in table 10.

**Table 10.** Risks associated in organic farming

| | | | Business risks | | | |
|---|---|---|---|---|---|---|
| | Agricultural/ Technological | | Communication risks | Non- agricultural | | |
| | Production risks | | | Managerial risks | Social risks | Profit risks |
| Yield risks | Specific primary production risks | | | Institutional and regulatory risks | Informational risks-information shortage, misinformation etc. | |
| | Climatic risks | Environmental risks and disasters-weather conditions-temperature, rain, sun etc. and climate change-extreme weather events such as frost, droughts, floods, storms etc. | | Management and organization risks-including human resources risks | Seasonality risks Research, development and knowledge transfer risks | |
| | Sanitary risks | Ecological risks-pests, diseases & weeds risks-epidemic diseases-GMO risks-pollution risks (land, water etc.) | | Market and marketing risks including price risks-inputs and outputs | Society risks | |
| | | Input risks including supply risks | | Administrative risks | Local development risks | |
| | Post- harvest production risks | Processing risks | | Financial & investments (land, facilities, equipment, technology and other) risks | Life standards and purchasing power risks | |
| | | Storage risks Sorting and packaging risks | | Unfair competition risks | Intentional or non-intentional violations risks-including theft, warfare, terrorism etc. | |
| | | Transport and distribution risks | | | | |
| | | | Price risks | | | |

*Source*: Terziev *et al*., 2015

## Government policies in promoting organic farming

The policy of Ministry of Agriculture tends to promote technically sound, economically viable, environment friendly and socially acceptable use of natural resources in favour of organic agriculture. The policy seeks to actualize the area and crop potential for strengthening rural economy, promoting value addition, increasing growth of agro-business and ensuring a fair standard of living for the farmers, agricultural workers and their families.

### *Area approach and crop identification*

For promotion of organic farming, identification of potential areas and crop is very important. The Government strategy is to promote organic farming for the crops having high market potential such as wheat, cotton, basmati rice, fruits, vegetables, spices, pulses, oilseeds etc. Depending on the potential areas, three priority zones have been identified.

*Category*-I: The very first or top priority areas for promotion of organic farming are the rainfed areas where fertilizer and agro-chemical consumption is already very low.

*Category-II*: The Category II areas are primarily under rainfed farming with little irrigation support.

*Category-III*: The last priority areas are those with moderate to heavy use of fertilizers and pesticides, mostly multiple cropped areas. Some efforts being made in high intensive cropping farming like Punjab, Haryana, etc. But it requires sufficient assess its acceptability in these areas.

### *National project on organic farming*

The Department of Agriculture Cooperation (DAC), Ministry of Agriculture, Government of India has launched a new Central Sector Scheme 'National Project on Organic Farming' with an outlay of Rs. 57.05 crore for production, promotion and market development of organic farming in the country during the 10thPlan on pilot basis. The project is operational since 1$^{st}$ October, 2004. The various components of the project are presented in table 11.

Table 11. National project on organic farming

(Rs. in lakh)

| Components | Xth Plan | |
|---|---|---|
| | Physical target | Financial outlay |
| Establishment of NCOF/RCOFs | - | 829.00 |
| Capacity building through service provider | 300 | 900.00 |
| Setting up of production units- | | |
| i) Fruit/vegetable compost units | 35 | 1400.00 |
| ii) Biofertilizer units | 30 | 600.00 |
| iii) Vermiculture hatcheries | 2000 | 300.00 |
| Training (on certification, input production, extension etc) | 400 | 258.00 |
| Field demonstration | 3920 | 896.00 |
| Market development and promotion | - | 522.00 |
| **Total** | | **5705.00** |

*Source:* Bhattacharyya and Chakraborty, 2005

### *Other projects of MOA in support of organic farming*

Apart from DAC project on organic farming, the projects like Technology mission on horticulture, Cotton scheme etc. is supporting organic farming. Besides, under Macro management programme, Government of India is supporting different State schemes on organic agriculture

### *FAO-DAC project on organic farming*

The Ministry of Agriculture has taken up the Technical Cooperation Programme (TCP) of FAO (TCP/IND/3003 -Development of Technical Capacity Base for the Promotion of Organic Agriculture in India) to overcome the knowledge gap by quickly producing necessary information taking into consideration the various ecological zones of the country, based on the current state of knowledge on organic crop production packages, input production and utilization and certification issues including legal and institutional aspects. The project is operational since April, 2005. The preparation of technical package of practice of different crops for different regions is under process.

## Organic farming on the internet

India has entered the age of information technology. The whole world is open for us to be browsed and to find out for ourselves the latest knowledge on the subject of our choice. There are as many as 1000,00,000 web sites referring to organic farming. Further scrutiny with revealed that there are as many as 5,61,00,000 sites listed under the head "organic farming in India" Since, we have to think about Agriculture for Tomorrow, it will be appropriate to familarise the readers about few of these web sites. the search engine Google One web site running into 18 pages and titled "Naya Krishi Andolan" (UBINIG)from Bangladesh is reproduced verbatim hereunder. A cursory reading will show that this Andolan has caught the imagination of the farmers in 16 districts as a result of which as many as 1,00,000 small farmers (average 5 family members) have now adopted organic farming to their advantage. There is no reason why the poor Indian farmers should not follow in the footsteps of his Bangladeshi brother. The authors are fully convinced that there is no substitute for organic farming and environmentally-sound practices and techniques.

Farida Akhtar is one of the founders of UBINIG, a Bangladeshi NGO which has set up one of the biggest community seed banks in the world. She narrared about development of UBINIG in Bangladesh as:

The term "UBINIG" is the Bangla acronym of "Policy Research for Development Alternative." In the early 1980s, there was an urgent need to look for development alternatives and translate them into the policy language in order to influence policy makers. At that time, the World Bank was pushing export-oriented policies and several sectors were being impacted. Shrimp farmers were being affected by large-scale shrimp operations and we were increasingly concerned about the environmental impacts of commercial shrimp culture. Weavers and handloom workers were increasingly losing their jobs as the garment industry relied completely on imported cloth. We began question whether the present approach was something that the people wanted. UBINIG was set up in 1984.

After severe floods in 1988, farmers of Tangail District approached us for help. They had lost everything in the floods and needed to start all over again. They asked us for fertilisers, since they thought the "modern" way of growing food was the only option. We started having discussions with the village women and finding out what the problems really were and what the farmers needed. Our approach built on the ideas of "ecological agriculture" that were sprouting up in Europe and India. So the Naya Krishi Andolan – new agriculture movement – was born. The farmers coined the term themselves. This meant following ecological principles to produce food in harmony with nature. Women were the first to respond as they had come face-to-face with excessive chemical use in agriculture. Midwives were especially concerned about miscarriages and other birth deformities. Then next the small poor farmers responded to Naya Krishi, since they could not afford to buy chemicals.

**Why the particular interest in agrobiodversity?**

Naya Krishi was not meant to be a quick fix technical solution. We had to do it, experience it and then keep it alive. For pest management and soil management, diversity was the obvious choice. As the farmers developed diverse cropping, they found many partner-plants coming up. The fish returned in the water and there were also other interesting results with cross-pollination. The birds were doing their job better! It was like a whole web of life being reactivated. Seed management is important to maintain diversity. There is a conscious policy to emphasise local seed varieties and farmers facilitate seed exchanges. Because of careful seed management, we have more than a thousand varieties of rice, 37 vegetable varieties and more than 40 varieties of chillies alone.

**How do you spread the word?**

The farmers who practise Naya Krishi talk about it to others and give demonstrations.

The Bangla people have an appreciation of the gastronomic qualities of food, and when that improves the new way of farming speaks for itself. In Bangla we say, we do not simply eat. We serve our body when we take food. The tongue is important not the teeth.

This taste and variety is made possible by the mixture of cropping.

**How is your approach to farming different?**

Naya Krishi is not input-based. Modern agriculture is all about high yields and fashion foods. Our agriculture is about nurturing the seed. The main capital is not cash, but farmers' knowledge. It is about restoring "culture" in agriculture. When you ask a farmer how many members there are in her/his family, the reply will be in "1 son, 2 daughters, 2 cows, 5 chickens"all in the same breath. Farming is about tending the whole family. It is about growing a particular kind of paddy because it makes good straw for your cow, even though it does not fetch a good price on the export market. Our approach to farming is more human, as opposed to motivated by greed, as modern agriculture is.

**How do the seed banks work?**

We do not call them "seed banks," but "community seed wealth centres." We have one in each of our regional centres. They work on the basis of give and take. We exchange with the farmers, keeping just enough for a sample. We encourage a decentralised seed system. We do not want to create dependency on UBINIG or its seed centres. So in every village there is a seed hut where seed preservation and seed storage takes place. Indivdual households also have their own collections.

**How many people are involved with the Naya Krishi Andolan?**

It is growing every day. Today spans over 16 districts and over 100,000 farming families (with an average of five members per family).

The Naya Krishi Andolan is renowned for the way it brings together people from a variety of cultures and religions. Why has it been so successful in doing so?

Bangladesh is 83% Muslim country. But our culture imbibes principles from Buddhism and Hinduism. This mix is reflected in – and is important to – our agriculture. Songs are an important part of our work and community life. As a daily ritual we start the day with spiritual songs, with songs of Krishna who herds the cowseven though we are Muslims. A villager in these parts is happy only because there is song. Modern agriculture pollutes this cultural environment and gives no room or reason for song.

### Why are women so important to the movement?

Modern agriculture disempowers women, making them a redundant part of the farming family. Naya Krishi thrives on women's knowledge. And even in a patriarchal society like ours, women's knowledge is being acknowledged in the decisions being made about which crop to grow at what time, in what way. Food security is not possible without women. Missing the crucial link between the seed and the woman is "bogus feminism".

### What is your role in the wider South Asian region?

We do not strive to set standards or be a model. And the South Asian region as a whole is rich in experience and new initiatives. There are several such examples from India, in the famed forest preservation of Himachal or the seed-saving work in Hyderabad. And there are important linkages to be made. That is why in 1996 in preparation of the World Food Summit we got together as groups from South Asia under SANFEC – South Asia Network for Food Ecology and Culture. Our emphasis is equally on food, ecology and culture to keep diversity alive.

SANFEC members are against the patenting of life. There is no compromise on this position. The organisations comprising the membership of SANFEC are all those who are actually working with the people. We regularly interact. The farmers from all the South Asian countries are happy to meet and exchange seeds. The two premiers of India and Pakistan may not be on talking terms, but are farmers are. We constantly hold biodiversity festivals and fairs together.

### What are the long term goals?

Our effort comes from the realization that so much has been lost; and so there is so much to bring back and there are so many ways to enhances our biodiversity. We want to establish our fight against the transnational corporations that are destroying agricultlure and show that another life is possible. Biodiversity and diversity alone has the answers.

## Conclusion

During the last four decades of the 20th Century, the global population has almost doubled itself from 3 to 6 billion and it is estimated to reach around 8 million during the year 2020. Food and nutritional security is therefore, a serious global concern. Neither conventional farming with inorganic alone nor organic farming only with the use organic input can face this challenge. The combination of inorganic and organic is certainly the best option to resolve this challenge unless the existing dietary system is changed.

The farmers of India had been practicing environment friendly agriculture for centuries till the introduction of the so-called 'green revolution' which was based on the conventional farming methods came into existence in the western countries. Still many resource-poor, marginal and small farmers, have not fully adopted the conventional farming and they follow more or less the traditional eco-friendly agriculture systems. They use local or own farm derived renewable resources and manage self-regulated ecological and biological processes. This has become vital to cultivate the acceptable levels of crop, livestock and human nutrition products and above all to safeguard both the crops and humans from pests and diseases through the use of locally available low-cost inputs. Such a situation is suitable for making the farming community aware of the organic farming methods to make the switch over less troublesome. A country like India can enjoy a number of benefits from the adoption of organic farming. The price premiums for the products, conservation of the natural resources in terms of improved soil fertility and water quality, prevention of soil erosion, preservation of natural and agro-biodiversity are major benefits of organic farming. Economic and social benefits like generation of rural employment, lower urban migration, improved household nutrition, local food security and reduced dependence on external inputs will be large gains in the Indian conditions. The control of environment pollution and the consequent increase in the quality of human life will be other contributions of organic farming. There is a good demand for organic products in the domestic and export market. The main rules and regulations for accreditation and certification of organic products are in place in India. Considering the above, it may be concluded that organic farming will progress tremendously in India.

The organic production is coming farmer's movement and consumer choice which cannot be ignored. All facilities should be necessarily provided to organic farmers in the form of appropriate package of practice, voluminous amounts of organic inputs and good domestic as well as export market. Organic farming should not be discouraged under any circumstances. The immediate task is to arrange availability of organic inputs and low cost certification process.

Therefore, it is recommended that Governments should develop appropriate policies for extensive adoption of organic agriculture by millions of individual farmer households. Specifically, Governments should encourage farmers to establish effective organizations to promote organic agriculture. In addition to this, they should support the development of alternatives to chemical pest control and chemical fertilizers. Meanwhile, local NGOs and all other related sectors should assist farmers to minimize the use of synthetic chemicals in farming. Actions should be taken to increase consumers' awareness of organic consumption. For this massive extension works and support from state government for capacity building of local NGO's are needed. More research need to be conducted on markets for organic products and operational mechanism for risk insurance should be secured through financial support during the conversion period and provision of incentives or subsidies for organic agricultural production.

# References


Abrol P and Palaniappan S P (1987) Green manure crops in irrigated and rainfed lowland rice based cropping system in South Asia. In: Green Manure In rice farming. Proc. of a Symposium on "Sustainable Agriculture", Los Banos, Philippines, pp-71.

Aggarwal PA and Goswami B (2003) Bed planting system for increasing water use efficiency in wheat (*Triticum aestivum*) grown in Inceptisol (Typic Ustochrept). *Nidian Journal of Agricultural Sciences* **73**:422-425.

Agromisa (2003) Agroforestry CTA. Wageningen Netherlands.

Aher S B, Sengupta B and Bhaveshanada S (2012) Organic agriculture: way towards sustainable development. *ARPN Journal of Science and Technology* **2**:318-324.

Alkhader AMF (2015) The impact of phosphorus fertilizers on heavy metals content of soils and vegetables grown on selected farms in Jordan. *Agrotechnology* **5**:137.

Alvares C (1999) *The Organic Farming Source Book*. The Other India Press, Mapusa, Goa, India.

Amadou Binta BA and Barbier B (2015) Economic and Environmental Performances of Organic Farming System Compared to Conventional Farming System: A Case Farm Model to Simulate the Horticultural Sector of the Niayes Region in Senegal. *Journal of Horticulture* **2**:152.

Anonymous (1999) Organic Agriculture, Item 8 of the Provisional Agenda. FAO Committee on Agriculture. Proceedings of the 15th Session, Rome.

APEDA (2013-14) National Programs for Organic Production (NPOP). www.apeda.gov.in.

Arora Y K and S C Mohan (1986) Agrihorticulture systems for watershed management. *Indian Journal of Soil Conservation* **14**(3):100-104.

Arshad S and Shafqat A (2012) Food Security Indicators, Distribution and Techniques for Agriculture Sustainability in Pakistan. *International Journal of Applied Science & Technology* **2**: 137-147.

Asiegbu I E and Oikeh S O (1993) Growth and yield of tomato to sources andrates of organic manure in ferralitic soils. *Bioresource Technology* **45**: 21-25.

Askegaard M, Olesen JE, Rasmussen IA, Kristensen K (2011) Nitrate leaching from organic arable crop rotations is mostly determined by autumn field management. *Agricultural Ecosystem Environment* **142** : 149–160. https://doi.org/10.1016/j.agee.2011.04.014

Atungwu JJ, Aiyelaagbe I O O, Sobowale PAS, Oni AO and Garba SH (2009) Integrating African Traditional Farming and Organic Agriculture. In: Organic Agriculture for Health, Wealth and Environmental Conservation. Proceedings of the fifth National Conference of Organic Agricultural Project Intertiary Institutions in Nigeria. FUTO. November, 15-19, 2009.

Aubert C (2007) Can organic farming mitigate the impact of agriculture on global warming? In: International Conference on Organic Agriculture and Food Security (FAO, Rome, Italy, 3-5 May 2007) pp:1-2.

Auobamiri and Ahmad GA (2014) Effects of conservation tillage inorganic brkshavrzy. *International Journal of Plant Animal and Environmental Sciences* **4**(4): 263-274.

Avasthe R, Singh R, Babu S and Kumar A (2020) Success Story on Integrated Organic Farming System in Sikkim. In: International Science Conference on "Organic Farming Research, Technologies and Extension" 31st October 2020, India (Online) pp 46-49.

Avery A (2007) Going organic. *Crops & Soils. American Society of Agronomy* **40**:8-12.

Bagchi A (2015) Organic farming practice for quality improvement of tea and its anti parkinsonism effect on health defense. *Journal of Physics Chemistry Biophysics* **5**:178.

Baker BP, Benbrook CM, Groth E and Benbrook KL (2002) Pesticide residues in conventional, integrated pest management (IPM)-grown and organic foods: Insights from three US data sets. *Food Additives and Contaminants* **19**(5):427–446.

Bakhetia DRC, Sukhwinder Kumar and Sandhu IS(1995) Cultural control of cutworm in Sunflower. *Journal Research Punjab Agricultural University* **32**(2): 137-141.

Balak Ram (2003) Impact of human activities on land use changes in arid Rajasthan: Retrospect and Prospects. In: Human impact on desert environment, eds. P. Narain, S. Kathju, A. Kar, M. P. Singh and Praveen-Kumar, 44-59. Jodhpur, India: Scientific Publishers.

Basak A B, Lee M W and Lee T S (2002) Inhibitive activity of cow urine and cow dung against *Sclerotinia sclerotiorum* of cucumber. *Mycobiology* **30**: 3.

Bengtsson J, Ahnström J and Weibull AC (2005) The effects of organic agriculture on biodiversity and abundance: a meta-analysis. *Journal of Applied Ecology* **42**: 261–269.

Berry PM, Stockdale EA, Sylvester-Bradley R, Philipps L, Smith KA, Lord EI (2003) N, P and K budgets for crop rotations on nine organic farms in the UK. *Soil Use and Management* **19**:112–118.

Bhadauria T and Ramakrishnan P S (1996) Role of earthworms in nitrogen cycling during the cropping phase of shifting agriculture (*Jhum*) in north-east India. *Biology and Fertility of Soils* **22**: 350-354.

Bhattacharya P and Chakraborty G (2005) Current status of organic farming in India and other countries. *Indian Journal of Fertilisers* **1**(9):111-123

Blaise D (2006) Yield, boll distribution and fibre quality of hybrid cotton (*Gossypium hirsutum* L.) as influenced by organic and modern methods of cultivation. *Journal of Agronomy and Crop Science* **192**:248-256.

Brar SK and Sarma SJ (2013) Bio-Hydrogen production: a promising strategy for organic waste management. *Hydrology: Current Research* **5**:002.

Brentrup F, Kusters J, Kuhlmann H and Lammel J (2001) Application of the life cycle assessment methodology to agricultural production: an example of sugar beet production with different forms of nitrogen fertiliser. *European Journal of Agronomy* **14**:221–233.

Buys J (1993) Conversion Towards organic agriculture in Russia - A preliminary study, Biological Codex Alimentarius commission (2001). Organically produced foods. Rome, Italy: FAO AND WHO 77pp.

Cederberg C (2002) Life cycle analysis (LCA) of animal production. Ph.D. Thesis, Department of Applied Environmental Science, G¨oteborg University

Chand H (2007) Ecofriendly methods of plant disease management. In: *Low Cost Biological Approaches for Sustainable and Organic Agriculture*. Regional Centre of Organic Farming, Hisar.

Chander M, Kumar S, Rathore RS, Mukherjee R, Kondaiah N and Pandey HN (2007) Organic vis-a-vis Conventional Livestock Production Potential in India. In: International Conference on Organic Agriculture and Food Security (FAO, Rome, Italy, 3-5 May 2007) pp: 48-49.

Chandra KK (2014) Growth, fruit yield and disease index of carica papaya inoculated with pseudomonas straita and inorganic fertilizers. *Journal of Biofertilizer Biopesticides* **5**:146.

Chandra S and Chauhan SK (2004) Prospectsoforganic farming in India. *Indian Farming* **52**, (2): 11–14.

Chandra R (2014) Sustainability through organic agro- biotechnology with special reference to Jammu & Kashmir scenario. *International Journal of Genetic Engineering and Biotechnology* **5** (2):169-178.

Chandrashekar (2010) Changing scenario of organic farming in India: An overview. *International NGO Journal* **5** (1): 034-039.

Chhonkar P K (2002) Organic farming myth and reality. In: Proceedings of the FAI Seminar on Fertilizer and Agriculture Meeting the Challenges, New Delhi, India.

Chriki-Adeeb R and Chriki A (2015) Bayesian phylogenetic analysis of rhizobia isolated from root-nodules of threetunisian wild legume species of the genus sulla. *Journal of Phylogeny Evolution Biology* **3**:149.

Coleman D C (1989) Agro-ecosystems and sustainable agriculture. *Ecology* **70:** 15-90.

Conacher J, Conacher A (1998) Organic farming and environment, with particular reference to Australia: a review. *Biological Agriculture and Horticulture* **16:** 145-171.

Condron LM, Cameron KC, Di HJ, Clough TJ, Forbes EA, McLaren RG (2000) A comparison of soil and environmental quality under organic and conventional farming systems in New Zealand. *New Zealand Journal of Agricultural Research* **43**:443–466.

Connor DJ and Minguez MI (2012) Evolution not revolution of farming systems will best feed and green the world. Global Food Security **1**:106–113. https://doi.org/10.1016/j.gfs.2012.10.004.

Cormack WF (2000) Report to Ministry of Agriculture, Fisheries and Food, London, UK. Retrieved March 3, 2011 from http://orgprints.org/8169.

Dahlin S, Kirchmann H, Kätterer T, Gunnarsson S, Bergström L (2005) Possibilities for improving nitrogen use from organic materials in agricultural cropping systems. *Ambio: A Journal of the Human Environment* **34**:288–295. https://doi.org/ 10.1579/0044-7447-34.4.288

Dai J (1999) Quality identification and logo systems for agricultural products in France. *World Agriculture* **243**: 11- 12. (in Chinese)

Das A, Layek J, Ramkrushna G, Babu S, Devi M T, Dey U, Suting D, Yadav G S, Lyngdoh D B Dand Prakash N (2019) Integrated organic farming system: an innovative approach for enhancing productivity and income of farmers in north eastern hill region of India. *Indian Journal of Agricultural Sciences* **89**(8): 1267–1272.

de Ponti T, Rijk B, van Ittersum MK (2012) The crop yield gap between organic and conventional agriculture. Agricultural Systems **108**:1–9. https:// doi.org/10.1016/j.agsy.2011.12.004.

Deivasigamani S (2013) Influence on certain herbicides for the control of water hyacinth (*Eichhornia Crassipes* (Mart.) Solms) and its impact on fish mortality. *Journal of Biofertilizers Biopesticides* **4**:138.

Densilin DM, Srinivasan S, Manju P, Sudha S (2011) Effect of individual and combined application of biofertilizers, inorganic fertilizer and vermicompost on the biochemical constituents of chilli (Ns - 1701). *Journal of Biofertilizers Biopesticides* **2**: 106.

Desai VR, Sabale RN and Raundal PV (2015) Integrated nitrogen management in wheatcoriander cropping system. *Journal of Maharasthra Agricultural Universities* **24** (3): 273–275.

Deshmukh M S and Babar N (2015) Present status and prospects of organic farming in India. *European Aca- demic Research* **3** (4): 4271-4287.

Dhaliwal G S, Sangha G S and Ralhan P K (2000) *Fundamentals of Environment Science.* Kalyani Publishers, Ludhiana pp 129.

Dhir R P (1997) Problems of desertification in the arid zone of Rajasthan, India. *Desertification Control Bulletin* **27**:45-52.

Diercks R (1986) Alternativen im Landbau: Eine kritische Gesamtbilanz. Ulmer, Stuttgart. pp. 280-291.

Dixit G and Gupta BR (2000) Effect of FYM, chemical and biofertilizers on yield and quality of rice and soil properties. *Journal of Indian Society Soil Science* **48**: 773-780.

Du X G, Wang H M (2001) *An Introduction to Organic Farming*. China Agricultural University Press, Beijing. (in Chinese)

Durgude A G, B P Role, V A Joshi and J D Patil (1996) Effect of different organic manures on yield, nutrient uptake and moisture utilization by rabi sorghum. *Indian Journal of Dryland Agriculture Research and Development* **11**(2): 90-92.

Edwards CA, Lal R, Madden P, Miller RH and House G (eds) (1990) *Sustainable Agricultural Systems*. Soil and Water Conservation Society, Ankeny, IA.

Ekoweb (2017) Ekologisk livsmedelsmarknad. ("The market for organic foods."). http://www.ekoweb.nu/?p=11247&m=2146. Accessed 14 Aug 2017

EU (2014) Commission Implementing Regulation (EU) No. 354/2014 amending and correcting Regulation (EC) No. 889/2008 laying down detailed rules for the implementation of Council Regulation (EC) No. 834/2007 on organic production and labelling of organic products with regard to organic production, labeling and control.

Evans J (1992) *Plantation Forestry in the Tropics* (second edition). Oxford University press Inc. New York. P6.

Eyhorn F, Ramakrishnan R and Mäder P(2007) The Viability of cotton-based organic farming systems in India. *International of Journal Agricultural Sustainability* **5**(1): 25–38.

Feber R E, Firbank LG, Johnson PJ and Macdonald DW (1997) The effects of organic farming on pest and non-pest butterfly abundance. *Agriculture, Ecosystems and Environment* **64**:133–139.

Fertiliser Association of India (FAI) (1998) *Fertiliser Statistics 1997-98*. New Delhi: The Association.

FiBL (2020) *The World of Organic Agriculture Statistics and Emerging Trends.* pp. 1-337. Research Institute of Organic Agriculture FiBL and IFOAM Organics International.

FiNieberg H and Oppermann N (2003) *The Profitability of Organic Farming in Europe.* pp 141–152 in OECD, ed. Organic agriculture. Sustainability, markets and policies. CABI Publishing, Wallingford, UK.

Food and Agriculture Organisation of the United Nations (FAO) / World Health Organisation (WHO) (2001) Guidelinesfor the Production, Processing, Labeling and Marketing of Organically Produced Foods (GL 32-1 999/2001). Codex Alimentarius. pp. 5-11.

Food and Agriculture Organisation of the United Nations (FAO) (2002) Organic agriculture, environment and food security. Environment and Natural Resources. pp. 6-48.

Food and Agriculture Organisation of the United Nations (FAO)/International Trade Centre (ITC)/Technical Centre for Agricultural and Rural Co-operation (CTA) (2001) World Markets for Organic Fruit and Vegetables, Rome. pp. 28-50.

Forschungsinstitut fuer Biologischen Landbau (FiBL (2017) Organic Agriculture Worldwide: Current Statistics. Helga Willer and Julia Lernoud, Research Institute of Organic Agriculture (FiBL), Frick, Switzerland.

Forschungsinstitut fuer Biologischen Landbau (FiBL 2002) Organic Farming in Europe - Provisional Statistics, Provisional Results of a FiBL Survey. October 2003. http://www.organic.aber.ac.uk/stats.shtm1

Foster C and Lampkin N (2000) Organic and In-conversion Land Area, Holdings, Livestock and Crop Production in Europe. Commission of the European Communities, Agriculture and Fisheries (FAIR). pp. 2-15. ftp://ftp.fao.org/codex/standard/organic/gl9932e. pdf

Francis CA, Flora CB and King LD (eds) (1990) *Sustainable Agriculture in Temperate Zones*. Wiley, New York, NY.

Frankenberger WT & Abdelmagid HM (1985) Kinetic parameters of nitrogen mineralisation rate of leguminous crops incorporated into soil. *Plant and Soil* **87**: 257-271

Frieben B (1997) Arten-und Biotopschutz durch Organischen Landbau. In Naturschutz durch okologischen Landbau. Deukalion, okologische Konzepte 95. Weiger, H. and H. Willer (eds.). pp73–92.

Gandhi M, Sangwan V, Kapoor KK and Dilbaghi N (1997) Composting of household wastes with and without earthworms. *Environment and Ecology* **15**(2): 432–434.

Gaur R B and Sharma R N (2010) Validation of some traditional wisdom of farming communities of semi arid region in Rajasthan, India-Regional review. *In: Proceeding of International Conference on Traditional Practices in Conservation Agriculture*. Choudhary S L, Khandelwal S K and Nene Y L (Eds.). September 18-20, 2010. pp 142-151.

Gauri P V S M (2005) National Programme for Organic Production. In: *Proceedings of National Seminar on National Policy on Promoting Organic Farming*, pp 44-48. Ghaziabad (India): National Centre of Organic Farming.

Ghai S and Thomas George (1989) Occurrence ofAzospirillum spp. in coconut-based farming systems. *Plant and Soil* **114**: 235 – 241

Gharaibeh SH (1998) Removal of selected heavy metals from aqueous solutions using processed solid residue of olive mill products. *Water Research* **32**:498-502.

Gholz H L (1987) Agroforesty: realities, possibilities, and potentials.(http://books.google.com/books7idY1YflgzthEsC&p^PA137 &Ipg= PA137 &dq ^Year+agrofores1r y+came+ mto+mgeria %3F&l&sig = pLhySxcjWsl sHueqfvd6vOsji3g&hlen).

Ghorab MA and Khalil MS (2016) The effect of pesticides pollution on our life and environment. *Journal of Pollution Effects and Control* **4**:159.

Ghosh S K (2000) Organic farming for sustainable development. *Kisan world* **27** (5) :14-17.

Giovannucci D (2005) International Fund for Agricultural Development (IFAD) Report No. 1664, July 2005. Rome.

Gliessman S R (1990) Agroecology: Researching the Ecological Basis for Sustainable Agriculture. pp. 3-10.

Graf S, Willer H (2001) Organic Agriculture in Europe-Current Status and Future Prospects of Organic Farming in Twenty- five European Countries. Stiftung Okologie & Landbau (SOEL). pp. 8-21.

Greene C R (2001) U.S. Organic farming emerges in the 90s: Adoption of certified systems. Agriculture Information Bulletin. pp. 770.

Gupta J P and A K Sharma (1996) Agrihorticulture in arid region for stability and sustainability. In: *Agroforestry systems for degraded lands*, eds. Singh P. *et al.*, 164-167. New Delhi: Oxford and IBH.

Gupta R D, Kher Deepak and Jalali V K (2005) Organic farming: concept and prospective in Jammu and Kashmir. *Journal of Research, SKUAST-J* **4:** 25-37.

Haas G, Wetterich F and K¨opke U (2001) Comparing intensive, extensified and organic grassland farming in southern Germany by process life cycle assessment. *Agriculture Ecosystem and Environment* **83**:43–53.

Haase T, Schuler C, Haase NU and Heb J (2007) Suitability of organic potatoes for industrial processing: effect of agronomical measures on selected quality parameters at harvest and after storage. *Potato Research* **50**:115-141.

Hansen B, Altroe HF and Kristensen ES (2001) Approaches to assess the environmental impact of organic farming with particular regard to Denmark. *Agriculture Ecosystem and Environment* **83**:11–26.

Herrmann G, Plakolm G (1991) Oekologischer Landbau, Grundwissenfuer die Praxis. Wien, Verlagsunion Agrar. pp.27-32.

Hill J K W (2016) Organic agriculture in india and participatory guarantee systems (PGS): a case study from West Bengal. *Jharkhand Journal of Development and Management Studies* **14**:7037-7055

Hoeppner JW, Entz MH, McConkey BG, Zentner RP and Nagy CN (2006) Energy use and efficiency in two Canadian organic and conventional crop production systems. *Renewable Agriculture and Food Systems* **21**(1):60-67.

Hole DG, Perkins AJ, Wilson JD, Alexander IH, Grice PV and Evans AD (2005) Does organic farming benefit biodiversity? *Biological Conservation* **122**:113–130.

Howard A (1940) *An Agricultural Testaments.* Oxford University Press.

Ibeawuchi I I, Obiefuna J C, Tom C T, Ihejirika G O, Omobvude S O (2015) Indigenous and current practices in organic agriculture in Nigeria: A review. *Agriculture, Forestry and Fisheries* **4**(6): 257-262.

Ibeawuchi I I and Ofoh M C (2000) Productivity of maize cassava/food legume mixtures in south-eastern Nigeria. *Nigeria Journal of Agricultural Rural Development* **1**(1): 19.

IFOAM (2005) The principles of organic agriculture. Preamble. http:// www.ifoam.bio/ sites/default/files/poa_english_web.pdf. Accessed 13 Jan 2016

IFOAM (2009) Organic Agriculture Worldwide: The main results of the FiBL – IFOAM Survey 2009 Helga Willer, Research Institute of Organic Agriculture FiBL, Frick, Switzerland Presented at BioFach Congress Nürnberg, February 20, 2009.

IFOAM (International Federation of Organic Agriculture Movements) (2002) Organic Agriculture and Food Security (Dossier 1, 2002).

IFOAM (International Federation of Organic Agriculture Movements) and Food and Agriculture Organisation of the United Nations (FAO) (2002) Conference Conclusions, IFOAM Conference on Organic Guarantee Systems- International Harmonization and Equivalence on Organic Agriculture. 17-19 February 2002, Nuremberg, Germany. pp. 11-18.

IFOAM EU Group (2010) Working for organic farming in Europe. IFOAM EU Group Annual Report. IFOAM, Bonn.

IFOAM (2004) The World of Organic Agriculture Statistics and emerging trends (Helga Willer and Minou Yussefi eds).pp 1 - 167.

IFPRI (International Food Policy Research Institute) (2002) Green Revolution: Curse or Blessing? Retrieved March 8, 2011 from http://www.ifpri.org/pubs/ib/ib11.pdf

Indian Horticulture Database (2014) National Horticulture Board, Ministry of Agriculture, Government of India: Gurgaon, India, 2015. Available online: http// www.nhb.gov.in (accessed on 25 October 2015).

International Trade Centre (ITC) (1999) Organic Food and Beverages: World Supply and Major European Markets. International Federation of Organic Agriculture Movements (IFOAM). pp. 46-71.

ISO/DIS (2000) Environmental Management – Life Cycle Assess- ment – Life Cycle Impact Assessment. ISO 14042.

Johan P (1999) *Anthropology of Food: The Social Dynamics of Food Security*. Cambridge, UK; Malden, a: Polity Press; Blackwell Publishers.

Jorden R, Muller A and Oudes A (2009) High carbon sequestration, low emission food secure farming. Organic Agriculture- a guide to climate change and Food Security.

Juma M A (2007) IFOAM's perspectives on organic agriculture, food security and sovereignty. In: Papers Submitted to the International Conference on Organic Agriculture and Food Security (FAO, Rome, Italy, 3-5 May 2007). pp 5-6.

Kahnt G (1986) Biologischer Pflanzenbau: Moeglichkeiten und Grenzen Biologischer Anbausysteme. Ulmer, Stuttgart. pp 19-36.

Kalyanan A (2007) *Fundamentals of Agriculture*. Vol. 2. Kaushal Publications and Distributors, Varanasi.

Karmegam N and Daniel T (2000) Effect of vermicompost on the growth and yield of green gram (*Phaseolus aureus* Rob.) *Tropical Agriculture* **76**(2), 143–146.

Kaur K and Toor MS (2015) Organic Farming: Status and Constraints. *Indian Journaal of Economic Development* **11**(1): 333-38.

Kellogg C E and Orvedai AC (1996) Potentially arable soils of the world and critical measures for their use. *Advances in Agronomy* **21:** 109-170.

Khan M A ed. (2002) *Watershed Management for Sustainable Agriculture*. Jodhpur (India): Agrobios.

Khanna B (2012) Organic farming attracts more Punjab peasant. *Hindustan Times* May 13, Bathinda.

Kotikal YK and Math M (2016) Insect and non-insect pests associated with drumstick, *Moringa oleifera* (Lamk.). *Entomology Ornithology Herpetology* **5**:180.

Kotschi J and Muller-Samann K (2004) The Role of Organic Agriculture in Mitigating Climate Change- Scoping Study. IFOAM, Bonn.

Kuepper, G. 2003. Manures for organic crop production. http: www.attra. mcat.org/

Kumar A and Mali P C (2010) Biochemical evidence of raw cow milk and *Gliocladium virens*induced protection against downy mildew of pearl millet. *In: Proceeding of International Conference on Traditional Practices in Conservation Agriculture*. Choudhary S L, Khandelwal S K and Nene Y L (Eds.). September 18-20, 2010. pp 151-159.

Kumar A S and Singh S C (2013) Adsorption and desorption behavior of chlorotriazine herbicides in the agricultural soils. *Journal Phylogenetics Evolutionary Biology* **4**:154.

Kumar S (2015) Biopesticide: An environment friendly pest management strategy. *Journal of Biofertilizer and Biopesticides* **6**: 127.

Kumar S and Singh A (2014) Biopesticides for integrated crop management: environmental and regulatory aspects. *Journal of Biofertilizer and Biopesticides* **5**:121.

Kumar V (1999) Effect of phosphate solubilising analogue resistant mutants of *Azotobacter chroococcum* on sorghum. *Indian Journal of Agricultural Sciences* **69**:198-200.

Kundu BS and Gaur AC (1984) Rice response to inoculation with $N_2$ fixing and P solubilizing microorganisms. *Plant and Soil* **79**:227-234.

Lammerts van, Bueren ET, Jones SS and Tamm L (2011) The need to breed crop varieties suitable for organic farming, using wheat, tomato and broccoli as examples: a review. *NJAS Wageningen Journal Life Sciences* **58**: 193–205. https://doi.org/ 10.1016/ j.njas. 2010.04.001.

Lampkin N (1992) The economic implication of conversion from conventional to organic fanning systems. Ph.D. Thesis, Department of Economics and Agricultural Economics, University of Wales, Aberystwyth.

Lampkin NH (1990) Estimating the impact of widespread conversion to organic farming on land use and physical output in the United Kingdom, in Lampkin, N.H./Padel, S. (ed.) Economics of Organic Farming. UK: CAB International.

Larson A, Ching A, Messner F and Messner H (2000) High quality, cost effective production of diverse horticultural crops grown organically: an individual case study in Northwest Missouri, U.S.A. Acta Hort. (ISHS) **536**:53-60.

Layek J, Das A, Krishnappa R, Patra S and Kandpal B K (2020) Success story on integrated organic farming system cluster in Meghalaya. In: International Science Conference on "Organic Farming Research, Technologies and Extension" 31st October 2020, India (Online) pp 50-52.

LEI (1990) Productie en atzet Van BD- en Eko-Produkten (production end economics or bio-dynamic food). National agro-economic institute of the whether lands. Band 2. LEI medelingen 425, Den Haeag. 89p.

Letourneau D and van Bruggen A (2006) *Crop Protection in Organic Agriculture. Organic Agriculture: A Global Perspective.* CSIRO Publishing, Collingwood.

Lin X J (2003) The seven development tendencies of the world's organic agriculture. *World Agriculture* **293**: 17-21. (in Chinese).

Lotter DW (2003) Organic agriculture. *Journal of Sustainable Agriculture* **21**: 59–128.

Lotter D W, Seidel R and Liebhardt W (2003) The performance of organic and conventional cropping systems in an extreme climate year. *American Journal of Alternative Agriculture* **18** (3):146-154.

Mader P, Fliebach A, Dubois D, Gunst L, Fried P and Niggli U (2002) Soil fertility and biodiversity in organic farming. *Science* **296** (5573):1694–1697.

Mahajan H S, Hirve N A and Deshmukh M R (2013) Effect of integrated nutrient management on seed yield and economics of sesame (*Sesamum indicum* L) in assured rainfall zone of North Maharashtra. *Journal of Oilseeds Research* **30**: 147-49.

Maheswarappa H P, Nanjappa H V and Hegde M R (1999) Influence of organic manures on yield of arrowroot, soil physico-chemical and biological properties when grown as intercrop in coconut garden. *Annals of Agricultural Research* **20**: 318-323.

Mahnke-Plesker S, Lorenz S and Brenndörfer E (2005) Installation of a monitoring system for fresh fruit and vegetables in the German natural food market. Bericht, Bundesverband Naturkost Naturwaren (BNN) Herstellung und Handel e.V., D-Köln. Retrieved February 4, 2011 from http://orgprints.org/6742.

Marinari S, Masciandaro G, Ceccanti B and Grego S (2000) Influence of organic and mineral fertilisers on soil biological and physical properties. *Bioresource Technology* **72**(1): 9-17.

Marinari S, Liburdi K, Masciandaro G, Ceccanti B and Grego S (2007) Humification-mineralization pyrolytic indices and carbon fractions of soil under organic and conventional management in central Italy. *Soil Tillage Research* **992** (2):10-17.

Marks B (2004) Diversity and classification of mycorrhizal associations. *Biological Reviews* **79:** 273-295.

Mattsson B (1999) Environmental life cycle assessment (LCA) of agricultural food production. Ph.D. Thesis, Department of Agricultural Engineering, Swedish University of Agricultural Sciences, Alnarp

May C (2001) New Zealand and the organic market. Inaugural OFA National Organics Conference 2001. Rural Industries Research and Development Corporation, Canberra. pp. 109- 110.

Maynard DN and Barker AV (1979) Regulation of nitrate accumulation in vegetables. *Acta Horticulturae* (ISHS) **93**:153- 162.

McCalla TM, Peterson JR & Lue-Hing C (1986) *Properties of Agricultural and Municipal Wastes*. In: ASA CSSA SSSA (Editors), Soils for Management of Organic Wastes and Waste Waters. ASA, Madison, pp. 11-46.

Meek D and Anderson C R (2020) Scale and the politics of the organic transition in Sikkim, India. *Agroecology Sustainable Food Systems* **44**: 653-72.

Mendoza TC (2002) Comparative productivity, profitability and energy use in organic, LEISA and conventional rice production in the Philippines. *Livestock Research for Rural Development* **14**(6): 70-81.

Meredith N (2008) Sustainable soils: reducing, mitigating, and adapting to climate change with organic agriculture. *Sustainable Development Law & Policy* **19-23**: 68-69.

Mitchell A and Edwards CA (1997) The production of vermicompost using *Eisenia fetida* from cattle manure. *Soil Biology and Biochemistry* **29**:3–4.

Mitra S and Devi H (2016) Organic Horticulture in India. *Horticulturae* **2**(4): 17. https://www.mdpi.com/2311-7524/2/4/17/htm#

Moghtader M, Salari H & Farahmand A (2011) Evaluation of the antifungal effects of rosemary oil and comparison with synthetic borneol and fungicide on the growth of *Aspergillus flavus*. *Journal of Ecology and the Natural Environment* **3**(6): 210-214.

Mourao I, Brito LM & Coutinho J (2008) Cultivating the-future based-on science, volume-1: organic crop- production," In Proceedings of the 2nd Scientific Conference of the International Society of Organic Agriculture Research ISOFAR, held at the 16th IFOAM Organic World Conference in Cooperation with the International Federation of Organic Agriculture-Movements IFOAM and the Consorzio ModenaBio, pp. 596-599.

Munda GC (2006) Problems and prospects of organic farming in NER. Paper presented in the Seminar on "Organic farming – Status and Road Map" during the North East Agri. Expo on March 06 at Dimapur, Nagaland.

Munda GC, Ghosh PK, Das Anup, Ngachan SV and Bujarbaruah K M (2007) *Advances in Organic Farming Technology in India.* ICAR Research Complex for NEH Region, Umaim-793 103, Meghalaya.

Muthana K D, Sharma S K and Harsh L N (1985) Study on silvi-pastoral system in arid zone. *My Forest* **21**(3):233-238.

Nagavallemma KP, Wani SP, Stephane Lacroix, Padmaja VV, Vineela C, Babu Rao M and Sahrawat KL. 2004. Vermicomposting: Recycling wastes into valuable organic fertilizer. Global Theme on Agrecosystems Report no. 8. Patancheru 502 324, Andhra Pradesh, India: International Crops Research Institute for the Semi-Arid Tropics.pp 20.

Namasivayam KR, Bharani RSA, Ansari MR (2013) Natural occurrence of potential fungal biopesticide nomuraea Rileyi (Farlow) samson associated with agriculture fields of tamil nadu, india and its compatibility with metallic nanoparticles. *Journal of Biofertilizers Biopesticides* **4**: 132.

National Project on Organic Farming (NPOF) (2005) *Organic Farming Newsletter* **1**(1): 9-18.

National Research Council (1989) *Board on Agriculture, Committee on the Role of Alternative Farming Methods in Modern Production Agriculture, Alternative Agriculture.* National Academy Press, Washington, DC.

Nemecek T, Huguenin-Elie O, Dubois D and Gaillard G (2005) Ökobilanzierung vonAnbausystemen im Schweizerischen Acker- und Futterbau, Zürich. pp:156.

Niggli U, Lockeretz W (1996) Development of research in organic agriculture. In: Stergaard T, ed, Proceedings of the 11th IFOAM International Scientijk Conference on Fundamentals of Organic Agriculture, 1, 11-15.

Niggli U, Schmidt J, Watson C *et al.* (2016) Organic knowledge network arable. State-of-the-art research results and best practices. Report D.3.1; http://www.ok-neable.eu/images/OK_Net_WP3_D3.1_final. pdf. Accessed 14 Aug 2017.

Ofoegbu RU, *et al.* (2015) Bioremediation of crude oil contaminated soil using organic and inorganic fertilizers. *Journal of Petroleum and Environmental Biotechnology* **6**:198.

Olesen JE, Hansen EM, Askegaard M, Rasmussen IA (2007) The value of catch crops and organic manures for spring barley in organic arable farming. *Field Crops Research* **100**(2–3):168–178. https://doi. org/10.1016/j.fcr.2006.07.001

Pacini C, Wossink A, Giesen G, Vzazana C, Huirne R (2002) Evaluation of sustainability of organic, integrated and conventional farming systems: a farm and field-scale analysis. *Agriculture, Ecosystems and Environment* **95**(1): 273-288.

Pagnanelli F *et al.*(2010) Development of new composite biosorbents from olive pomace wastes. *Applied Surface Science* **256**:5492-5497.

Pandit NP, Ahmad N and Maheshwari SK (2012) Vermicomposting biotechnology: An eco-loving approach for recycling of solid organic wastes into valuable biofertilizers. *Journal of Biofertilizers and Biopesticides* **2**: 113.

Panwar N, Ramesh P, Bahadur A and Ramana S (2010) Influence of organic, chemical, and integrated management practices on soil organic carbon and soil nutrient status under semi-arid tropical conditions in central India. *Communications in Soil Science and Plant Analysis* **41**(9): 1073-1083.

Partap T (2006) *India Organic Pathway: Making Way for Itself.* Occasional Paper. International Competence Centre for Organic Agriculture, Bangalore. Retrieved March 4, 2011 from www.iccoa.org pub.

Partap T (2010) *Organic Farming Potentials for Green Productivity, Ecological Services and Sustainable Rural Development.* In: Organic Agriculture and Agribusiness: Innovation and Fundamentals. Asian Productivity Organization, Hirakawacho, (Partap, T. and Saeed, M. (Eds.)) Chiyoda-ku, Tokyo, Japan.

Partap T and Vaidya CS (2009) *Organic Farmers Speak on Economics and Beyond.* West View Press, New Delhi.

PAU (2020-21) *Package of Practices for Crops of Punjab (Rabi).* Punjab Agricultural University, Ludhiana.

PAU (2021) *Package of Practices for Crops of Punjab (Kharif).* Punjab Agricultural University, Ludhiana.

Pavan Kumar P, Satyanarayana SDV (2014) Soil microbial exploration for the efficient exploitation of unknown culturable PGPR for geographically similar crop lands. *Journal of Biofertilizers and Biopesticides* **5**: 118.

Payraudeau S and van der WerfHMG (2005) Environmental impact assessment for a farming region: a review of methods. *Agriculture, Ecosystems and Environment* **107**:1–19.

Pimentel D, Acquay H, Biltonen M, Rice P, Silva M, Nelson J, Lipner V, Giordano S, Horowitz A and D'Amore M (1992) Environmental and economic cost of pesticide use. *Bioscience* **42**: 750–760.

Ponisio LC, M'Gonigle LK, Mace KC, Palomino J, deValpine P, Kremen C (2015) *Diversification Practices Reduce Organic to Conventional Yield Gap.* Proc. Royal Society ofBiological Sciences 282:20141396. https://doi.org/10.1098/ rspb.2014.1396.

Prasad D, Prashar K S and Singh R P (1981) Quarter Century of Agronomic Research in India. Silver Jubliee Publication, Indian Society of Agronomy, IARI, New Delhi. In: *Principles of Agricultural Ecology*, Dhaliwal G S and Kler D S (Eds). Himalaya Publishing House.

Prasad K and Gill M S (2009) Developments and strategies perspective for organic farming in India. *Indian Journal of Agronomy* **54**(2): 184-192.

Prasad R and Power J F (1991) Crop residue management. *Advances in Soil Science* **15**: 205-251.

Pratap T (2006) Indian organic pathways: making way for itself. The international competence of organic agriculture. Accessed at www.iccoa.org.

Pretty JN (1995) *Regeneration Agriculture: Policies and Practices for Sustainability and Self-Reliance London, Washington*. National Academy Press, Action Aidand Vikas.

Pretty JN, Noble AD, Bossio D, Dixon J, Hine RE, Penning de Vris FWT and Morison JIL (2006) Resource conserving agriculture increases yields in developing countries. *Environmental Science Technology* **40**(4):1114-1119.

Purohit M L (1998) Impact of recent spurt in human and livestock population on India arid ecosystem. In: *Desertification Control in the Arid Ecosystem of India,* eds. S. Singh and A. Kar, pp 198-204. Bikaner (India): Agrobotanical Publishers.

Rajeshwar Rao BR (1999) Medicinal plants for dry areas. In: *Sustainable Alternate Land Use Systems for Drylands*, eds. R. P. Singh and M. Osman, 139-156. Dehradun (India): Oriental Enterprises.

Ramangouda H P (2002) *Evaluation of insect pest management components in soybean ecosystem.* Ph.D. Thesis. Himachal Pradesh Krishi Vishvadyala, Palampur . pp 84-88.

Ramesh P, Panwar NR, Singh AB, Ramana S, Yadav SK, Shrivastava R and Subba Rao A (2010) Status of organic farming in India. Current Science India **98**(9):1190-1194.

Rao AU, Murthy KMD, Sridhar TV, Krishnam RS and Adi Lakshmi D (2014) Studies on performance of organic farming and chemical farming in rainy season rice. *International Journal of Applied Power Engineering* **4**: 1-5.

Rao Kishore VSK, Supe R, Menon MK and Partap T (2006) Market for organic foods in India: Consumer perceptions and market potential. Bangalore: International Competence Centre for Organic Agriculture.

Rathore V S, Tanwar S P S, Kumar Pand Yadav O P (2019) Integrated Farming System: Key to sustainability in arid and semi-arid regions. *Indian Journal of Agricultural Sciences* **89**(2): 181–92.

Reganold JP and Wachter JM (2016) Organic agriculture in the twenty-first century. *Nature Plants* 15221. https://doi.org/10.1038/NPLANTS. 2015.221

Rigby D, Young T and Burton M (2001) The development and prospects for organic farming in the UK. *Food Policy* **26**: 599-613.

Rothschild M (1998) The Butterfly Gard eners by Miriam Rothschild and elive farell, Great Britain, pp 128-130.

Roychowdhury R, Banerjee U, Sofkova S and Tah J (2015) Organic farming for crop improvement and sustainable agriculture in the era of climate change. *Journal of Biological Sciences* **13**(2): 50-65. doi: 10.3844/ ojbsci.2013.50.65

Sane SA and Mehta SK (2015) Isolation and Evaluation of Rock Phosphate Solubilizing Fungi as Potential Bio- fertilizer. *Journal of Biofertilzers and Biopesticides* **6**:156.

Sangha K K (2014) Modern agricultural practices and analysis of socio-economic and ecological impacts of development in agriculture sector, Punjab, India - A review. *Indian Journal of Agricultural Research* **48**(5): 331-341.

Sannaveerrappaanavar V T (2003) Biopesticides in pest management. In: Organic Inputs in Organic Farming, ed Veeresh G K. Association for Promotion of Organic Farming. UAS, Alumni Building, Bellary Road, Bangalore. pp 87-89.

Sarma J S, Brar SK (2015) Industrial production of bacillus thuringiensis based bio-insecticide: which way forward. *Journal of Biofertilzers and Biopesticides* **6:** 126.

Sarwar, Ghulam & Hussain, Nazir & Schmeisky, H & Suhammad, S & Ibrahim, M & Ahmad, S. (2008) Efficiency of various organic residues for enhancing rice-wheat production under normal soil conditions. *Pakistan Journal of Botany* **40:** 2113.

Saxsena K B, Ravishankar K, Vijaya Kumar R, Sreejith K Pand Srivastava R K (2010) Vegetable pigeonpea-a high protein food for all ages. Information Bulletin No. 83, ICRISAT. pp 31.

Scialabba E N (2003) Organic agriculture: The challenges of sustaining food production while enhancing biodiversity. United Nations. The matic Group meeting on wild life, biodiversity and organic agriculture. Anakara, Turkey, 15-16 April 2003, FAO Corporate documents repository.

Setboonsarng, Sununtar (2008) Can Ethical Trade Certification Contribute to the Attainment of the Millennium Development Goals? A Review of Organic and Fair-trade Certification. ADBI Discussion Paper 115. Tokyo: Asian Development Bank Institute. Available: http://www.adbi.org/discussion-paper/2008/08/25/2675.organic.fairtrade.certification/

Seufert V, Ramankutty N (2017) Many shades of gray—the context- dependent performance of organic agriculture. *Science Advances* **3**(3): e1602638. https://doi.org/10.1126/sciadv.1602638

Seufert V, Ramankutty N, Foley J A (2012) Comparing the yields of organic and conventional agriculture. *Nature* **485**:229–232. https:// doi.org/10.1038/nature11069.

Sharma A K (2000) Sustainability through simplicity: Rainwater conservation in the arid zone of India. *Indian Water Resource Quarterly* **48**(1): 5-10.

Sharma K and Pradhan S (2011) Organic farming: problems and prospects. *Yojana* **55:** 68-70.

Sharma A K (2001) *A Handbook of Organic Farming*. Jodhpur (India): Agrobios.

Sharma A K and Goyal R K (2000) Addition in tradition: Agroforestry in the arid zone of India. *LEISA India* **2**(3):19-20.

Shenbagavalli S and Mahimairaja S (2012) Characterization and effect of biochar on nitrogen and carbon dynamics in soil. *International Journal Advanced Biological Research* **2**:249-255.

Sheng D Z, Xie J R, Zhang Q W (1995) Natural agriculture in Japan. *World Agriculture* **195**: 9- 11. (in Chinese)

Shepherd M, Pearce B, Cormack B, Philipps L, Cuttle S, Bhogal A, Costigan P and Unwin R (2003) An assessment of the environmental impacts of organic farming. ADAS, Wolverhampton.

Sidhu H S (2002) Crisis in agrarian economy in Punjab- Some urgent steps. *Economic and Political Weekly* **37**: 3132-3138.

Sikkim Becomes India's First Organic State. (2016) Avail- able online: http:// www.thehindu.com/news/national/ sikkim-becomes-indias-first-organic state/ article8107170.ece (accessed on 14 January 2016)

Sims GK (2014) Bioavailability in biodegradation and function of herbicides. *Journal of Bioremediation and Biodegradation* **5**:144.

Singh AU and Sharma K (2016) Pests of mushroom. *Advances in Crop Science and Technology* **6**(4):213. |

Singh S S (2001) *Crop Management.* Kalyani Publishers, New Delhi.

Smith J L and Elliott L F (1990) Tillage and residue management effects on soil organic matter dynamics in semiarid regions. *Advances in Soil Science* **13**: 69-88.

Sofia P K, Prasad R, and Vijay V K (2006) Organic farming- tradition reinvented. *Indian Journal of Traditional Knowledge* **5**(1): 139–142.

Soil Association (2001) *The Organic Food and Farming Report 2000*. Soil Association Publications.

Speiser B, Tamm L, Amsler T, Lambion J, Bertrand C, Hermansen A, Ruissen MA, Haaland P, Zarb J, Santos J, Shotton P, Wilcockson S, Juntharathep P, Ghorbani R, Leifert C (2006) Improvement of late blight management in organic potato production systems in Europe: field tests with more resistant potato varieties and copper-based fungicides. *Biological Agriculture and Horticulture* **23**:393–412. https://doi.org/ 10.1080/ 01448765.2006.9755339.

Sreerag RS, Jayaprakas CA (2015) Management of two major sucking pests using neem oil formulation. *Journal of Biofertilzers and Biopesticides* **6**: 147.

Stiftung Oekologie and Landbau (SOEL) (2002) Organic Agriculture Worldwide 2001 / Statistics and Future Prospect.

Stockdale EA, Lampkin NH and Hovi M (2001). Agronomic and environmental implications of organic farming systems. *Advances in Agronomy* **70**:261–327.

Stockdale EA, Shepherd MA, Fortune S and Cuttle SP (2002) Soil fertility in organic farming systems – fundamentally different? *Soil Use Management* **18**:301–308.

Stoll G (2002) *Asia and the International Context. Organic agriculture and rural poverty alleviation: Potential and best practices in Asia*. Bangkok: United Nations Economic and Social Commission for Asia and the Pacific.

Stolton S (2002) *Biodiversity and Organic Agriculture*. IFOAM, Tholey-Theley, Germany.

Stolze M, Piorr A, Haring A and Dabbert S (2000) *The Environmental Impacts of Organic agriculture in Europe– Organic agriculture in Europe: Economics and Policy*, vol. 6. University of Hohenheim, Stuttgart.

Stonehouse D P, Clark E A and Ogini Y A (2001) Organic and conventional dairy farm comparisons in Ontario, Canada. *Biological Agriculture and Horticulture* **19**(2):115–125.

Subrahmanyeswari B and Chander M (2008) Animal husbandry practices of organic farmers: An appraisal. *Veterinary World* **1**(10): 303-305.

Sudadi Sumarno (2014) Azolla - based organic farming: low biotechnology for high rice productivity. *The International Journal of Plant, Animal and Environmental Sciences* **4**(2): 425-429.

Sujatha V, Mosha K, Subbaiah G, Prasuna Rani P (2014) Residual soil fertility and productivity of rice (*Oryza sativa* L.) As influenced by different organic sources of nitrogen. *The International Journal of Plant, Animal and Environmental Sciences* **4**(1): 266-269.

Sunitha ND, Giraddi RS, Kulkarni KA and Lingappa S (1997) Evaluation methods of vermicomposting under open field conditions. *Karnataka Journal of Agricultural Sciences* **10**(4): 987–990.

Terziev V, Arabska E and Radovic V (2015) *Risk management in Organic Farming*. 15th International Academic Conference, Rome.

Teshome Z and Kibret K (2015) Characterization of soil management groups of metahara sugar estate in terms of their physical and hydraulic properties. *Advances in Crop Science and Technology* **3**:159.

Tewari H C and Tewari P (2007) Off Season Organic Vegetables: A Potential Source of Household Food Security. In: Papers Submitted to the International Conference on Organic Agriculture and Food Security (FAO, Rome, Italy, 3-5 May 2007) pp:45-47.

Thakore Y (2006) The biopesticide market for global agricultural use. *Industrial Biotechnology* **2** (3): 194-208.

Thorup-Kristensen K, Magid J and Jensen LS (2003) Catch crops and green manures as biological tools in nitrogen management in temperate zones. Advances in Agronomy **79**: 227-302.

Tilman D, Cassman KG, Matson PA, Naylor R, Polasky S (2002) Agricultural sustainability and intensive production practices. *Nature* **418:**671.

Tom C T and Asiegbu I E (2002) Evaluation of short duration pigeon peacultivars in intercrop with FARZ-7 maize in derived savanna of Nigeria. *International Journals of Agricultural Science* **3**(1)**:** 138-143.

TP Organics (2010) *European Technology Platform for Organic Food and Farming Research (TP Organics)Technology Platform Organics (TP Organics)*, c/o IFOAM EU Group, Rue du Commerce 124, 1000 Brussels, Belgium

Tybrik K, Alroe H F and Frederiksen P (2004) Nature and quality in organic farming: A conceptual analysis of consideration and criteria in a European context. *Journal of Agricultural and Environmental Ethics* **17**(3):249–274.

Ul-Haq S, Hasan SS, Dhar A, Mital V, Sahaf KA (2014) Antifungal properties of phytoextracts of certain medicinal plants against leaf spot disease of mulberry, *Morus* spp. *Journal of Plant Pathology and Microbiology* **5** (2): 224.

United States Department of Agriculture (USDA) (1980) Report and Recommendations on Organic Farming. U.S. Department of Agriculture, US Government Priting Office, Washington D C. pp. 94.

Van Lenteren JC (2012) The state of commercial augmentative biological control: plenty of natural enemies, but a frustrating lack of uptake. *Bio Control* **57**(1):1–20. https://doi.org/10.1007/s10526-011-9395-1.

VanMansvelt J D and Mulder J A (1993) European feature for sustainable development: a contribution to the dialogue. *Landscape and Urban Planning* 27: 67-90.

Velimirov A (2006) Europäische Sommerakademie für Biolandwirtschaft, Lednice na Moravé, CZ, Retrieved March 9, 2011 from http://orgprints.org/9163.

Venkataraman G S (1961) The role of blue-green algae in agriculture. Scientific Culture **27:** 9–13.

Venkataraman G S (1972) *Algal Biofertilizers and Rice Cultivation*. Today and Tomorrow's Printers, Faridabad (Haryana).pp 75.

Verma R S, Motiwale M P, Chauhan R S and Tiwari R K (1987) Studies on intercropping spices and tobacoo with autumn sugarcane. *Indian Sugar* **31**: 451-456.

Wai O K (2000) *Humus Consultancy, Malaysia*. Revised Edition, October 2000.

Wani S A, Chand S, Najar G R and Teli M A (2013) Organic Farming: As a Climate Change Adaptation and Mitigation Strategy. *Current Agriculture Research Journal* **1**(1): 45-50.

Watson CA, Bengtsson H, Ebbesvik M, Løes A-K, Myrbeck A, Salomon E, *et al*, (2002) A review of farm-scale nutrient budgets for organic farms as a tool for management of soil fertility. *Soil Use Management* **18**:264–273.

Williams AG, Audsley E and Sandars DL (2006) Determining the environmental burdens and resource use in the production of agricultural and horticultural commodities. Main Report. Defra Research Project IS0205, Cranfield University and Defra, Bedford.

Wynen E (1994) Economics of organic farming in Australia. In: *The Economics of Organic Farming: An International Perspective*. (Lampkin N H, and Padel S (eds.)). CAB International, Wallingford. pp185–199.

Xu HL (2000) Effects of a microbial inoculant and organic fertilizers on the growth, photosynthesis and yield of sweet corn. *Journal Crop Production* **3**(1):183-214.

Yadav A K(2010) Status of organic agriculture in India. *Organic Farming Newsletter* **6**(4): 3-10.

Yadav RH and Vijayakumari B (2004) Impact of vermicompost on biochemical characters of Chilli (*Capsicum annum*). *Journal of Ecotoxicology and Environmental Monitoring* **14**: 51-56.

Yu H F, Dai JY (1995) The trend of thoughts in modem agriculture. *World Agriculture* **200**:9-11. (in Chinese)

Yussefi M, Willer H (2003) The World of Organic Agriculture- Statistics and Future Prospects. International Federation of Organic Agriculture Movements (IFOAM). www.soel.de/inhalte/ publikation/s/s-74. pdf

Zajonc I and Sidor V (1990) Use of some wastes for vermicompost preparation and their influence on growth and reproduction of the earthworm Eisenia fetida. Pol'nohospodars-tvo (*CSFR*) **36**(8):742–752.

Zia M S (1993) Soil fertility evaluation and management for flooded lowland-rice soils of Pakistan. Ph.D. Thesis, Dissertation Kyoto University, Japan.

Ziesemer J (2007) Energy use in organic food systems. Natural Resources Management and Environment Department Food and Agriculture Organization of the United Nations, Rome.

Znaor D (1994) Ecological Agriculture: Analysis of the most commonly criticized aspects. Why shall Estonian Agriculture be Ecological? Proceedings of the International Conference held in Tartu June 13, 1994; Centre of Ecological Engineering Tartu.

Znaor D (1997) What future for sustainable agriculture? *Danube watch* **3**(2): 2-3.

Znaor D (2002) Contribution of organic agriculture to macroeconomic and environmental performance of the countries with economic in transition. *Vagos Research Papers* **53**(6):41-46. Lithuanias University of Agriculture, Akademija.